AF817431

Antibiotics and Microbial Transformations

Editors

Surendar S. Lamba, Ph.D.
Professor of Pharmacognosy and
Medicinal Chemistry
College of Pharmacy
The Florida Agricultural and Mechanical University
Tallahassee, Florida

Charles A. Walker, Ph.D.
Chancellor
University of Arkansas
Pine Bluff, Arkansas

CRC Press, Inc.
Boca Raton, Florida

Library of Congress Cataloging-in-Publication Data

Antibiotics and microbial transformations.

Bibliography: p.
Includes index.
1. Microbiological synthesis. 2. Antibiotics--
Metabolism. 3. Drug resistance in micro-organisms.
I. Lamba, Surendar S., 1934- . II. Walker,
Charles Andrew, 1935- . [DNLM: 1. Antibiotics--
metabolism. 2. Bacteria--metabolism. 3. Biotransforma-
tion. 4. Drug Resistance, Microbial. QW 52 A629]
QR88.A55 1987 615′.329 86-29908
ISBN 0-8493-6408-6

This book represents information obtained from authentic and highly regarded sources. Reprinted material is quoted with permission, and sources are indicated. A wide variety of references are listed. Every reasonable effort has been made to give reliable data and information, but the author and the publisher cannot assume responsibility for the validity of all materials or for the consequences of their use.

All rights reserved. This book, or any parts thereof, may not be reproduced in any form without written consent from the publisher.

Direct all inquiries to CRC Press, Inc., 2000 Corporate Blvd., N.W., Boca Raton, Florida, 33431.

© 1987 by CRC Press, Inc.

International Standard Book Number 0-8493-6408-6

Library of Congress Card Number 86-29908
Printed in the United States

PREFACE

The Florida Agricultural and Mechanical University, College of Pharmacy and Pharmaceutical Sciences, has been hosting an annual clinical symposium since 1976. The tenth in the series was entitled, ''Antibiotics and Microbial Transformations'', and was held on the University campus February 28 to March 3, 1985. The chapters presented in this monograph are derived from lectures presented at this symposium. The Symposium was attended by practicing pharmacists, physicians, graduate and Doctor of Pharmacy students, and some paramedical personnel. The seminarians selected were outstanding scientists, researchers, and practioners in their respective areas.

The topics chosen for the symposium were not intended to cover the entire spectrum of antibiotics and/or microbial transformation; rather, they were selected to provide an appreciation to those in practice as health professionals on a wide range of antibiotic activities, including selected research aspects. Some of the topics discussed dealt with antibiotic formulations; bioavailability and pharmacokinetics of selected antibiotics; noscomial and urinary tract infections; and selected microbial conversions. In recent years, microbial systems have been described which metabolize selected drugs by pathways known to operate in mammals. Selected scientists addressed the issue of how microorganisms, which offer comparative ease of manipulation, may be employed to study the formation of drug metabolites as they occur in mammalian systems and also, how to prepare such products in amounts sufficient for biological evaluation, structure determination, and the development of analytical methods.

Briefly, the monograph provides succinct and integrated information on selected antibiotics and microbial transformations.

S. S. Lamba, Ph.D.
C. A. Walker, Ph.D.
Tallahassee, Florida
February, 1986

THE EDITORS

Surendar S. Lamba, Ph.D., was born in India and received the B.S., B. Pharm., and M. Pharm. degrees from that country. In 1960, he came to the U.S. and pursued graduate work at the University of Nebraska, where he received the M.S. degree (1963), and the University of Colorado, where he completed requirements for the Ph.D. degree in 1966. Subsequently, he assumed a faculty position in the College of Pharmacy, Florida A&M University, Tallahassee, Florida and attained a professorial rank in 1967.

Dr. Lamba has travelled widely and has held two visiting professorships, one at the University of Panama, Republic of Panama (1977 to 1978) and the other at the University of Benin, Nigeria (1981 to 1982).

Dr. Lamba is an active member of the American Pharmaceutical Association, the American Society of Pharmacognosy, the Academy of Pharmaceutical Sciences, the New York Academy of Sciences, the American Association for the Advancement of Science, the American Association of Colleges of Pharmacy; and two honor societies, namely, Sigma Xi and Rho Chi. He has authored and co-authored over 25 scientific publications and presented papers at several national and international scientific meetings.

In 1970, he was nominated one of the Outstanding Educators of America; and he is included in *American Men of Science, International Men of Achievement, and Personalities of the South.*

Charles A. Walker, Ph.D., is Chancellor and Professor of Pharmacology and Toxicology at the University of Arkansas at Pine Bluff.

Dr. Walker is a graduate of the Department of Agriculture at Arkansas AM&N College, College Heights (1957) and holds a master's degree in biochemistry and nutrition from Washington State University and a Ph.D. in neuropharmacology from the Loyola University Stritch School of Medicine in Chicago, Illinois. His academic experience includes service as Professor and Chairman of the Department of Physiology and Pharmacology in the School of Veterinary Medicine at Tuskegee Institute. He served for twelve years as Dean of the College of Pharmacy and Professor of Pharmacology at Florida A&M University in Tallahassee.

Dr. Walker's major research interest has been in the area of circadian rhythms of biogenic amines in the central nervous system and drug responsiveness. In 1978, he sponsored the first International Symposium on Chronopharmacology and Chronotherapeutics, in which fifteen countries participated. He holds membership in several scientific organizations and societies, including the National Advisory Board of the National Aeronautics and Space Administration.

Dr. Walker is the author of numerous scientific articles, books, and book chapters in his area of research expertise.

CONTRIBUTORS

Walter G. Chambliss, Ph.D.
Department of Process Engineering
Bristol-Myers Company
Industrial Division
Syracuse, New York

Joyce I. Ciadella
Senior Research Microbiologist
Department of Infectious Diseases
Research
The Upjohn Company
Kalamazoo, Michigan

David W. Elrod
Research Chemist
Research Laboratories
The Upjohn Company
Kalamazoo, Company

Patrick J. Davis, Ph.D.
Associate Professor of
Medicinal and Natural
Product Chemistry
College of Pharmacy
University of Texas
Austin, Texas

Zafar H. Israili, Ph.D.
Associate Professor of Medicine
Emory University and
Research Pharmacologist
Veterans Administration
Medical Center
Atlanta, Georgia

Surendar S. Lamba, Ph.D.
Professor of Pharmacognosy
and Medicinal Chemistry
College of Pharmacy
Florida A&M University
Tallahassee, Florida

Hugh N. Lunan, Pharm.D.
Associate Professor of
Clinical Pharmacy
College of Pharmacy
Florida A&M University
Tallahassee, Florida

Parshotam L. Madan, Ph.D.
Chairman and Professor
Department of Pharmaceutics
St. John's University
Jamaica, New York

Vincent Marshall, Ph.D.
Senior Research Scientist
Department of Infectious Diseases
Research
The Upjohn Company
Kalamazoo, Michigan

C. Martin Parkinson, Jr.,
Assosciate Scientist
Analytical Research &
Development Department
Bristol-Myer Company
Industrial Division
Syracuse, New York

Tarun R. Patel, Ph.D.
Associate Director, Process Engineering
Bristol-Myers U.S. Pharmaceutical and
Nutritional Group
Evansville, Indiana

Larry W. Robertson, Ph.D.
Professor
Division of Medicinal
Chemistry and Pharmacognosy
College of Pharmacy
Ohio State University
Columbus, Ohio

Ronald J. Sax
Associate Scientist
Analytical Research &
Development Department
Bristol-Myers Company
Industrial Division
Syracuse, New York

Sampat M. Singhvi, Ph.D.
Research Group Leader
Department of Drug Metabolism
Squibb Institute of Medical Research
Princeton, New Jersey

Michael Thompson, Pharm.D.
Assistant Professor of
Clinical Pharmacy
College of Pharmacy
Florida A&M University
Tallahassee, Florida

Joseph Tricome, Ph.D.
Manager, Analytical Research &
Development
Microscopy and Chemistry
Laboratories
Bristol-Myers Company
Industrial Division
Syracuse, New York

Paul F. Wiley, Ph.D.
Distinguished Scientist
Reseach Laboratories
The Upjohn Company
Kalamazoo, Michigan

TABLE OF CONTENTS

Dr. Surendar S. Lamba, dedicates this monograph to his loving wife, Betty, and son, Sunjai, for their encouragement and support throughout the preparation of this manuscript.

Introduction and Microbial Transformations

Chapter 1

INTRODUCTION

S. S. Lamba

Antibiotics are generally characterized as microbial products which inhibit the growth of specific microorganisms in low (< 1 mg/mℓ) concentrations. They vary greatly in their antimicrobial spectra, chemical structures, biosynthesis, and mode of action. They belong to a large group of secondary metabolites which also include certain aliphatic and aromatic compounds, peptides, toxins, terpenoids, phenazines, alkaloids, sterols, prophyrins, and others. They are among the most extensively used compounds for the treatment of a wide range of bacterial and fungal infections, and some are effective in certain neoplastic diseases. They are of significant value in veterinary medicine and in the control of plant pathogens. They are also used in subinhibitory levels in animal production as feed additives, since they cause substantial increases in the weight gain rates and feed efficiency in livestock and poultry. Antibiotics are manufactured in rather large quantities in the U.S. Reports indicate that the production of antibiotics in recent years was only surpassed by the production of central nervous system depressants and stimulants, gastrointestinal agents, and therapeutic nutrients and vitamins. Most antibiotics are produced by fermentation, and some are also used as substrates for the manufacture of clinically superior semisynthetic compounds.

In view of the continuously developing antibiotic-resistant populations of infectious bacteria and the need to control diseases which do not respond to the existing antibiotics, a great deal of effort has been directed to screen for new compounds which would be effective in such situations. In time, these screens were expanded, modified, and supplemented by additional ones which are of clinical and agricultural interest: antiviral and antitumor, anticoccidial, anthelmintic, antitrichomonal, and insecticidal, as well as others which detect low molecular-weight enzyme inhibitors and immunomodifiers. As a result of these efforts, several new antibiotic substances have been isolated.

Since the beginning of the antibiotic era in the early 1940s, screening of various soil microorganisms has been the main method for the discovery of new antibiotics. Of those produced by fungi, only penicillins, cephalosporin C, and griseofulvin (and fusidic acid) are clinically valuable. Some of those produced by unicellular bacteria are of similar use: bacitracins, gramicidin, colistin, polymyxin, pyrrolnitrin, tyrocidine, and tyrothricin. However, most of the thousands of antibiotics found to date are products of microorganisms belonging to the genus *Streptomyces*. A number of them are produced on an industrial scale; adriamycin and daunorubicin, chloramphenicol, erythromycin, kanamycin, lincomycin, neomycin, novobiocin, spectinomycin, tetracylines, tobramycin, tylosin, or vancomycin. Two others of considerable practical importance are produced by closely related organisms: gentamicin by *Micromonospora purpurea* and rifamycin by *Nocardia mediterranei.* Because of the nature of their structures, only chloramphenicol (produced by *S. venezuelae)* and pyrrolnitrin (produced by two pseudomonads) are now manufactured only by chemical synthesis.

The original methods for the discovery of new antibiotics were fairly simple and straightforward. They consisted of random isolation of soil microorganisms; growing them on agar plates and in tubes, bottles, or in shaken flasks; testing samples for their ability to inhibit selected microorganisms; and carrying out preliminary identification of the active compound and isolating it for further evaluation. In time, various modifications and changes were added. Organisms were isolated from samples collected in ecologically widely different places where microbial flora varies with the nature of and changes in the environmental factors (humidity, temperature, nutrients, pH, presence of other living organisms). New techniques were added to direct soil enrichment and to improve culture selection (use of specific inhibitors including antibiotics, chemicals, and mutagens). Mechanical devices were used for inoculation, culture transfer, and multipoint sample applications. New methods were also applied to preserve the desired property unchanged on long-term storage and to maintain the culture for reliable reproduction in large-scale fermentations (deep freeze, lyophilization, liquid N_2).

If the data indicated that the new culture produces a new antibiotic, the organism is grown under different nutritional and physiological conditions to allow the maximum expression of the new genome. The antibiotic production may be monitored biologically and also by physicochemical methods (paper, thin-layer, and high-performance liquid chromatography). Further work is then directed toward preparing sufficient amounts for further examination. The identification is generally carried out by comparing antimicrobial spectra, chromatographic behavior, solubility, extractibility, and physicochemical properties of the new compound with those of known antibiotics. The compound is also evaluated in animals with experimentally induced infections and also for toxicity, absorption, distribution, metabolism, and excretion. Chemical work may also be carried out to modify the molecule in an attempt to improve its desirable properties.

As in the case of other secondary metabolites, the synthesis of antibiotics is usually initiated in the late growth phase (idiophase) when the enzymes involved are derepressed and become functional. And since such enzymes (synthetases) are not as substrate specific as those of the primary metabolism, it is a common occurrence that an antibiotic producer will synthesize a mixture of chemically related antibiotics rather than a single antibiotic entity. Their synthesis can be regulated to a different degree by metabolic manipulation or by mutation. Thus if a component of such a mixture becomes of interest and its amounts are small and difficult to separate from the other components, conditions may be found under which the metabolic flow toward the synthesis of the desired component is increased and the production of the others reduced or completely suppressed. Such improvements have been made by means of strains selected both from original and mutated populations. They have also been made by precursor feeding, metabolic inhibition, mutasynthesis, genetic manipulation, and biotransformation.

The oldest and best-known example of precursor feeding is the incorporation of phenylacetic and phenoxyacetic acids into benzyl and phenoxymethyl penicillins (penicillins G and V), respectively. These are not only outstanding therapeutic agents themselves, but are also unique substrates for enzymatic hydrolysis to 6-aminopenicillanic acid, a starting material for chemical preparation of semisynthetic penicillins. Utilizing broad specificity of antibiotic synthetases, the corresponding analogues have been prepared in a similar way from a number of other antibiotics, namely, actinomycin, bleomycin, celesticitin, gramicidin, lincomycin, novobiocin, polyoxin, pyrrolnitrin, quinomycin, and tyrocidine. Some of the more recent methods used for the modification and/or preparation of antibiotic analogues are briefly discussed below.

Antibiotics have also been modified by means of metabolic inhibitors. Thus, fermentations which normally produce chlortetracycline, streptomycin, or lincomycin can be made to yield N-demethylated analogues of these respective antibiotics by incorporating methylation inhibitors, such as D-methionine, ethionine, aminopterin, etc., in the fermentation media. Another good example of metabolic inhibition is by cerulenin, which is an inhibitor of fatty acid synthesis. Therefore, if cerulenin is included in the picromycin-producing fermentation medium, it is capable of inhibiting the synthesis of the aglycone portion of the 14-membered macrolide. When the aglycone of a 16-membered marcolide (tylosin) was added preformed to the medium, it could be incorporated into a new antibiotic. This method is referred to as "hybrid biosynthesis", and has also been used similarly to modify other macrolide antibiotics, such as leucomycin, spiramycin, carbomycin, etc.

Another method of bringing about such modification employs antibiotic-blocked mutants, which have been invaluable in another area of antibiotic research, namely, in the elucidation of biosynthesic sequences of these compounds. It was first envisioned by Professor Birch in 1963, and has since been used to prepare many new analogues of antibiotics of interest: aminocyclitols (butirosin, gentamicin, kanamycin, neomycin, ribostamycin, sisomicin, streptomycin); anthracyclines (adriamycin); beta-lactams (cephalosparin); macrolides (erythromycin, mycinamicin, oleandomycin); and coumarins (coumeramycin, novobiocin). Of potentially practical interest is the mutasynthetic formation of amikacin, a broad-spectrum semisynthetic aminocyclitol.

As is the case with many natural products, antibiotics are known to undergo microbial modifications. This approach has also been an effective means for the preparation of analogues of members of several classes of antibiotics. It is an intriguing area of antibiotic research and is discussed in Chapter Two by Dr. V. P. Marshall.

The study of microbial transformations of drugs is making both direct and indirect contributions to the complete understanding of the mammalian metabolism of drugs. In recent years, it has been suggested that microbial systems can be defined which mimic metabolic systems of mammals in the production of drug metabolites. This fascinating area of research, which is making significant contributions to the understanding of human drug metabolism is specifically discussed in this volume by Drs. Patrick J. Davis (Chapter 3) and Larry W. Robertson (Chapter 4).

Chapter 2

MICROBIAL TRANSFORMATIONS OF ANTIBIOTICS

Vincent P. Marshall, Joyce I. Cialdella, David W. Elrod, and Paul F. Wiley

TABLE OF CONTENTS

I. INTRODUCTION

Microorganisms and their enzymes have been effective agents in the synthesis of analogues of members of various classes of antibiotics.[175,176,252,253,279,281,282,284,297] In this review of these transformation systems, the aim is to report as completely as possible all microbial transformations of antibiotics. The antibiotic classes covered include the aminocyclitol, macrolide, β-lactam, peptide, actinomycin, anthracycline, lincosaminide, nucleoside, griseofulvin, chloramphenicol, novobiocin, and aureolic acid, as well as other minor classes. Antibiotic transformations in the context of this review are considered both as metabolic alterations of completely synthesized antibiotics (direct transformations) and as other manipulations involving their biosyntheses (indirect transformations). The direct transformations are reported on the basis of reaction type. These divisions include conversions by oxidation, reduction, acylation, deacylation, hydrolysis, alkylation, dealkylation, phosphorylation, and nucleotidylylation, as well as other transformations by miscellaneous reactions. The indirect transformations include precursor-directed biosynthesis, interrupted biosynthesis, mutasynthesis, and genetic recombination.

Some compounds mentioned as substrates are not antibiotics, but are closely related to them and are included. In many cases which are included, it was not shown by isolation that a new compound had been formed, but only that the substrate had been changed. These cases were decided on an individual basis, and some similar ones were excluded on the basis of personal judgment as to whether or not a transformation had occurred. The reference list is not exhaustive, although (as mentioned earlier) representative coverage of antibiotic transformation by microorganisms was attempted. Some transformations were discussed in more than one publication and some of these publications were not included in the references.

The structures presented in Figures 1 and 2 are representative of the different classes of antibiotics which have been modified using reactions mediated by microorganisms. They illustrate general structures and give the usual numbering systems. The two macrolides, leucomycin A and erythromycin, are representative of 16- and 14-membered ring macrolides, respectively. In the case of the anthracyclines the two compounds represented, aclacinomycin A and daunomycin, show the substantial variation in structure that exists in this family. The structure of gentamicin C_{1a}, which is representative of most aminocyclitols, and the structure of streptomycin, which differs considerably from most others of this family, are shown. β-Lactams are represented by the penicillin and cephalosporin types. Toyocamycin is the only nucleoside structure shown, but several quite different ones are reported in the various tables.

II. BIOTRANSFORMATIONS OF ANTIBIOTICS

A. Biotransformations of β-Lactam Antibiotics

The β-lactam antibiotics have been the most valuable substrates for the preparation of clinically useful antibiotic analogues through means involving microbial or enzymic biotransformations. Of these transformations, the deacylation of the penicillins at position 6 and some cephalosporins at position 7 have been of prime importance. These reaction products, which include 6-aminopenicillanic acid (6-APA) and 7-aminocephalosporanic acid (7-ACA) or 7-amino-3-deacetoxycephalosporanic acid (7-ADCA) can be reacylated to form a wide variety of β-lactam antibiotics (Table 1). In addition to these reactions involving penicillin and cephalosporin type substrates, enzyme-catalyzed deacylation of nocardicin C to 3-aminonocardicinic acid has been reported.[146] Other useful biotransformations employing cephalosporins and 7-aminocephem compounds as substrates include deacylation and reacylation at position 3 (Table 1).

All of the β-lactam antibiotic biotransformations mentioned result in formation of compounds which retain the β-lactam ring structure. However, the most thoroughly investigated

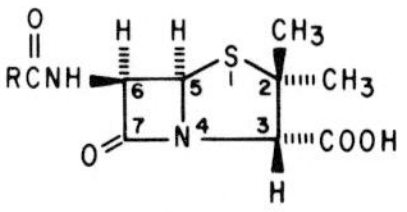

PENICILLIN TYPE
β-LACTAM ANTIBIOTIC

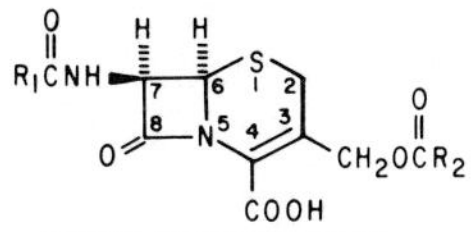

CEPHALOSPORIN TYPE
β-LACTAM ANTIBIOTIC

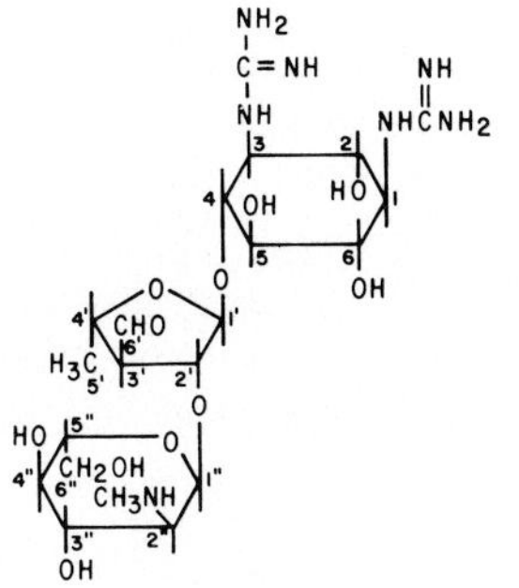

STREPTOMYCIN

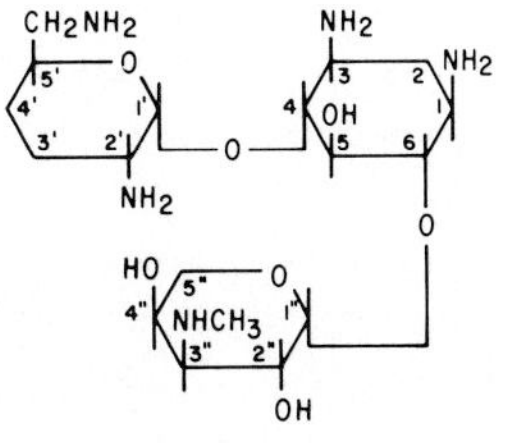

GENTAMICIN C_{1a}

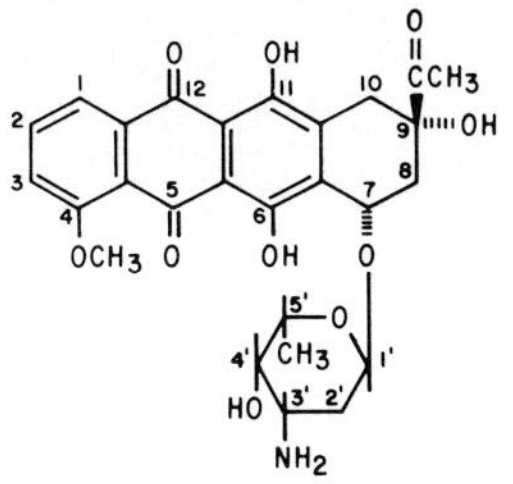

DAUNOMYCIN

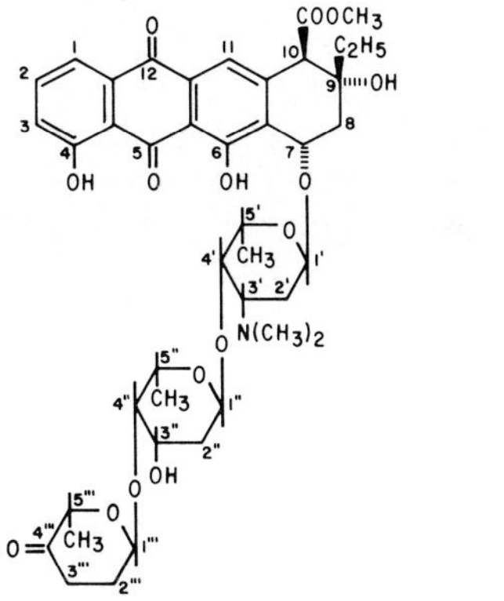

ACLACINOMYCIN A

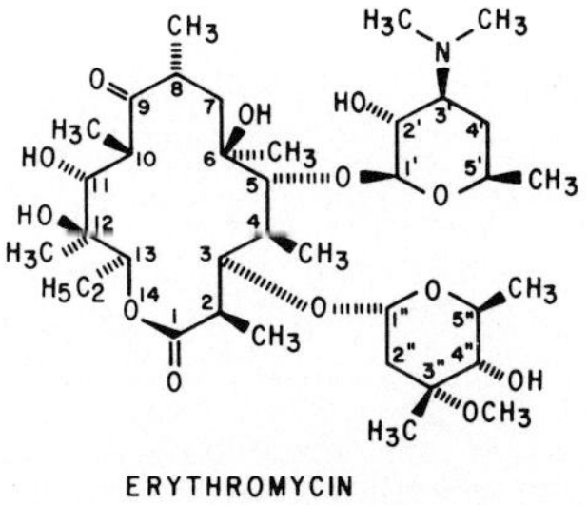

ERYTHROMYCIN

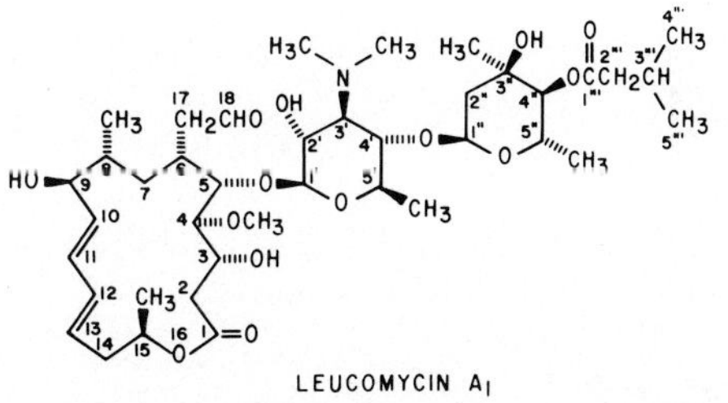

LEUCOMYCIN A_1

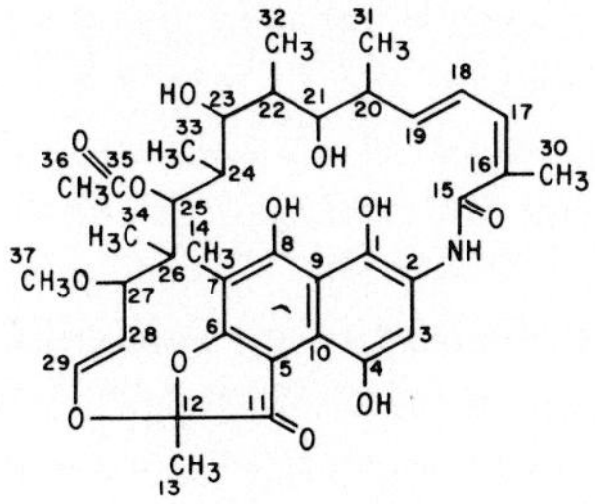

RIFAMYCIN SV

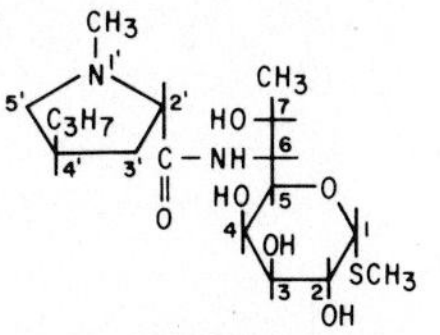

LINCOMYCIN

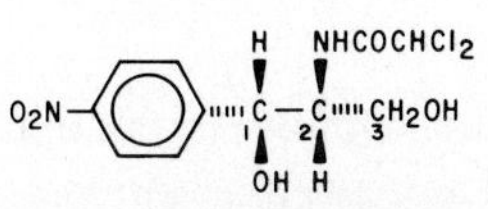

CHLORAMPHENICOL

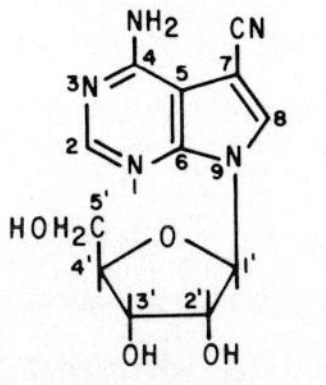

TOYOCAMYCIN

FIGURE 1. Antibiotic structures, Part 1.

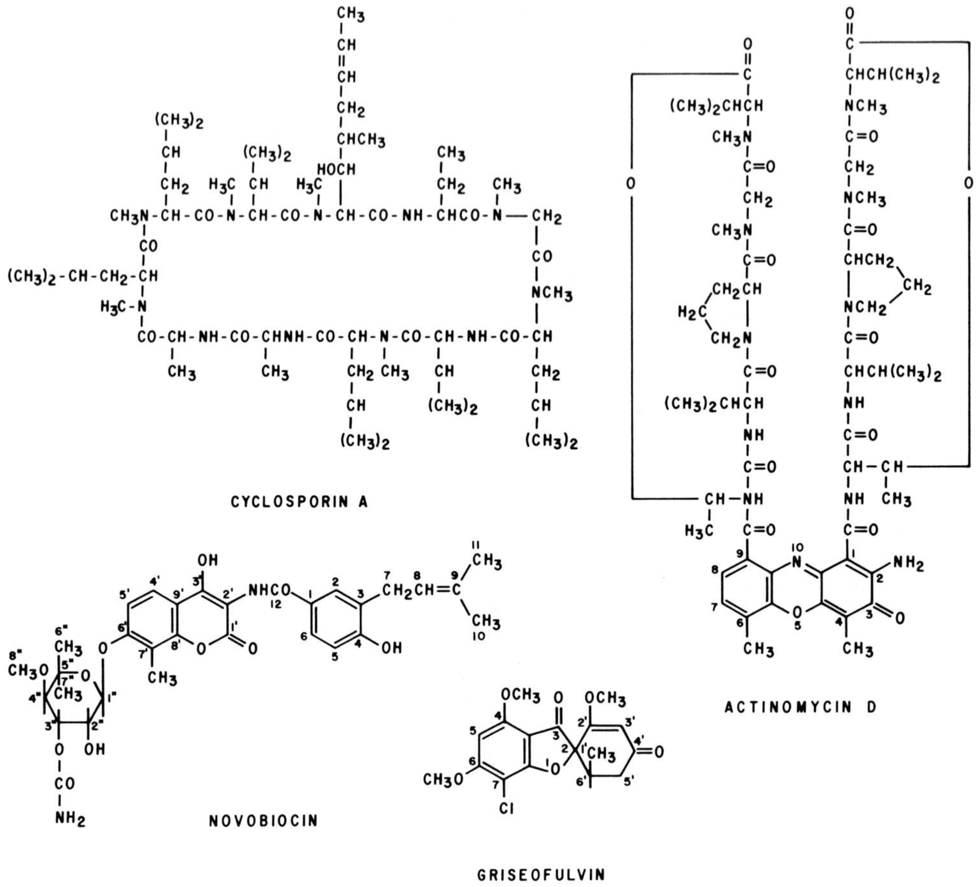

FIGURE 2. Antibiotic structures, Part 2.

biotransformation involving this class of antibiotic is β-lactam ring cleavage. As these ring cleavage products lack antibiotic activity, they have been only superficially summarized in Table 1.

In general, penicillin deacylation at position 6 and the reacylation of 6-APA are catalyzed by penicillin acylases, with their directional equilibria controlled by reaction pH.[91] Deacylation of the penicillins occurs at somewhat alkaline pH (7.5 to 8.5) while acylation occurs at more acidic pH (4 to 5). A notable example of an alternate mode of 6-APA acylation involves a reaction catalyzed by an enzyme from *Penicillium chrysogenum* which requires acyl-CoA.[81]

Penicillin acylase activity has been observed in a variety of microorganisms, with three general types of enzyme being recognized.[164] Enzymes of broad bacterial origin typically recognize penicillin G while those of fungal and actinomycete origin typically recognize penicillin V. A third type of enzyme termed ampicillin acylase occurs in some bacterial types including actinomycetes. As there are exceptions to these relationships, these enzymic categories are not an absolute means of classification.

6-APA can be produced alternatively as a secondary metabolite through its fermentation production in the absence of phenylacetic acid.[286] Biological production of 7-ACA appears to be more difficult than 6-APA. Unlike the latter, 7-ACA has never been detected as a secondary metabolite in a major producing organism.[164] In addition, the application of cephalosporin deacylation to the preparation of 7-ACA has been limited by rather narrow

Table 1
BIOTRANSFORMATIONS OF β-LACTAM ANTIBIOTICS

Substrate	Transformation product	Biological agent	Ref.
Deacylation			
Penicillin G	6-Aminopenicillanic acid (6-APA) and phenylacetic acid	*Escherichia coli*	268
		Alcaligenes faecalis	33
		Penicillium chrysogenum and *Aspergillus oryzae*	272
Penicillin V	6-APA and phenoxyacetic acid	*Cephalosporium sp.*	34
		Erwinia aroideae	358
		Pseudomonas acidovorans	163
Penicillins G and V	6-APA and respective acyl groups	*Fusarium semitectum*	28
Ampicillin	6-APA and phenylglycine	Pseudomonad	201, 220
Cephalosporin C	7-Aminocephalosporanic acid (7-ACA) and aminoadipic acid	*Aspergillus* and *Alternaria* spp.	208
		Pseudomonad	83
Cephalosporin G	7-ACA and phenylacetic acid	Streptomycete and *Achromobacter* sp.	365
7-Phenoxyacetamidedeacetoxy-cephalosporanic acid	7-Amino-3-deacetoxy-cephalosporanic acid (7-ADCA) and phenoxyacetic acid	*E. aroideae*	66
7-Phenylacetamidedeacetoxy-cephalosporanic acid	7-ADCA and phenylacetic acid	*Bacillus megaterium*	73
		E. coli	213
Glutaryl-7ACA	7-ACA and glutaric acid	Pseudomonad and *Comomonas* sp.	178, 298
Phenoxyacetylcephalosporin and phenoxymercaptoacetyl-cephalosporin	7-ACA and respective acyl groups	*Nocardia* and *Proteus* spp.	104
Nocardicin C	3-Aminonocardicinic acid	*Pseudomonas schuylkilliensis*	146
Acylation			
6-APA and phenylacetic acid	Penicillin G	*E. coli*	124, 268
		A. faecalis	33
		Kluyvera citrophila	200
6-APA and phenoxyacetic acid	Penicillin V	*A. faecalis*	33
6-APA and derivatives of phenylglycine	Ampicillin	*K. citrophila*	200
6-APA and *p*-hydroxy-phenyl-glycine ester	Amoxycillin	*E. coli*	171
6-APA and *p*-aminophenyl-acetic acid	*p*-Aminobenzylpenicillin	*K. citrophila*	200
6-APA and phenylmalonic acid ethyl ester	α-Carboxybenzylpenicillin	*K. citrophila*	200
6-APA and aminocyclohexane carbonic acid methyl ester	Aminocyclohexylpenicillin	*K. citrophila*	200
6-APA and phenoxyacetylglu-tamic acid	Phenoxymethylpenicillin	*E. coli*	124
6-APA and *O*-Chlorophenylacetylanaline	*O*-Chlorobenzylpenicillin	*E. coli*	124
6-APA and 23 carboxylic acids	23 Penicillins	*E. coli*	39
6-APA and 4 Acyl-CoA derivatives	4 Penicillins	*P. chrysogenum* enzyme	81
7-ACA, 7-ADCA, and esters of organic acids	Cephalexin and Cephaloglycine	7 Species of microorganisms	320
7-ADCA and 1-(1*H*)-tetra-zolylacetic acid	7-[1-(1*H*)-tetrazolylacetamido]-desacetoxycephalosporanic acid	*K. citrophila*	302a
7-ACA, 7-ADCA and 14 esters of organic acids	20 Cephalosporins	*K. citrophila*	302a

Table 1 (continued)
BIOTRANSFORMATIONS OF β-LACTAM ANTIBIOTICS

Substrate	Transformation product	Biological agent	Ref.
Deacetylation (position 3)			
Cephalosporin C	Deacetylcephalosporin C	*Bacillus subtilis*	1, 148
		Rhodotorula rubra	85
		Rhodosporidium toruloides	310
		Aureobasidium pullulans	106
		Bacillus cereus	51
Cephalothin	Deacetylcephalothin	*E. coli*	207
Acetylation (position 3)			
Deacetylcephalosporin C and Acetyl-CoA	Cephalosporin C	*Cephalosporium acremonium* enzyme	74
Ring expansion			
Penicillin N	Deacetoxycephalosporin C	*C. acremonium*	388, 397
β-Lactam ring cleavage			
Penicillin G	β-Lactam ring cleavage product	*E. coli*	4
		B. subtilis	255
Penicillins G, V, N, and ampicillin	Ring cleavage products	*B. cereus*	149
Penicillin G and 6-APA	Ring cleavage products	*Bacillus licheniformis*	256
Penicillins G, V, and pheoxyethylpenicillin	Ring cleavage products	*Staphylococcus aureus*	260
Penicillins G, V, ampicillin, and carbenicillin	Ring cleavage products	*E. coli*	261
Penicillins G, V, and ampicillin	Ring cleavage products	Streptomycete and *Streptomyces albus*	110
Penicillin G, ampicillin, and carbenicillin	Ring cleavage products	*Pseudomonas aeruginosa*	319
Cephalosporin C, cephalothin, cephaloridine, and cephaloglycine	Ring cleavage products	*B. cereus*	149
Cephalosporin C	Ring cleavage products	*B. licheniformis*	256
Cephalosporin C	Ring cleavage products	*S. aureus*	260
Cephaloridine and cephalothin	Ring cleavage products	*E. coli*	261
Cephaloridine, cephalothin, and cephalosporin C	Ring cleavage products	*Enterobacter cloacae*	94
Cephaloglycine and cephalothin	Ring cleavage products	Streptomycetes	110
Cephaloridine	Ring cleavage products	*P. aeruginosa*	34
Sulfoxidation			
7-Phenoxy- and 7-Phenylacetamido-3-deacetoxycephalosporanic acids, and cephalexin	Correponding (R)- and (S)-sulfoxides	*Aphyllophorales* spp.	337
Interrupted biosynthesis (blocked mutants)			
Endogenous precursor	Epithienamycins A and B	Mutants of *S. griseus* (defective oxidation)	212
Endogenous precursor	Antibiotic C-19393E_5	Mutants of *S. griseus* (defective methylation)	212

enzymic substrate specificities. However, 7-ADCA can be produced on a commercial scale through the enzymic deacylation of ring-expanded penicillin G or V.[66,73] Further, glutaryl-7ACA can be deacylated to form 7-ACA through the action of acylases derived from *Pseudomonas* or *Comomonas*.[178,298] Glutaryl-7ACA can be efficiently derived from cephalosporin C through the use of D-amino acid oxidase.[65,182] Although direct biological deacylation of cephalosporin C is not suitable for commercial production of 7-ACA, this reaction has been observed.[83,208] Numbered structures of typical β-lactam antibiotics are presented in Figure 1.

B. Biotransformations of Aminocyclitol Antibiotics

The aminocyclitols are a large family of antibiotics characterized by a cyclohexane ring usually substituted by two amino groups in a 1,3-relationship and by several hydroxyl groups. Deoxystreptamine is the most characteristic nucleus. At least one amino sugar, and commonly more than one, is attached to the carbocyclic system through oxygen. Sugars are usually attached at either 4,5 or 4,6 with one sugar most frequently being a pentose. The numbered structures of streptomycin and gentamicin C_{1a} are shown in Figure 1.

For the most part, microbial modifications of aminocyclitols have been studied as a result of microbial inactivations rather than in a planned attempt to improve activity by modification. The usual reactions involved have been either N- or O-acetylation, phosphorylation, or nucleotidylylation. An excellent review of these aspects of aminocyclitol microbial modification is available.[353] No useful biotransformations of the clinically important aminocyclitols have been reported. However, development of mutasynthesis was the result of a designed approach to the preparation of new aminocyclitols.[299] A number of recent reviews detail this approach.[37,86,233,263-267] Mutasynthesis involves mutation of an aminocyclitol-producing organism to give a mutant requiring the addition of the aminocyclohexane moiety to complete antibiotic biosynthesis. Addition to the mutant fermentation of aminocyclohexanes which are not produced by the parent organism gives new antibiotics. These transformations are listed in Table 2. A number of indirect transformations have been reported as a consequence of biosynthetic studies on the aminocyclitols.

C. Biotransformations of Macrolide Antibiotics

The macrolide antibiotics are a numerous group of antibiotics characterized by a large lactone ring. The original term applied only to a group of antibiotics whose aglycone is a 12-, 14-, or 16-membered ring, to which is attached amino sugars and frequently neutral sugars, usually at C-3 or C-4′. In some cases only a neutral sugar is present. This is the only group of macrolide antibiotics which has been subjected to extensive microbial modification, and it is the group that is discussed in this review with the exception of the lankacidins and the lankacidinols. Microbial conversions discussed are not limited to chemical changes brought about in antibiotics themselves, but the discussion is extended to compounds derived from fermentations and chemical or microbial conversions of substrates which are closely related to macrolides. In addition, the preparation of antibiotics by microbial modification of antibiotic precursors such as the aglycones is reviewed. The structures of erythromycin and leucomycin A_1 are shown in Figure 1 as representing 14- and 16-membered ring types, respectively, and indicate the numbering system.

Most reported microbial conversions were done in studying the biosyntheses of macrolide antibiotics rather than in attempts to prepare better ones and most have involved the 16-member lactone macrolides as substrates. The chemical reactions brought about by microorganisms are usually rather simple ones and most are reactions normally occurring in the biosynthesis of this family of antibiotics. However, specificities are obtained which are difficult to obtain chemically. For example, conversions are known in which oxidation occurs at a secondary hydroxyl group in the presence of other such groups, an aldehyde carbonyl, and double bonds. The biological agents most frequently used have been mutants of species which form macrolide antibiotics or enzymes isolated from such organisms, but a very large number of other species and genera have been used. In those cases in which a chemical substance is used to prevent formation of a part of an antibiotic, the standard producing organisms usually have been used.

The two most common microbial conversions are acylation and deacylation (Tables 3a and 3b). As several subfamilies of macrolide antibiotics differ among themselves only in the acyl groups present, interconversion between members has been carried out by microbial means. Reported acylations have been for the most part at the 3 and 4′ positions for basic

Table 2
BIOTRANSFORMATIONS OF AMINOCYCLITOL ANTIBIOTICS

Substrate	Transformation product	Biological agent	Ref.
Transglycosidation			
Dihydrostreptosylstreptidine-6-phosphate	Dihydrostreptomycin-6-phosphate	*Streptomyces griseus*	138
Neamine	5-Glucosylneamine	Clarase®	61
	Neomycin	*Streptomyces fradiae*	63
		Streptomyces rimosus	250
Kanamycin A	Glucosylkanamycin A	Clarase®	61, 62
Gentamicin	Glucosylgentamicin	Clarase®	61
Paromomycin	Glucosylparomomycin	Clarase®	61
Validoxylamine A	Validamycins A and D	*Rhodotorula glutinis*	114
		Rhodotorula marina	114
		Rhodotorula lactosa	114
		Rhodotorula rubra	114
	β-D-Galactosylvalidoxylamine A	*R. marina*	115
	Di-β-D-Galactosylvalidoxylamine A	*R. marina*	115
	β-D-Galactosylvalidamycin A	*R. marina*	115
	Validoxylamine-α-D-glucosides	*R. lactosa*	116
	Validoxylamine-β-D-glucosides	*R. marina*	116
Hydrolysis			
Mannosidostreptomycin	Streptomycin	*Streptomyces griseus*	251
N-p-Nitrophenylvalidamine	5-Hydroxymethyl-2,6-dihydroxy-2-cyclohexene-1-one	*Flavobacterium saccharophilium*	15
Validamycin A	Validoxylamine A	*Pseudomonas denitrificans*	113, 114
		Erwinia aroidea	113, 114
		Corynebacterium aquaticum	113, 114
		Oospora destructa	113, 114
		Rhizopus stolonifer	113, 114
		Syncephalastrum sp.	113, 114
	3-Keto-1-epivalidatol	*Flavobacterium saccharophilium*	117, 118
	4-Keto-1-epivalidatol	*F. saccharophilium*	117, 118
	1,3,4-Epivalidatol	*F. saccharophilium*	117, 118
	1-Epivalidatol	*F. saccharophilium*	117, 118
	1,3-Epivalidatol	*F. saccharophilium*	117, 118
Validamycin B	Hydroxyvalidamine	*Pseudomonas denitrificans*	113
Validamycins C, E, and F	Validamycin A	*Endomycopsis fibuliger*	114
		Endomycopsis chodatii	114
		Endomyces decipiens	114
		Candida intermedia	114
Validamycin D	Validoxylamine	*E. fibuliger*	114
		E. chodatii	114
		E. decipiens	114
		C. intermedia	114
		P. denitrificans	114
		Pseudomonas maltophilia	114
		Pseudomonas melanogenum	114
N-Acetylation		*Candida* sp.	114
Amikacin	6′-*N*-Acetylamikacin	*Streptomyces kanamyceticus*	89
		Pseudomonas aeruginosa	127
		Staphylococcus aureus	160

Table 2 (continued)
BIOTRANSFORMATIONS OF AMINOCYCLITOL ANTIBIOTICS

Substrate	Transformation product	Biological agent	Ref.
		Klebsiella pneumoniae	187
		Serratia marcescens	190
		Serratia proteamaculans	190
		Serratia sp.	190
Astromicin	6′-*N*-Acetylastromicin	*Staphyococcus aureus*	231
Butirosin A	2′-*N*-Acetylbutirosin A	*Streptomyces tenebrarius*	386
	6′-*N*-Acetylbutirosin A	*S. aureus*	160
		P. aeruginosa	381
Butirosin B	6′-*N*-Acetylbutirosin B	*S. aureus*	160, 381
Butirosin	2′-*N*-Acetylbutirosin	*Providentia* sp.	383
Dibekacin	2′-*N*-Acetyldibekacin	*S. tenebrarius*	386
	6′-*N*-Acetyldibekacin	*Mycobacterium tuberculosis*	189
3′,4′-Dideoxykanamycin B	6′-*N*-Acetyl-3′,4′-dideoxykanamycin B	*P. aeruginosa*	127, 128, 379
Fortimicin A	1-*N*-Acetylfortimicin A	*E. coli*	274
Gentamicin A	6′-*N*-Acetylgentamicin A	*S. kanamyceticus*	22
Gentamicin B	6′-*N*-Acetylgentamicin B	*S. aureus*	160
Gentamicin C_1	3-*N*-Acetylgentamicin C_1	*P. aeruginosa*	23, 29
		E. coli	352
	6′-*N*-Acetylgentamicin C_1	*S. kanamyceticus*	89
		P. aeruginosa	381
	2′-*N*-Acetylgentamicin C_1	*Providentia* sp.	383
	Acetylgentamicin C_1	*P. aeruginosa*	188
Gentamicin C_{1a}	6′-*N*-Acetylgentamicin C_{1a}	*E. coli*	20
		S. kanamyceticus	89
		P. aeruginosa	89, 381
		Moraxella sp.	158
		S. aureus	160
		K. pneumoniae	187
	2′-*N*-Acetylgentamicin C_{1a}	*Streptomyces spectabilis*	22
		Providentia sp.	32, 383
		S. kanamyceticus	89
	3-*N*-Acetylgentamicin C_{1a}	*P. aeruginosa*	29
		E. coli	157, 352
	Acetylgentamicin C_{1a}	*P. aeruginosa*	188
Gentamicin C_2	6′-*N*-Acetylgentamicin C_2	*K. pneumoniae*	187
		P. aeruginosa	188, 381
Gentamicin	3-*N*-Acetylgentamicin	*E. coli*	375
	6-*N*-Acetylgentamicin	*P. aeruginosa*	257
6′-*N*-Acetylgentamicin C_{1a}	3,6′-Di-*N*-acetylgentamicin C_{1a}	*P. aeruginosa*	29
Kanamycin A	2′-*N*-Acetylkanamycin A	*S. tenebrarius*	386
	3-*N*-Acetylkanamycin A	*P. aeruginosa*	23
	6′-*N*-Acetylkanamycin A	*E. coli*	20, 345
		S. kanamyceticus	22, 89
		P. aeruginosa	89, 127, 128, 381, 385
		Moraxella sp.	158
		S. aureus	160
		K. pneumoniae	187
		M. tuberculosis	189
Kanamycin B	2′-*N*-Acetylkanamycin B	*S. tenebrarius*	386
		Providentia sp.	383
	3-*N*-Acetylkanamycin B	*P. aeruginosa*	23
	6′-*N*-Acetylkanamycin B	*E. coli*	20
		S. kanamyceticus	22, 89

Table 2 (continued)
BIOTRANSFORMATIONS OF AMINOCYCLITOL ANTIBIOTICS

Substrate	Transformation product	Biological agent	Ref.
		P. aeruginosa	89, 128, 381
		Moraxella sp.	160
		S. aureus	160
		K. pneumoniae	187
		M. tuberculosis	189
Kanamycin C	2′-*N*-Acetylkanamycin C	*Providentia* sp.	383
	6′-*N*-Acetylkanamycin C	*S. kanamyceticus*	22, 89
		P. aeruginosa	89, 381
Lividomycin A	2′-*N*-Acetyllividomycin A	*Providentia* sp.	383
		S. tenebrarius	386
6′-*N*-Methylkanamycin B	6′-*N*-Acetyl-6′-*N*-methylkanamycin B	*P. aeruginosa*	128, 381
Neamine	2′-*N*-Acetylneamine	*S. tenebrarius*	386
	6′-*N*-Acetylneamine	*Moraxella* sp.	158
		S. aureus	160
Neomycin	6′-*N*-Acetylneomycin	*Moraxella* sp.	158
Neomycin B	2′-*N*-Acetylneomycin B	*S. tenebrarius*	386
	3-*N*-Acetylneomycin B	*P. aeruginosa*	23
		Streptomyces fradiae	262
	6′-*N*-Acetylneomycin B	*E. coli*	20
		S. kanamyceticus	22, 89
		P. aeruginosa	89
		S. aureus	160
Neomycin C	3-*N*-Acetylneomycin C	*S. fradiae*	262
	6′-*N*-Acetylneomycin C	*S. aureus*	160
Paromamine	6′-*N*-Acetylparomamine	*P. aeruginosa*	381
Paromomycin	2′-*N*-Acetylparomomycin	*S. tenebrarius*	386
	3-*N*-Acetylparomomycin	*E. coli*	157
		P. aeruginosa	23
	6′-*N*-Acetylparomomycin	*S. kanamyceticus*	22
Ribostamycin	2′-*N*-Acetylribostamycin	*Providentia* sp.	383
		S. tenebrarius	386
	3-*N*-Acetylribostamycin	*Streptomyces ribosidificus*	144
	6′-*N*-Acetylribostamycin	*S. aureus*	160
		P. aeruginosa	381, 385
Seldomycin factor 5	3-*N*-Acetylseldomycin factor 5	*E. coli*	274
Sisomicin	2′-*N*-Acetylsisomicin	*Providentia* sp.	32
	3-*N*-Acetylsisomicin	*P. aeruginosa*	23, 29
	6′-*N*-Acetylsisomicin	*S. kanamyceticus*	22, 89
		P. aeruginosa	89, 217, 218
		Moraxella sp.	158
		K. pneumoniae	187
Tobramycin	2′-*N*-Acetyltobramycin	*Providentia* sp.	32
	3-*N*-Acetyltobramycin	*P. aeruginosa*	23
		E. coli	157
	6′-*N*-Acetyltobramycin	*S. kanamyceticus*	22, 89
		P. aeruginosa	89, 381
		Moraxella sp.	158
		Enterobacter cloacae	187
		K. pneumoniae	187
		M. tuberculosis	189
Phosphorylation			
Amikacin	2″-*O*-Amikacin phosphate	*S. aureus*	160
	3′-*O*-Amikacin phosphate	*S. aureus*	42
Butirosin	3′-*O*-Butirosin phosphate	*E. coli*	30
		P. aeruginosa	30, 354

Table 2 (continued)
BIOTRANSFORMATIONS OF AMINOCYCLITOL ANTIBIOTICS

Substrate	Transformation product	Biological agent	Ref.
Butirosin A	3′-*O*-Butirosin A phosphate	*E. coli*	380
		P. aeruginosa	179
		Bacillus circulans	180
Dihydrostreptomycin	3″-*O*-Dihydrostreptomycin phosphate	*P. aeruginosa*	125, 140, 191
		E. coli	347
		S. griseus	381
	3′-(2-Hydroxymethyl-*O*-dihydrostreptomycin phosphate	*Streptomyces bikiniensis*	185, 364
		Streptomyces sp.	363
3″-*O*-Dihydrostreptomycin phosphate	3″,6-*O*-Dihydrostreptomycin diphosphate	*S. bikiniensis*	364
Gentamicin A	2″-*O*-Gentamicin A phosphate	*S. aureus*	160
Gentamicin B	2″-*O*-Gentamicin B phosphate	*S. aureus*	160
Gentamicin C_1	2″-*O*-Gentamicin C_1 phosphate	*S. aureus*	160
Gentamicin C_{1a}	2″-*O*-Gentamicin C_{1a} phosphate	*S. aureus*	160
Gentamicin C_2	2″-*O*-Gentamicin C_2 phosphate	*S. aureus*	160
3′-(2-Hydroxymethyl-*O*-)-dihydrostreptomycin phosphate	6-*O*-3′-(2-Hydroxymethyl-*O*-)dihydrostreptomycin diphosphate	*S. bikiniensis*	364
Kanamycin A	3′-*O*-Kanamycin A phosphate	*P. aeruginosa*	30, 54, 179, 349, 354
		E. coli	30, 147, 191, 347, 380
		S. aureus	55
	2″-*O*-Kanamycin A phosphate	*S. aureus*	160
Kanamycin B	3′-*O*-Kanamycin B phosphate	*P. aeruginosa*	54, 56, 236, 354
	2″-*O*-Kanamycin B phosphate	*S. aureus*	160
Kanamycin C	3′-*O*-Kanamycin C phosphate	*P. aeruginosa*	54, 56
Lividomycin A	5″-*O*-Lividomycin A phosphate	*S. aureus*	141
		P. aeruginosa	179, 354
Neamine	3′-*O*-Neamine phosphate	*E. coli*	380
Nebramine	3′-*O*-Nebramine phosphate	*P. aeruginosa*	236
Neomycin	3′-*O*-Neomycin phosphate	*P. aeruginosa*	54
Neomycin B	3′-*O*-Neomycin B phosphate	*E. coli*	30
		P. aeruginosa	30, 236
Paromamine	3′-*O*-Paromamine phosphate	*E. coli*	147
		P. aeruginosa	166, 191, 347
Paromomycin	3′-*O*-Paromomycin phosphate	*P. aeruginosa*	236
Ribostamycin	3′-*O*-Ribostamycin phosphate	*P. aeruginosa*	236
		E. coli	353, 380
	5″-*O*-Ribostamycin phosphate	*P. aeruginosa*	130

Table 2 (continued)
BIOTRANSFORMATIONS OF AMINOCYCLITOL ANTIBIOTICS

Substrate	Transformation product	Biological agent	Ref.
Sisomicin	2″-*O*-Sisomicin phosphate	*S. aureus*	160
Streptomycin	6-*O*-Streptomycin phosphate	*S. griseus*	186, 204, 205, 210, 211
		S. tenebrarius	386
	3′-*O*-Streptomycin phosphate	*E. coli*	237
	Streptomycin phosphate	*P. aeruginosa*	54, 139
		S. bikiniensis	185
	3″-*O*-Streptomycin phosphate	*S. griseus*	364
Tobramycin	2″-*O*-Tobramycin phosphate	*S. aureus*	160
Nucleotidylation			
Amikacin	4′-*O*-Adenylylamikacin	*S. aureus*	159, 160
Butirosin A	4′-*O*-Adenylylbutirosin A	*S. aureus*	159, 160
		Bacillus brevis	304
Butirosin B	4′-*O*-Adenylylbutirosin B	*S. aureus*	159, 160
3′,4′-Dideoxykanamycin B	2″-*O*-Adenylyl-3′,4′-dideoxykanamycin B	*E. coli*	192, 377
	2″-*O*-Guanosylyl-3′4′-dideoxykanamycin B	*E. coli*	378
	2″-*O*-Inosylyl-3′4′-dideoxykanamycin B	*E. coli*	378
Gentamicin A	4′-*O*-Adenylylgentamicin A	*S. aureus*	159, 160
	O-Adenylylgentamicin A	*E. coli*	21
Gentamicin B	4′-*O*-Adenylylgentamicin B	*S. aureus*	159, 160
Gentamicin C_1	*O*-Adenylylgentamicin C_1	*E. coli*	21, 309
		P. aeruginosa	112
Gentamicin C_{1a}	*O*-Adenylylgentamicin C_{1a}	*E. coli*	21, 309
		P. aeruginosa	112
Gentamicin C_2	*O*-Adenylylgentamicin C_2	*E. coli*	21, 309
		P. aeruginosa	112
Kanamycin A	*O*-Adenylylkanamycin A	*E. coli*	21, 225, 309
		P. aeruginosa	112
	4′-*O*-Adenylylkanamycin A	*S. aureus*	159, 160
		B. brevis	303
Kanamycin B	*O*-Adenylylkanamycin B	*P. aeruginosa*	112
		E. coli	309
	4′-*O*-Adenylylkanamycin B	*S. aureus*	159, 160
Lividomycin	4′-*O*-Adenylyllividomycin	*S. aureus*	159, 160
Neamine	4′-*O*-Adenylylneamine	*B. brevis*	303
Neomycin B	4′-*O*-Adenylylneomycin B	*S. aureus*	159, 160
Neomycin C	4′-*O*-Adenylylneomycin C	*S. aureus*	159, 160
Paromamine	*O*-Adenylylparomamine	*E. coli*	225
Paromomycin	4′-*O*-Adenylylparomomycin	*S. aureus*	159, 160
Ribostamycin	4′-*O*-Adenylylribostamycin	*S. aureus*	159, 160
Sisomicin	*O*-Adenylylsisomicin	*P. aeruginosa*	112
Spectinomycin	9-*O*-Adenylylspectinomycin	*E. coli*	19
	O-Adenylylspectinomycin	*S. aureus*	126
Streptomycin	3″-*O*-Adenylylstreptomycin	*E. coli*	19, 225, 322, 348, 382
	O-Adenylylstreptomycin	*S. aureus*	126
Tobramycin	4′-*O*-Adenylyltobramycin	*P. aeruginosa*	112
		S. aureus	159, 160
		E. coli	309
Xylostatin	4′-*O*-Adenylylxylostatin	*B. brevis*	303
Methylation			
Gentamicin C_{1a}	Sagamicin (Gentamicin C_{2b})	*Micromonospora sagamiensis*	120

Table 2 (continued)
BIOTRANSFORMATIONS OF AMINOCYCLITOL ANTIBIOTICS

Substrate	Transformation product	Biological agent	Ref.
		Micromonospora purpurea	332
Gentamicin C_2	Gentamicin C_1	*M. purpurea*	332
Sisomicin	6′-*N*-Methylsisomicin	*Micromonospora zionensis*	121
Verdamicin	6′-*N*-Methylverdamicin	*M. zionensis*	121
Mutasynthesis			
Streptidine	New product	*B. circulans*	35
Streptamine	2-Hydroxybutirosins	*B. circulans*	35, 79, 333
2,5-Dideoxystreptamine	5-Deoxybutirosamine	*B. circulans*	333
3′,4′-Dideoxyneamine	3′,4′-Dideoxybutirosins A and B	*B. circulans*	323, 324, 325, 328
6′-*N*-Methylgentamine C_{1a}	3′,4′-Dideoxy-6′-*N*-methylbutirosins A and B	*B. circulans*	323, 326
6′-*N*-Methylneamine	6′-*N*-Methylbutirosins A and B	*B. circulans*	323, 327
Gentamine C_2	3′,4′-Dideoxy-6′-*C*-methylbutirosins A and B	*B. circulans*	323, 327
S-11-P	Butirosin	*B. circulans*	76
Scyllo-inosose	2-Hydroxybutirosins	*B. circulans*	79
Scyllo-inosamine-2	2-Hydroxybutirosins	*B. circulans*	79
3′-Chloro-3′-deoxyneamine	3′-Chloro-3′-deoxybutirosin A	*Bacillus botulinus*	96
Streptamine	2-Hydroxygentamicins C_1, C_{1a}, and C_2	*Micromonospora purpurea*	269
2,5-Dideoxystreptamine	5-Deoxygentamicins C_1, C_{1a}, and C_2	*M. purpurea*	269
3-Epistreptamine	New products	*M. purpurea*	269
2,4,6/3,5-Pentahydroxylcyclohexanone	2-Hydroxygentamicins C_1, C_{1a}, and C_2	*M. purpurea*	47, 48
2,6-Dideoxystreptamine	6-Deoxyneomycins A and B	*S. fradiae*	36, 37
Streptamine	2-Hydroxyneomycins A and B (Hybrimicins A_1 and A_2)	*S. fradiae*	299, 300, 301
2-Epistreptamine	2-Epihydroxyneomycins A and B (Hybrimicins B_1 and B_2)	*S. fradiae*	299, 300
Streptamine	Hybrimicins C_1 and C_2	*S. rimosus*	300, 302
2,6-Dideoxystreptamine	6-Deoxyparomomycins I and II	*S. rimosus*	36, 37
Streptamine	2-Hydroxysisomicin (Mu 1)	*Micromonospora inyoensis*	45, 46, 330
	3″-*N*-Demethyl-3″-*N*-Acetyl-2-hydroxysisomicin (Mu 1a)	*M. inyoensis*	45, 46
	3″-*N*-Demethyl-2-hydroxysisomicin (Mu 1b)	*M. inyoensis*	45, 46
2,5-Dideoxystreptamine	5-Deoxysisomicin (Mu 2)	*M. inyoensis*	46, 330
	5-Deoxygentamicin A (Mu 2a)	*M. inyoensis*	46, 330
2-Epistreptamine	2-Epihydroxysisomicin (Mu 4)	*M. inyoensis*	45, 46
5-Amino-2,5-dideoxystreptamine	5-Amino-5-deoxysisomicin (Mu 5)	*M. inyoensis*	45, 46
5-Epi-2-deoxystreptamine	5-Episisomicin (Mu 6)	*M. inyoensis*	362
3-*N*-Methyl-2-deoxystreptamine	3-*N*-methylsisomicin (Mu 7)	*M. inyoensis*	45, 46
1-*N*-Methyl-2,5-dideoxystreptamine	1-*N*-Methyl-5-deoxysisomicin (Mu 7)	*M. inyoensis*	45, 46

Table 2 (continued)
BIOTRANSFORMATIONS OF AMINOCYCLITOL ANTIBIOTICS

Substrate	Transformation product	Biological agent	Ref.
5-Epifluoro-2-deoxystreptamine	5-Epifluorosisomicin (Mu 8)	*M. inyoensis*	45, 46
Neamine	Ribostamycin	*Streptomyces ribosidificus*	143
Streptamine	2-Hydroxyribostamycin	*S. ribosidificus*	143
2-Epistreptamine	2-Epihydroxyribostamycin	*S. ribosidificus*	143
1-*N*-Methyldeoxystreptamine	1-*N*-Methylribostamycin	*S. ribosidificus*	143
Gentamine C_{1a}	3′,4′-Dideoxyribostamycin	*S. ribosidificus*	143
2-Epistreptamine	6′-Deamino-6′-hydroxy-2-epihydroxykanamycin A	*S. kanamyceticus*	143
1-*N*-Methylstreptamine	6′-Deamino-6′-hydroxy-1-*N*-methylkanamycin A	*S. kanamyceticus*	143
2-Deoxystreptidine	Streptomutin A	*S. griseus*	193
Kanamycin B	Combimicins B_1 and B_2	*Micromonospora* sp.	219
Kanamycin A	Combimicin A_2	*Micromonospora* sp.	219
Streptamine	2-Hydroxysagamicin	*M. sagamiensis*	133
Multiple conversions			
Sisomicin	Gentamicin C_{1a} and Sagamicin	*M. sagamiensis*	121
Verdamicin	Gentamicin C_{2a}, C_2, C_1	*M. sagamiensis*	121
Gentamicin C_{1a}	Sagamicin	*M. sagamiensis*	120
2-Deoxystreptamine, Paromamine, Gentamicin A, Gentamicin X_2	Gentamicin C_{1a}, C_2, C_1 and Sagamicin	*M. sagamiensis*	120
Gentamicin C_{2a}, Antibiotic G418, Antibiotic JI-20B	Gentamicin C_2 and C_1	*M. sagamiensis*	120
Antibiotic JI-20A	Gentamicin C_{1a} and Sagamicin	*M. sagamiensis*	120
Gentamicin C_2	Gentamicin C_1	*M. sagamiensis*	120
2-Deoxystreptamine	2-Deoxy-*scyllo*-inosamine	*M. sagamiensis*	119a
2-Deoxystreptamine	Sagamicin	*M. sagamiensis*	119a
2-Deoxystreptamine, paromamine, Gentamicin A, Gentamicin X_2, and Gentamicin C_2	Gentamicin C_1	*M. sagamiensis*	215
2-Deoxystreptamine, paromamine, Neamine, Ribostamycin, and Xylostatin	Butirosins A and B	*B. circulans*	79, 324, 325, 328
Butirosin B	Butirosin A	*B. circulans*	324, 325, 328
2-Deoxy-*scyllo*-inosose	Butirosins A and B	*B. circulans*	79
2-Deoxy-*scyllo*-inosamine	Butirosins A and B	*B. circulans*	79
Kanamycin A	Amikacin	*B. circulans*	31
1-Deamino-1-hydroxy-2-deoxystreptamine	Butirosin	*B. circulans*	75
No substrate	1-Deamino-1-hydroxyxylostatin	*B. circulans*	75
Sisomicin	Gentamicin C_{2b}	Micromonospora rhodorangea	156
Gentamicin A	Gentamicins C_1, C_{1a} and C_2	*M. purpurea*	332
Gentamicin X_2	Gentamicins C_1, C_{1a}, and C_2	*M. purpurea*	332
Antibiotic JI-20A	Gentamicins C_1, C_{1a}, and C_2	*M. purpurea*	332
Antibiotic JI-20B	Gentamicins C_1 and C_2	*M. purpurea*	332
Antibiotic G-418	Gentamicins C_1 and C_2	*M. purpurea*	332
2-Deoxystreptamine, Neamine, paromamine	Sisomicin	*M. inyoensis*	331

Table 2 (continued)
BIOTRANSFORMATIONS OF AMINOCYCLITOL ANTIBIOTICS

Substrate	Transformation product	Biological agent	Ref.
Gentamicin A_1	Antibiotic 66-40D	*M. inyoensis*	331
Gentamicin A_2 and X_2	Sisomicin	*M. inyoensis*	331
Antibiotic JI-20A	Sisomicin	*M. inyoensis*	331
Antibiotic 66-40B	Sisomicin	*M. inyoensis*	331
Ribostamycin	3-*N*-Carboxymethylribostamycin	*M. inyoensis*	142

Table 3a
BIOTRANSFORMATIONS OF MACROLIDE ANTIBIOTICS

Substrate	Transformation product	Biological agent	Ref.
Acylation			
3-*O*-Acetyltylosin	3-*O*-Acetyl-4″-O-butyryltylosin	*Streptomyces thermotolerans*	224
3-*O*-Propionyltylosin	3-*O*-Propionyl-4″-*O*-butyryltylosin	*S. thermotolerans*	224
	3-*O*-Propionyl-4″-*O*-isovaleryltylosin	*S. thermotolerans*	224
4″-*O*-Depropionylmaridomycin III	Maridomycin V	*Streptomyces* sp.	355
5-*O*-(4′,6′-Dideoxy-3′-*C*-acetyl-β-D-xylohexopyranosyl)-platenolide II	3-*O*-Acetyl-5-*O*-(4′,6′-dideoxy-3′-*C*-acetyl-β-D-xylohexopyranosyl)-platenolide II	*Streptomyces hygroscopicus*	88
	3-*O*-Propionyl-5-*O*-(4′,6′-dideoxy-3′-*C*-acetyl-β-D-xylohexopyranosyl)-platenolide II	*S. hygroscopicus*	88
9-Dihydrojuvenimicin B	3-*O*-Acetyl-9-dihydrojuvenimicin B_1	*Streptomyces ambofaciens*	271
Forocidin I	Foricidin III	*S. ambofaciens*	241
Lankacidin C + formate	14-*O*-Formyllankacidin C	*Aspergillus niger, Aspergillus sojae,* and *Trametes sanguinea*	70
Lankacidin C + acetate	Lankacidin A	*A. niger, A. sojae, and T. sanguinea*	70
Lankacidin C + propionate	14-*O*-Propionyllankacidin C	*A. niger, A. sojae, and T. sanguinea*	70
Lankacidin C	8-*O*-Acetyllankacidin C	*Bacillus megaterium*	198
Lankacidin C + butyrate	8-*O*-Butyryllankacidin C	*B. megaterium*	198
Lankacidin C + isobutyrate	8-*O*-Isobutyryllankacidin C	*B. megaterium*	198
Lankacidin C + valerate	8-*O*-Valeryllankacidin C	*B. megaterium*	198
Lankacidin C + isovalerate	8-*O*-Isovaleryllankacidin C	*B. megaterium*	198
Lankacidinol + acetate	Lankacidinol A, 2′-*O*-acetyllankacidinol A	*Streptomyces rochei*	92
Lankacidinol + propionate	14-*O*-Propionyllankacidinol	*S. rochei*	92
Leucomycin A_1	Leucomycin A_3	Macrolide-3-acyltransferase	221
		Streptomyces kitasatoensis	135, 136, 238, 239
Leucomycin A_1	Leucomycin A_3	*S. thermotolerans*	222
Leucomycin U	Leucomycin A_3	*S. thermotolerans*	222
Leucomycin U	Leucomycin A_3	*S. kitasatoensis*	135
Leucomycin V	Leucomycin A_3	*S. kitasatoensis*	135
Leucomycin V	Leucomycin U	*S. kitasatoensis*	135
M-4365G_2	3-*O*-Acetyl M-4365G_2	*S. kitasatoensis*	271
Neospiramycin I	Neospiramycin III	*S. ambofaciens*	135, 241
Spiramycin I	Spiramycin II	*S. ambofaciens*	240

Table 3a (continued)
BIOTRANSFORMATIONS OF MACROLIDE ANTIBIOTICS

Substrate	Transformation product	Biological agent	Ref.
Spiramycin I	Spiramycin III	*S. ambofaciens*	241
Tylosin	3-*O*-Acetyltylosin	*S. thermotolerans*	224
	3-*O*-Propionyltylosin	*S. thermotolerans*	224
	4″-*O*-Butyryltylosin	*S. thermotolerans*	224
	4″-*O*-Isovaleryltylosin	*S. thermotolerans*	224
	3-*O*-Acetyl-4″-*O*-isovaleryltylosin	*S. thermotolerans*	224
Deacylation			
8-*O*-Acetyllankacidin A	8-*O*-Acetyllankacidin C	*S. rochei*	92
3″-*O*-Acetylleucomycin A_5	Leucomycin V	*Nocardia* sp.	340
8-*O*-Benzoyllankacidin A	8-*O*-Benzoyllankacidin C	*S. rochei*	92
8-*O*-Butyryllankacidin A	8-*O*-Butyryllankacidin C	*S. rochei*	92
Carbomycin A	4″-*O*-Deisovalerylcarbomycin A	Members of 8 genera	334
Carbomycin B	4″-*O*-Deisovalerylcarbomycin B	Members of 8 genera	334
8-*O*-Crotonyllankacidin A	8-*O*-Crotonyllankacidin C	*S. rochei*	92
8,14-Di-*O*-acetyllankacidin C	8-*O*-Acetyllankacidin C	*A. niger, A. sojae,* and *T. sanguinea*	70
8,14-Di-*O*-butyryllankacidin C	8-*O*-Butyrllankacidin C	*B. megaterium*	198
8,14-Di-*O*-propionyllankacidin C	8-*O*-Propionyllankacidin C	*A. niger, A. sojae,* and *T. sanguinea*	70
Lankacidin A	Lankacidinol	*A. niger, A. sojae,* and *T. sanguinea*	70
Lankacidin A	Lankacidin C	*S. rochei*	95
Lankacidinol A	Lankacidinol	*A. niger, A. sojae, T. Sanguinea*	70, 71
Leucomycin A_1	Leucomycin V	Members of 8 genera	334
Leucomycin A_1, A_3-A_9	Products A and B	*Mucor spinescens*	306
Leucomycin A_3	Leucomycin U	*B. megaterium*	194, 222
Leucomycin A_3	Leucomycin U	Members of 8 genera	334
Leucomycin A_3	Leucomycin A_1	*S. kitasatoensis*	136
Leucomycin A_3	Leucomycin A_1	*Bacillus subtilis*	222
Leucomycin A_4	Leucomycin A_5	*B. subtilis*	223
Leucomycin A_5	Leucomycin V	*Actinoplanes missouriensis*	308
Maridomycin III	4″-*O*-Depropionylmaridomycin III	*B. megaterium*	194
		Streptomyces pristinaespiralis	196
		Streptomyces lavendulae	295, 296, 356
Maridomycin V	4″-*O*-Depropionylmaridomycin III	*Streptomyces olivaceus*	195
8-*O*-Nicotinyllankacidin A	8-*O*-Nicotinoyllankacidin C	*S. rochei*	92
Niddamycin	4″-*O*-Deisovalerylniddamycin	Members of 8 genera	334
9-*O*-Propionyl-4″-*O*-depropionyl-maridomycin III	4″-*O*-Depropionylmaridomycin III	*S. pristinaespiralis*	196
9-*O*-Propionylmaridomycin III	9-*O*-propionyl-4″-*O*-depropionyl-maridomycin III	*S. olivaceus*	195
9-*O*-Propionylmaridomycin V	9-*O*-propionyl-4″-*O*-depropionyl-maridomycin III	*S. olivaceus*	195
8-*O*-Propionyllankacidin A	8-*O*-Acetyllankacidin C	*S. rochei*	92
9-*O*-Propionylmaridomycin III	9-*O*-Propiony-4″-*O*-depropionyl-maridomycin III	*B. megaterium*	194
		S. pristinaespiralis	196
		S. olivaceus	196
SF-837	SF-837 M_1	*M. spinescens*	202
SF-387 and acylated analogues	Deacylated compounds	Members of 11 genera	203
8-*O*-Valeryllankacidin A	8-*O*-Valeryllankacidin C	*S. rochei*	92
2′,8,14-Tri-*O*-acetyllankacidinol	8-*O*-Acetyllankacidinol, 2′,8-Di-*O*-acetyllankacidinol	*S. rochei*	92

Table 3a (continued)
BIOTRANSFORMATIONS OF MACROLIDE ANTIBIOTICS

Substrate	Transformation product	Biological agent	Ref.
Oxidation			
Carbomycin B	Carbomycin A	*S. hygroscopicus*	318
Lankacidinol A	Lankacidin A	*A. niger, A. sojae,* and *T. sanguinea*	70
Lankamycin	15-Deoxy-15-oxolankamycin	*Streptomyces erythreus*	84
Leucomycin A_3	3‴-Hydroxyleucomycin A_3	*S. olivaceus*	195
		Streptomyces piomogenes	222
Leucomycin A_3	Carbomycin B	*S. hygroscopicus*	318
Leucomycin A_3	Maridomycin II	*S. hygroscopicus*	318
M-4365G_2	Juvenimicin B_1	*S. ambofaciens*	271
Maridomycin I	3‴-Hydroxymaridomycin I	*S. olivaceus*	195
Maridomycin II	Carbomycin A	*S. hygroscopicus*	318
Midecamycin A_1	Midecamycin A_3	*Streptomyces mycarofaciens*	181
Narbomycin	Picromycin	*Streptomyces zaomyceticus*	167
Reduction			
Carbomycin A	Carbomycin AP1	*Streptomyces halstedii*	77
Carbomycin A	Maridomycin II	*S. hygroscopicus*	318
Carbomycin B	Leucomycin A_3	*S. hygroscopicus*	318
Leucomycin A_3	18-Dihydroleucomycin A_3	*Nocardia gardnei*	222
Leucomycin A_3	18-Dihydroleucomycin A_3	*Nocardia mexicana*	197
Maridomycin III	18-Dihydromaridomycin III	*Nocardia mexicana*	197
Maridomycin III	18-Dihydromaridomycin III	*Streptomyces* sp.	296
		S. lavendulae	295, 356
Midecamycin A_3	Midecamycin A_1	*S. mycarofaciens*	181
9-Propionylmaridomycin III	18-Dihydro-9-*O*-propionylmaridomycin III	*N. mexicana*	197
Tylosin	Relomycin	*S. ambofaciens*	64
		S. hygroscopicus	64
		Streptomyces griseospiralis	64
		Nocardia corrallina	64
		Streptomyces fradiae	17
Tylosin	9,20-Tetrahydrotylosin	*S. ambofaciens*	242
Glycosylation			
5-*O*-Dedesosaminyl-(8S)-8-fluoroerythromycin B	(8S)-8-Fluoroerythromycin B	*S. erythreus*	338
5-*O*-Dedesosaminyl-(8S)-8-fluoroerythromycin D	(8S)-8-Fluoroerythromycin D	*S. erythreus*	338
4′-*O*-Demycarosyl-20-deoxyrelomycin	20-Deoxyrelomycin	*S. fradiae*	246
Forocidin I	Neospiramycin III	*S. ambofaciens*	241
5-*O*-Mycaminosyltylonolide	Tylosin	*S. fradiae*	243, 244, 246
23-*O*-Mycinosyl-20-deoxyrelonolide	20-Deoxyrelomycin	*S. fradiae*	246
23-*O*-Mycinosylrelonolide	23-Mycinosyl-5-*O*-mycaminosylrelonolide	*Micromonospora rosaria*	80
Methylation			
Erythromycin C	Erythromycin A	*S. erythreus*	40
Erythromycin D	Erythromycin B	*S. aureus*	169
Macrocin	Tylosin	*S. fradiae*	17
Demethylation			
Lankamycin	3′-*O*-Demethyllankamycin	*S. erythreus*	84
Other			
Erythromycin A	Erythronolide, erythrolosamine, and 2 unknown products	*Pseudomonas* sp.	67
Maridomycin III	Products-A_1, A_2, A_3, A_4	*Streptomyces* sp.	296
Maridomycin III	Maridomycins-S_1 and S_2	*Streptomyces marcescens*	357

Table 3a (continued)
BIOTRANSFORMATIONS OF MACROLIDE ANTIBIOTICS

Substrate	Transformation product	Biological agent	Ref.
Other (continued)			
Erythromycin A	Erythromycin A-2′-phosphate	*Streptomyces coelicolor*	57
Directed biosynthesis			
Erythronolide A	Erythromycin A	*S. erythreus*	314
Erythronolide A oxime	3-*O*-Oleandrosyl-5-*O*-desosaminylerythronolide oxime	*Streptomyces antibioticus*	161
Erythronolide B	3-*O*-Oleandrosyl-5-*O*-desosaminyl-15-hydroxyerythromycin B	*S. antibioticus*	313
	3-*O*-Oleandrosyl-5-*O*-desosaminylerythronolide B	*S. antibioticus*	313
	3-*O*-Oleandrosyl-5-*O*-desosaminyl-(8S)-8-hydroxyerythronol ide B	*S. antibioticus*	313
	3-*O*-Oleandrosyl-5-*O*-desosaminyl-(8S)-8,19-epoxyerythronolide B	*S. antibioticus*	313
(8S)-8-Fluoroerythronolide A	(8S)-8-Fluoroerythromycin A, (8S)-8-Fluoroerythromycin C	*S. erythreus*	338
(8S)-8-Fluoroerythronolide A	3-*O*-Oleandrosyl-3-*O*-descladinosyl-(8S)-8-fluoroerythromycin A	*S. antibioticus*	339
(8S)-8-Fluoroerythronolide B	(8S)-8-Fluoroerythromycin B, (8S)-8-Fluoroerythromycin D	*S. erythreus*	338
(8S)-8-Fluoroerythronolide B	3-*O*-Oleandrosyl-3-*O*-descladinosyl-(8S)-8-fluroerythromycin B	*S. antibioticus*	339
23-Hydroxyprotylonolide	20-Deoxyrelomycin	*S. fradiae*	246
Narbonolide	Narbomycin, Picromycin	*Streptomyces narbonensis*	167
Narbonolide	5-*O*-Mycaminosylnarbonolide, 5-*O*-Mycaminosyl-9-dihydronarbonolide	*Streptomyces platensis*	168
Platenolide I	Turimycin	*S. hygroscopicus*	87
Platenolide II	Turimycin	*S. hygroscopicus*	87
Platenolide I	3-*O*-Propionyl-5-*O*-mycaminosylplatenolide I, 3-*O*-Propionyl-5-*O*-mycaminoslylplatenolide II, 4″-*O*-Desmycarosylplatenomycins, Platenomycins A_1, B_1, C_2	*S. platensis*	78
Platenolide III	3-*O*-Propionyl-5-*O*-mycaminosyplatenolide II, 4′-*O*-Desmycarosylplatenomycins, Platenomycins A_1, B_1, C_2	*S. platensis*	78
Protylonolide	M-4365G	*Streptomyces* sp.	244
Protylonolide	M-4365G	*Streptomyces* sp.	244
Protylonolide	Tylosin	*S. fradiae*	244, 245
Protylonolide	4′-*O*-Propionyl-5-*O*-mycarosylprotylonolide	*S. kitasatoensis*	271
Protylonolide	Chimeramycins A, B	*S. ambofaciens*	247
Tylactone	Tylosin, relomycin	*S. fradiae*	17

macrolides, which are the usual positions acylated in naturally occurring macrolides. However, in the case of tylosin, which is not acylated, microbial acylations have also been at positions 3 and 4′.[224] These acylations could be directed by adding various amino acids to the fermentations to introduce specific acyl groups. Lankacidin C, which is a 17-membered lactone antibiotic having no sugars attached, has been acylated at position 8 with various acyl groups by adding the corresponding esters to direct acylation.[198] Kitao et al.[136] have reported that a *Streptomyces kitasatoensis* macrolide acylase is induced by glucose and is repressed by butyrate. As is the case with all types of microbial conversions of macrolides, various species of *Streptomyces* have been the organisms most frequently utilized.

Table 3b
MULTIPLE BIOTRANSFORMATIONS OF MACROLIDE ANTIBIOTICS

Substrate	Transformation product	Type of transformation	Biological agent	Ref.
3″-Acetylleucomycin A_5	Leucomycin A_1	Acylation, deacylation	*Nocardia* sp.	340
5-*O*-Desosaminylmycinamicin IV	5-*O*-Dedesosaminyl-5-*O*-mycosaminyl-10,11-dihydromycinamicin IV	Glycosylation, reduction	*S. fradiae*	162
Deepoxycirramycin B_1	Tylosin	Glycosylation, O-methylation	*S. fradiae*	246
20-Deoxy-5-*O*-mycaminosylrelonide	Tylosin	Oxidation, glycosylation	*S. fradiae*	246
2′,8-Di-*O*-acetyllankacidin A	8-*O*-Acetyllankacidin A	Deacylation, oxidation	*S. rochei*	92
5-*O*-(4′,6′-Dideoxy-3′-*C*-acetyl-β-D-xylohexopyranosyl)platenolide II	3-*O*-acetyl-14-hydroxy-5-*O*-(4′,6′-dideoxy-3′-*C*-acetyl-β-D-xylohexopyranosyl)-platenolide II	Acylation, oxidation	*Streptomyces hygroscopicus*	87
	3-*O*-Propionyl-14-hydroxy-5-*O*-(4′,6′-dideoxy-3′-*C*-acetyl-β-3′-D-xylohexopyranosyl)-platenolide II	Acylation, oxidation	*S. hygroscopicus*	87
Forocidin I	Neospiramycin III, spiramycin III	Acylation, glycosylation	*S. ambofaciens*	241
Lankacidinol	Lankacidin A	Acylation, oxidation	*S. rochei*	92
Lankamycin	4″-*O*-Deacetyl-15-deoxy-15-oxolankamycin	Deacylation, oxidation	*S. erythreus*	84
Maridomycin III	4″-*O*-Depropionyl-18-dihydromaridomycin III	Deacylation, reduction	*Streptomyces* sp.	295, 296, 356
5-*O*-Mycaminosylprotylonolide	Tylosin	Glycosylation, *O*-methylation, oxidation	*S. fradiae*	246
Neospiramycin I	Spiramycin III	Acylation, glycosylation	*S. ambofaciens*	134
2′,8,14-Tri-*O*-acetyllankacidonol	8-*O*-Acetyllankacidin	Deacylation, oxidation	*S. rochei*	92

Deacylations (Tables 3a and 3b) have been reported even more frequently than acylations, with deacylations of lankacidins and leucomycins accounting for about two thirds of those reported. Harada et al.[92] and Fugono et al.[70] have selectively deacylated a series of lankacidins diacylated at positions 8 and 14. Deacylation occurs only at position 14, which is deacylation at the position at which acetyl occurs in lankacidin A. Deacylation of SF-837 by a deacylase from *Mucor spinescens* immobilized by attachment to glass beads has been reported.[202] Although deacylation with *Streptomyces* species has been carried out most frequently, about 20 other genera have been found to bring about this reaction.

Reported oxidations (Tables 3a and 3b) have not been numerous and most have been oxidations of C-9 hydroxyl to a carbonyl or a C-18 or C-20 hydroxymethyl to an aldehyde. However, a number of different types of oxidation have been found to occur, and many of these occur with great specificity. Suzuki et al.[318] have reported oxidation of 12,13-olefinic bonds to the corresponding epoxides with no attack on the neighboring double bond. Oxidation occurs at saturated carbon atoms to introduce a hydroxyl group or a carbonyl. Examples of this are oxidation of the isovaleryl groups in maridomycin I and leucomycin A_3 to introduce a hydroxyl at C-3″ and oxidation of narbomycin at C-12 to form picromycin.[194,167] Matsuhashi et al.[181] have found a requirement for NADP in the oxidation of the 9-hydroxyl group of midecamycin A to form carbonyl in a process catalyzed by a *Streptomyces mycarofaciens* enzyme. The oxidation of the 9-hydroxyl group of maridomycin II to give carbomycin A is pH dependent, with oxidation occurring at pH 8.5 and the reverse reaction occurring at pH 6.3.[318]

The reductions (Tables 3a and 3b) reported have almost all involved reduction of the C-9 carbonyl group,[318] or the aldehyde groups present at C-18 or C-20,[64,318] or both in 16-membered macrolides.[242] A recent report discusses enzymic reduction of the C-12 to C-13 epoxide in carbomycin A, with formation of carbomycin A Pl in which there is unsaturation at C-11 to C-12 and a hydroxyl at C-13.[77] This transformation required anaerobic conditions and was promoted by NADH, NADPH, and FMN. A total of 10 species of *Streptomyces* and species of 7 other genera also mediated the reaction, and it occurred with several other macrolides.

A few cases of glycosylation (Tables 3a and 3b) have been reported, with glycosylation occurring in all cases on a macrolide already containing a sugar.[241,244,246,338] A very few instances of O methylation and O demethylation (Tables 3a and 3b) are known and always involve mutants of macrolide-producing organisms. A number of multiple conversions involving macrolides are mentioned in the literature (Table 3b), but in all cases, these involved reactions already discussed.

Perhaps the most interesting microbial conversions, although not strictly antibiotic conversions, have been those carried out using aglycones of macrolide antibiotics (Tables 3a and 3b). Most of this work has involved studies of the biosynthesis of known antibiotics, but recently, modified and promising new antibiotics have been prepared. The earliest work in this field was reported by Maezawa et al.,[167,168] who converted narbonolide (the aglycone of narbomycin) to narbomycin, picromycin, and analogues of these compounds. Omura et al.[244,246] have used protylonolide as a precursor in fermentations containing cerulenin to prevent aglycone formation by strains of *Streptomyces fradiae* and *S. kitasatoensis* during the fermentation production of antibiotic M-4365G, tylosin, and various analogues of these antibiotics. Toscano et al.[338] used chemically synthesized (8S)-8-fluoroerythronolides A and B as substrates for a *Streptomyces erythreus* mutant which converted them to (8S)-8-fluoroerythromycins A, B, C, and D. The resulting products are quite active in vitro and a great deal more acid stable than naturally occurring erythromycins.

D. Biotransformations of Anthracycline Antibiotics and Analogues

Microbial and enzymic biotransformations of anthracycline antibiotics and their analogues have been achieved by direct means including oxidation, reduction, acylation, and alkylation, as well as by other indirect means of transformation. The latter category includes precursor-directed biosynthesis, interrupted biosynthesis, and genetic recombination. Specific examples of these direct and indirect transformations are cited in Tables 4a and 4b.

Reported biological oxidations of these compounds are of two general types. The first is oxidation by removal of hydrogen atoms. An example of this type of biotransformation is the conversion of aclacinomycin A to Y.[389] The other type of oxidative biotransformation is oxygenation. Specific examples of oxygenation include the conversions of alkavinone to ϵ-rhodomycinone,[392] daunomycin to adriamycin,[228] and auramycinone to 11-hydroxy-auramycinone.[100]

The reported biological reductions of these antibiotics are also of two types. The first of these is carbonyl reduction and the second is reductive cleavage of carbon-oxygen bonds. Reduction of the C-13 ketonic carbonyl of daunomycin and its analogues has been the most frequently reported of the transformations by carbonyl reduction (Table 4a). Another frequently reported microbial transformation is the reduction of the C-10 ketonic carbonyl of steffimycin-related compounds (Table 4a). Table 4a also cites examples of reductive cleavage of carbon-oxygen bonds usually at position 7 of anthracyclines and anthracyclinones.

Table 4a
DIRECT BIOTRANSFORMATIONS OF ANTHRACYCLINE ANTIBIOTICS AND ANALOGUES

Substrate	Transformation product	Biological agent	Ref.
Oxidation including oxygenation			
Aclacinomycin A	Aclacinomycin Y	*Streptomyces galilaeus*	389
Aklavinone	ϵ-Rhodomycinone	*Streptomyces coeruleorubidus*	24, 392
Auramycinone	11-Hydroxyauramycinone	*S. coeruleorubidus*	98
Daunomycin	Adriamycin	*Streptomyces peucetius var. caesius*	228
ϵ-Pyrromycinone	ϵ-Isorhodomycinone	*S. coeruleorubidus*	390
Carbonyl reduction			
Daunomycin	Daunomycinol (13-dihydrodaunomycin)	*Corynebacterium simplex*	206
Daunomycin	Daunomycinol	*Corynebacterium equi*	16
Daunomycin	Daunomycinol	*Mucor spinosus*	174
N-Acetyldaunomycin	*N*-Acetyldaunomycinol	*C. equi*	16
Daunomycinone	Daunomycinol aglycone (13-dihydrodaunomycinone)	*Streptomyces aureofaciens*	119
Daunomycinone	Daunomycinol aglycone	*S. coeruleorubidus* and *S. galilaeus*	24
7-Deoxyadriamycinone	7-Deoxyadriamycinol aglycone	*Streptomyces steffisburgensis*	174
7-Deoxydaunomycinone	7-Deoxydaunomycinol aglycone	*S. steffisburgensis*	173
ϵ-Isorhodomycinone	1-Hydroxy-13-dihydrodaunomycin	*S. coeruleorubidus*	390
Rubeomycin A	Rubeomycin B	*Rhodotorula glutinis*	216
Steffimycin	10-Dihydrosteffimycin	*Actinoplanes utahensis*	374
Steffimycin B	10-Dihydrosteffimycin B	*A. utahensis* and *Chaetomium* sp.	374
Steffimycinone	Steffimycinol (10-dihydrosteffimycinone	*S. peucetius* var. *caesius* and *Streptomyces nogalater*	373
Reductive carbon-oxygen bond cleavage			
Aclacinomycin B	Aclacinomycin Y	*S. galilaeus*	98
Adriamycin	7-Deoxyadriamycinone	*S. steffisburgensis*	174
Aklavinone	7-Deoxyaklavinone	*S. coeruleorubidus*	24
Cinerubin A	ζ-Pyrromycinone	*Aeromonas hydrophilia*	172
Daunomycin	7-Deoxydaunomycinone	*A. hydrophila*	172, 372
Daunomycin	7-Deoxydaunomycinone	*S. steffisburgensis*	173
Daunomycinol	7-Deoxydaunomycinol aglycone	*S. steffisburgensis*	174
Daunomycinol aglycone	7-Deoxydaunomycinol aglycone	*S. coeruleorubidus* and *S. galilaeus*	24
Nogalamycin	7-Deoxynogalarol	*A. hydrophila*	172, 372
Nogalamycin	7-Deoxynogalarol	*Streptomyces nogalater*	259
Steffimycin	7-Deoxysteffimycinone	*A. hydrophila, Escherichia coli* and *Citrobacter freundii*	172, 372
Steffimycin	7-Deoxysteffimycinone	*A. hydrophila*	165
Acylation			
Daunomycin	*N*-Acetyldaunomycin	*Bacillus cereus* var. *mycoides*	90
Daunomycinol	*N*-Acetyldaunomycinol	*B. cereus* var. *mycoides*	90
1-Hydroxy-13-dihydrodaunomycin	*N*-Formyl-1-hydroxy-13-dihydrodaunomycin	*S. coeruleorubidus*	390
Alkylation			
Carminomycin I	Daunomycin	*S. peucetius* var. *caesius*	228
Daunosaminylaklavinone	Aklavin	*S. coeruleorubidus*	226, 229
11-Hydroxyauramycinone	Feudomycin D	*S. coeruleorubidus*	98

Other transformations reported to occur among these compounds are acylation and alkylation (Table 4a). An example of acylation is the N-acetylation of daunomycin and daunomycinol by *Bacillus subtilis* var. *mycoides*.[90] The microbial alkylations which have been reported were performed primarily using streptomycetes and are biosynthetic in nature.

In addition to the direct transformations cited, other indirect anthracycline biotransfor-

Table 4b
INDIRECT BIOTRANSFORMATIONS OF ANTHRACYCLINE ANTIBIOTICS: INTERRUPTED BIOSYNTHESIS

Means of interruption	Antibiotic produced	Biological agent	Ref.
Interrupted biosynthesis			
Mutation: defective oxidoreductase	Aclacinomycin analogue MA-144-N1	*Streptomyces galilaeus* mutant	184, 393
Mutation: defective	Aclacinomycin analogues MA-144-U1, U2, U5-U9, and S1	*S. galilaeus* mutant	184, 393
Mutation: defective glycosidation	Aklavinone	*S. galilaeus* mutant	184, 393
Mutation: multiple defects	2-Hydroxyaklavinone and 2-hydroxy-7-deoxyaklavinone	*S. galilaeus* mutant	184, 336, 393
Mutation: oxygenation	Adriamycin	*Streptomyces peucetius* var. *caesius*	53
Mutation: multiple defects	Feudomycins A and B	*Streptomyces coeruleorubidus* mutant	227
Mutation: multiple defects	Feudomycinones C and D	*S. coeruleorubidus* mutant	227
Mutation: multiple defects	Maggiemycin and anhydromaggiemycin	Mutant streptomycete	249
Ethionine	Carminomycinone	*S. coeruleorubidus*	25
Precursor-directed biosynthesis			
Aklavinone	Glycosides of aklavinone	*S. coeruleorubidus*	24
Aklavinone	Aclacinomycin A	*S. galilaeus*	226
Aklavinone and ϵ-Rhodomycinone	Daunomycin	*S. coeruleorubidus*	391
9 Anthracyclinones	16 Anthracyclines	*S. galilaeus*	99
4-*O*-Methylaklavinone	4-*O*-Methylaclacinomycin A	*S. galilaeus*	226
10-Decarbomethoxyaklavinone	10-Decarbomethoxyaclacinomycin A	*S. galilaeus*	226
2-Hydroxyaklavinone	2-Hydroxyaclacinomycin A	*S. galilaeus*	393
Aklanoic acid	ϵ-Rhodomycinone	Streptomycete	361
Carminomycinone or 13-Dihydrocarminomycinone	Trisarubicinol	*S. galilaeus*	183, 226
Daunomycinone	Glycosides of daunomycinone	*S. coeruleorubidus*	24
Daunomycinol aglycone	Glycosides of daunomycinol Aglycone	*S. coeruleorubidus*	24
ϵ-Pyrromycinone	Glycosides of ϵ-pyrromycinone	*S. coeruleorubidus*	24
ϵ-Pyrromycinone	Cinerubins A and B	*S. galilaeus*	100
ϵ-Pyrromycinone	Galirubins A and B	*S. galilaeus*	359
ϵ-Pyrromycinone or ϵ-isorhodomycinone	1-Hydroxy-13-dihydrodaunomycin and *N*-formyl-1-hydroxy-13-dihydrodaunomycin	*S. coeruleorubidus*	390
ϵ-Isorhodomycinone	Glycosides of ϵ-isorhodomycinone	*S. coeruleorubidus*	24
ϵ-Isorhodomycinone and ϵ-pyrromycinone	Anthracycline glycoside	*S. galilaeus*	183, 226
β-Isorhodomycinone and α-rhodomycinone	12 Hybrid anthracyclines	*S. coeruleorubidus* and *S. galilaeus*	395
Rhodomycinones	Cinerulosyldeoxyfucosylrhodosaminylrhodomycinone	*S. galilaeus*	183, 226
ϵ-Rhodomycinone	Glycosides of rhodomycinone	*S. coeruleorubidus*	24

mations have been reported to occur through biosynthetic manipulation (Table 4b). One means involves interruption of antibiotic biosynthesis. Biosynthetic interruption can be achieved through genetic mutation or through chemically mediated inhibition. Specific examples of interrupted biosynthesis are cited in Table 4b. Another type of indirect transformation is termed precursor-directed biosynthesis. Because of tolerances in the substrate specificities

Table 5
BIOTRANSFORMATIONS OF LINCOSAMINIDE ANTIBIOTICS

Substrate	Transformation product	Biological agent	Ref.
Sulfoxidation			
Lincomycin	Lincomycin sulfoxide	*Streptomyces lincolnensis*	6
Clindamycin	Clindamycin sulfoxide	*Streptomyces armentosus*	7
Phosphorylation			
Lincomycin	Lincomycin-3-phosphate	*Streptomyces rochei*	8
Celesticetin	Celesticetin-3-phosphate	*S. rochei*	8
Clindamycin	Clindamycin-3-phosphate	*Streptomyces coelicolor*	38
4′-Depropyl-4′-ethyllincomycin	4′-Depropyl-4′-ethyllincomycin-3-phosphate	*S. rochei*	8
S-Demethyl-S-ethyllincomycin	S-Demethyl-S-ethyllincomycin-3-phosphate	*S. rochei*	8
1′-Demethyllincomycin	1′-Demethyllincomycin-3-phosphate	*S. rochei*	8
Nucleotidylylation			
Clindamycin	Clindamycin-3-cytidine-5′-phosphate	*S. coelicolor*	10, 14
Clindamycin	Clindamycin-3-adenosine-5′-phosphate	*S. coelicolor*	10, 14
Clindamycin	Clindamycin-3-uridine-5′-phosphate	*S. coelicolor*	10, 14
Clindamycin	Clindamycin-3-guanosine-5′-phosphate	*S. coelicolor*	10, 14
Demethylation			
Celesticetin	*N*-Demethylcelesticetin	*Streptomyces caelestis*	13
Celesticetin	*N*-Demethyl-7-*O*-demethyl-celesticetin	*S. caelestis*	13
Celesticetin	7-*O*-Demethylcelesticetin	*S. caelestis*	12
Clindamycin	*N*-Demethylclindamycin	*Streptomyces punipalus*	7
Lincomycin	1-Demethylthio-1-hydroxylincomycin	*S. lincolnenis*	6
Hydroxymethylation			
N-Demethylclindamycin	*N*-Demethyl-*N*-hydroxy-methylclindamycin	*S. lincolnensis*	11

of certain biosynthetic enzymes, it is often possible for antibiotic-producing microorganisms to incorporate precursor analogues into the structures of various antibiotics. Specific examples involving precursor-directed biosynthesis of anthracyclines and anthracyclinones are listed in Table 4b.

Genetic recombination is a novel approach in the discovery of anthracyclines through genetic manipulation. In one of these studies, *Streptomyces violaceus* subsp. *iremyceticus* was prepared through hybridizations between different streptomycetes blocked in anthracycline biosynthesis.[105] This hybrid organism produced a novel anthracycline termed iremycin. Although the application of genetic recombination to the biosynthesis of antibiotics is in an early stage of development, this technology appears to be extremely promising. The numbered structures of typical anthracycline antibiotics are presented in Figure 1.

E. Biotransformations of Lincomycin-Related Antibiotics

The lincomycin family of antibiotics is a very small one, and only six compounds have been reported as substrates for microbial conversions (Table 5). Only five types of conversions have been reported. These are oxidations of sulfur, phosphorylation, nucleotidylylation, demethylation, and hydroxymethylation. The organisms used were all streptomycetes. The numbered structure of lincomycin is shown in Figure 1.

Table 6
BIOTRANSFORMATIONS OF NUCLEOSIDE ANTIBIOTICS

Substrate	Transformation product	Biological agent	Ref.
Deamination			
Formycin	Formycin B	*Escherichia coli*	346
		Nocardia interforma	344
		7 Streptomycetes	344
Blasticidin S	Deaminohydroxyblasticidin S	*Aspergillus terreus*	384
Oxidation			
Formycin B	Oxoformycin	*N. interforma*	275
		Pseudomonas fluorescenes	275
		Streptomyces kasugaensis	275
Amination			
Formycin B	Formycin	*Streptomyces* sp.	214
		N. interforma	275
		Xanthomonas oryzae	275
Hydrolysis			
Cordicepin	Adenine and 3-deoxy-3-aminoribose	*Bacillus subtilis*	270
Toyocamycin	Sangivamycin	*Streptomyces rimosus*	315, 343
Isomerization			
Showdomycin	Isoshowdomycin	*Streptomyces* sp.	248
Phosphorylation			
Formycin B	5′-*O*-Formycin phosphate	*N. interforma*	275
Tubercidin	5′-*O*-Tubercidin phosphate	*Serratia marcescens*	304
Mutasynthesis			
Pyrimidine Bases	Nikkomycins J_T, J_H, Z_T, Z_H	*Streptomyces tendae*	50

F. Biotransformations of Nucleoside Antibiotics

The nucleoside antibiotics make up a large family in which there are two classes, the N nucleosides, exemplified by psicofuranine, and the C nucleosides, exemplified by formycin. Several different conversions such as amination, deamination, oxidation, etc. (Table 6), have been carried out using antibiotics of this type as substrates, but only a few such antibiotics have been used. A wide variety of streptomycetes and organisms of other genera have been reported to mediate nucleoside conversions. One case of mutasynthesis has been reported in which a *Streptomyces tendae* mutant unable to form pyrimidines was fermented in the presence of 5-substituted uracil analogues, forming four new nikkomycins.[50] Figure 1 shows the numbered structure of toyocamycin.

G. Biotransformations of Actinomycins and Peptide Antibiotics

Microbial conversion of this type of antibiotic has been achieved primarily through peptide bond hydrolysis and by precursor-directed or -controlled biosynthesis. The only other direct transformations reported involve acylation and deacylation. The reactions are summarized in Table 7.

Transformations involving peptide bond hydrolysis were performed using whole cells, cell-free extracts, and purified enzymes. The transformation products include amino acids, linear peptides, circular peptides, or various combinations of these products. In the case of the hydrolysis of bleomycin B_2, the reaction products were bleomycinic acid and agmatine. The former is useful in the preparation of semisynthetic bleomycins. The only hydrolysis of actinomycin reported was accomplished using fermentations of *Actinoplanes missouriensis* and yielded actinomycin monolactone and actinomycinic acid. An inducible lactonase which catalyzed the reaction was eventually isolated from this organism. Actinomycin monolactone has been demonstrated to occur as a biosynthetic product of actinomycin-producing organisms.[254]

Table 7
BIOTRANSFORMATIONS OF ACTINOMYCINS AND PEPTIDE ANTIBIOTICS

Substrate	Transformation product	Biological agent	Ref.
Acylation			
Virginiamycin (M factor)	*O*-Acetyl-M factor	*Staphylococcus aureus*	52
Deacylation			
Amphomycin	Deacylamphomycin	*Corynebacterium* sp.	368
Polymixin E	Deacylpolymixin E	*Pseudomonas* sp.	305
Polymixin S	Deacylpolymixin S	*Pseudomonas* sp.	305
Hydrolysis			
Actinomycin D	Actinomycin monolactone and actinomycinic acid	*Actinoplanes missouriensis*	101
Bleomycin B_2	Bleomycinic acid and agmatine	*Fusarium anguioides*	351
Colistin A or B	Cyclodepsipeptide and linear peptide	*Bacillus polymyxa* var. *colistinus*	107
Echinomycin, etamycin, staphylomycin, stendomycin, and vernamycin Bα	Complete linear peptides	*A. missouriensis*	101
Gramacidin S	Linear peptide	*Bacillus subtilis*	396
Polymyxin E	Amino acids and peptide	*B. polymyxa*	367
Polymyxin E	Cyclic and linear peptides	*Bacillus colistinus*	108
Polymyxin, circulin, tyrocidin	Constituent amino acids	*Pseudomonas* sp.	366
Precursor-directed biosynthesis			
Endogenous actinomycin precursor (EAP) + *cis*-4—methylproline	Actinomycin K_{1c}, actinomycin K_{2c}	*Streptomyces parvullus*	123
EAP + Trans-4-methylproline	Actinomycin L, actinomycin K_2t	*S. parvullus*	123
EAP + 4-Methylproline	3 novel actinomycins	*Streptomyces antibioticus*	387
EAP + DL-Pipecolic acid	Actinomycin Pip 2, actinomycin Pip 1α, actinomycin Pip 1β	*S. antibioticus*	68
EAP + Azetidine-1-carboxylic Acid	Azetomycins I, II, III	*S. antibioticus*	69
EAP + Sarcosine	Novel actinomycins	*Streptomyces* sp.	276
EAP + D-isoleucine	2 Novel *N*-methylisoleucine containing actinomycins	*Streptomyces chrysomallus*	122
EAP + norvaline	Actinomycin AuNv	Auranthin-producing organism	258
EAP + glycine	Actinomycin AuGl	Auranthin-producing organism	258
Endogenous bleomycin precursor (EBP) + amines	8 Novel bleomycins	*Streptomyces verticillius*	72
EBP + amines (especially diamines and triamines)	42 Bleomycins	*S. verticillius*	350
Endogenous polymyxin B precursor + L-isoleucine	Polymyxin B_1	*B. polymyxa*	376
Triostin A + methionine	Echinomycin (quinomycin A)	*Streptomyces echinatus*	41
Triostin C + methionine	Quinomycin C	*S. echinatus*	41
bis-Quinolinetriostin A	bis-Quinolinechinomycin	*S. echinatus*	41
bis-6-Bromoquinooxalinetriostin A	bis-6-Bromoquinoxalin-echinomycin	*S. echinatus*	41
Endogenous viridogrisein precursor (EVP) + L-proline	Neoviridogrisein II	*Streptomyces griseoviridus*	234
EVP + L-methionine	Neoviridogrisein VII	*S. griseoviridus*	235
EVP + L-α-amino-*n*-butyric acid	Neoviridogrisein I and II	*S. griseoviridus*	235

Biomodification of the actinomycins has been achieved primarily through precursor-directed or -controlled biosynthesis. The precursors utilized are related to the amino acid moieties of the actinomycins and include branched chain amino acids, norvaline, proline, proline analogs, pipecolic acid, sarcosine, and glycine, as well as azetidine-2-carboxylic acid. Precursor addition resulted in either enhanced formation of actinomycins normally

Table 8
BIOTRANSFORMATIONS OF GRISEOFULVIN AND ANALOGUES

Substrate	Transformation product	Biological agent	Ref.
Griseofulvin	5′-Hydroxygriseofulvin	*Streptomyces cinereocrocatus, Cephalosporium curtipes*	5, 26
	6′-Hydroxymethylgriseofulvin	*C. curtipes*	26
	Griseofulvic acid	*Botrytis allii,*	27
	6-Demethylgriseofulvin	*Cercospora melonis*	27
	4-Demethylgriseofulvin	*Microsporium canis*	27
Dehydrogriseofulvin	Griseofulvin	*S. cinereocrocatus*	5
Griseophenone A	Dehydrogriseofulvin	*Trametes gibbose, Fomes robustus, Trametes heteromorpha, Trametes sanguinea Coriolus fibula, Coriolus hirsutus, Gonodema lucidium, Gloseoporium lacticolor*	232

produced in trace amounts or synthesis of novel actinomycins. These conversions are summarized in Table 7.

Several studies also have been done utilizing precursor-directed biosynthesis for transformation of other peptide antibiotics. Addition of various amines to cultures of *Streptomyces verticillius* has resulted in the synthesis of numerous bleomycin analogues. In one such study, Umezawa reported the synthesis of 42 semisynthetic bleomycins.[350] Polymyxin B, triostin A and C, and viridogrisein have also been successfully transformed by addition of various amino acids to fermentations of the producing organism (Table 7). The structure of an actinomycin, numbered in the typical manner, is shown in Figure 2. The structure of a peptide antibiotic is also shown.

H. Biotransformations of Griseofulvin and Analogues

Modification of griseofulvin and its analogues by microbial means has been very limited. The fact that analogues can be prepared relatively easily by chemical synthesis in greater variety and the relatively low market potential for griseofulvin analogues probably account for this. Only a few direct transformations have been reported. Hydroxylation and demethylation of griseofulvin at different sites have been accomplished using a variety of microorganisms. The other transformations reported involve the oxidation or reduction of two analogues, dehydrogriseofulvin and griseophenone A (Table 8). The structure of griseofulvin is shown in Figure 2.

I. Biotransformations of Chloramphenicol

Chloramphenicol is a relatively simple antibiotic whose structure is shown in Figure 1. The most common microbial transformation of this compound has been acylation at the C-1 or C-3 position. This type of transformation has been carried out using streptomycetes, *Flavobacterium* sp., *Escherichia coli,* and *Staphylococcus aureus.* The latter two organisms appear to form only the 1- and 3-*O*-acetylchloramphenicol or the 1,3-diacetyl compound. Since chloramphenicol is inactivated by this reaction, these transformations are mainly mechanisms of resistance for these bacteria. In addition to the acetyl group, the streptomycetes are capable of adding a variety of other fatty acids to the C-3 position (Table 9). Relatively few other transformations have been found. Those reported include hydrolysis, oxidation, and reduction. These kinds of conversions were all performed by each of the following organisms: *B. subtilis, Bacillus mycoides, Proteus vulgaris, E. coli,* and *Streptomyces* strain 135 (Table 9).

Table 9
BIOTRANSFORMATIONS OF CHORAMPHENICOL

Transformation product	Biological agent	Ref.
Acylation		
1-*O*-Acetylchloramphenicol	*Streptomyces coelicolor*	9
	Streptomyces griseus	60
	Flavobacterium sp.	316
Monoacetylchloramphenicol	*Staphylococcus aureus*	209
	S. coelicolor	209
3-*O*-Acetylchloramphenicol	*S. coelicolor*	60
	S. griseus	60
	Escherichia coli	209
	Streptomyces fradiae	209
	S. coelicolor	209
	E. coli	293
	S. aureus	294
	E. coli	294
	Flavobacterium sp.	316
	E. coli	317
3-*O*-Propionylchloramphenicol	*S. coelicolor*	9
	S. griseus	60
3-*O*-Butyrylchloramphenicol	*S. griseus*	60
3-*O*-Isobutyrylchloramphenicol	*S. griseus*	60
	S. coelicolor	9
3-*O*-Valerylchloramphenicol	*S. griseus*	60
3-*O*-Isovalerylchloramphenicol	*S. griseus*	60
1,3-Di-*O*-acetylchloramphenicol	*E. coli*	293
	S. aureus	294
	E. coli	294
	E. coli	317
Hydrolysis		
1-(*p*-Nitrophenyl)-2-aminopropan-1,3-diol	*Streptomyces* sp., *Bacillus subtilis*, *Bacillus mycoides*, *Proteus vulgaris*, and *E. coli*	170, 309, 310
1-(*p*-Nitrophenyl)-2-acetamidopropan-1,3-diol	*Streptomyces* sp.	170
Reduction		
1-(*p*-Aminophenyl)-2-dichloroacetamido-propan-1,3-diol	*Pseudomonas aeruginosa*	59
3-(*p*-aminophenyl)-serine	*B. subtilis*	312
	B. mycoides	312
	P. vulgaris	312
	E. coli	312
p-Aminobenzaldehyde	*B. mycoides*	312
p-Aminobenzoic acid	*P. vulgaris*	312
	E. coli	312
Other transformations		
α-Amino-β-hydroxy-*p*-nitropropiophenone	*E. coli*	312
α-Dichloroacetamido-β-hydroxy-*p*-nitropropiophenone	*B. subtilis*	312
	B. mycoides	312
	P. vulgaris	312
N-Dichloroacetyl-β-(4-Nitrophenyl)-serine	*Flavobacterium* sp.	316
p-Nitrophenylserine	*E. coli*, *Flavobacterium* sp.	312
p-Nitrophenylserinol	*Flavobacterium* sp.	316
p-Nitrobenzoic Acid	*Streptomyces* sp.	170
	B. mycoides	312
	E. coli	312

Table 9 (continued)
BIOTRANSFORMATIONS OF CHORAMPHENICOL

Transformation product	Biological agent	Ref.
p-Nitrobenzaldehyde	*B. subtilis*	312
	B. mycoides	312
	P. vulgaris	312
	E. coli	312
p-Nitrobenzylalcohol ethanolamine	*Streptomyces* sp.	170
	B. subtilis	312
	B. mycoides	312
	P. vulgaris	312
	E. coli	312

J. Biotransformations of Ansamycins

Microbial transformations of ansamycins almost exclusively have involved *Nocardia mediterranei.* As this organism is a producer of the rifamycins, many of these citations have involved indirect transformations of these antibiotics. However, several instances of direct transformation by this and other organisms have been reported.

Among the reported indirect means of transformation is genetic recombination. Traxler and co-workers used a strain of *N. mediterranei* derived from genetic recombination for the synthesis of novel ansamycins.[341] In similar studies, Schupp et al. produced new rifamycins using recombinant strain R-21 which was derived from a cross between *N. mediterranei* mutants.[278]

Another means of indirect transformation which has been employed in synthesis of ansamycins is interrupted biosynthesis. This technology involves the use of either mutants or chemical inhibitors. Mutants of *N. mediterranei* blocked in methylation have been important in the synthesis of 27-demethylrifamycin SV.[153] Chemical preparation of this compound from rifamycin SV is not practical. Therefore, this mutant technology provided an important structural modification allowing the synthesis of an entire series of semisynthetic rifamycins. Mutants of rifamycin B-producing *N. mediterranei* blocked in acylation also have been isolated.[152] As a result of this genetic block, these organisms produced rifamycin SV directly as a secondary metabolite. In another instance of interrupted biosynthesis, barbital addition to fermentations of *N. mediterranei* functioned in promoting almost exclusive synthesis of rifamycin B.[155] In the absence of barbital, other rifamycins including rifamycins A, C, D, and E were produced. Several citations involving precursor-directed biosyntheses of these compounds are listed in Table 10.

In addition to these indirect transformations, other instances of direct transformation of ansamycins have been reported. These mostly have involved the rifamycins. In these studies, *N. mediterranei* usually has been reported as the transforming agent. However, the abilities of various other actinomycetes, *E. coli,* and a species of *Monocillium* also have been reported to cause these transformations. These direct transformations have involved oxidation, reduction, hydrolysis, and acylation, as well as demethylation, and are listed in Table 10. The structure of a typical ansamycin numbered according to the usual convention is presented in Figure 1.

K. Biotransformations of Miscellaneous Antibiotics

The final group of antibiotics covered represents no unified structural family and incorporates a great variety of transformations. Microbial biotransformations of this group have been performed through acylation, deacylation, amination, deamination, demethylation, glucosylation, hydrolysis, hydroxylation, isomerization, oxidation, and reduction. In addi-

Table 10
BIOTRANSFORMATIONS OF ANSAMYCINS

Substrate	Transformation product	Biological agent	Ref.
Oxidation			
Ansamitocin	15-Hydroxyansamitocin P3	6 Streptomycetes,	199
		several actinomycetes	109
Rifamycin B	Rifamycin Y	*Nocardia mediterranei*	150
Rifamycin B	Rifamycin O	*Monocillium* sp.	287
Reduction			
Rifamycin S	Rifamycin SV	*N. mediterranei* and *E. coli*	82
Rifamycin S	Rifamycin L	*N. mediterranei*	151
Rifamycin S	Rifamycin G	*N. mediterranei*	154
Rifamycin O	Rifamycin B	*N. mediterranei*	82
Hydrolysis			
Rifamycin O	Rifamycin S	*N. mediterranei*	82
Rifamycin L	Rifamycin SV	*N. mediterranei*	82
Acylation			
Rifamycin S	Rifamycin L	*N. mediterranei*	151
Rifamycin SV	Rifamycin L	*N. mediterranei*	82
Demethylation			
Ansamitocin P3	*N*-Demethylansamitocin P3	3 Streptomycetes	199
	20-*O*-Demethylansamitocin P3	7 Streptomycetes	199
Precursor-directed biosynthesis			
Rifamycin W	Rifamycins S and B	*N. mediterranei*	371
Rifamycin S	Thiazorifamycins	*N. mediterranei*	43
Rifamycin S	Rifamycins B and G	*N. mediterranei*	154
Rifamycin S	Rifamycins B and L	*N. mediterranei*	82, 154
Isoleucine, propionate, propionaldehyde, or *n*-propyl alcohol	Ansamitosin P-2	*Actinosynnema pretiosum*	93
Valine, isobutyrate, isobutyraldehyde, or isobutyl alcohol	Ansamitosin P-3	*A. pretiosum*	93
Leucine, isovalerate, isovaleraldehyde, or isoamyl alcohol	Ansamitosin P-4	*A. pretiosum*	93

tion, many general oxygenation type conversions have been reported primarily using fusidic acid and mycophenolic acid. A few examples also exist of indirect transformations involving precursor-directed biosynthesis, such as with the indolmycins and the paulomycins. All of these conversions are summarized in Table 11.

Table 11
BIOTRANSFORMATIONS OF MISCELLANEOUS ANTIBIOTICS

Substrate	Transformation product	Biological agent	Ref.
Acylation and deacylation			
Cycloheximide	Cycloheximide acetate	*Cunninghamella blakesleeana*	103
Chromomycin A_2	Deisobutyrylchromomycin A_2	*Whetzelinia sclerotiorum*	277
Olivomycin A	Deisobutyrylolivomycin A	*W. sclerotiorum*	277
Amination and deamination			
Antibiotic A23187	Demethylamino-A23187(cezomycin)	*Streptomyces chartreusis*	49
Ketomycin	3-Cyclohexeneglycine	*Bacillus subtilis*	129
Demethylation			
A23187 Methyl Ester	*N*-Demethyl A23187 methyl ester	*S. chartreusis*	2
Glucosylation			
Mycophenolic Acid	7-O-(β-D-glucopyranosyl)mycophenolic acid	*Streptomyces candidus* and *Streptomyces aureofaciens*	3
Hydrolysis			
Antimycin A	Blastamycic acid	*Aspergillus ochraceus*	307
Mikamycin B	Mikamycin B acid	*Streptomyces mikakaensis*	131, 132
Antibiotic ML-236B	ML-236A	*Penicillium citrinum*, 8 Other fungi	291
Novobiocin	Novenamine,3-(3-methylbuten-2-yl)-4-hydroxybenzoic acid	Nonmotile bacterium NRRL B3652	280
Hydroxylation			
Antibiotic A23187 Methyl Ester	16-Hydroxy-A23187 methyl ester	*S. chartreusis*	2
Chlorobiocin	11-Hydroxychlorobiocin	*Sebekia benihana*	285
Compactin (ML-236B)	3α-Hydroxycompactin	*Syncephalastrum nigricans* (3 types), *Syncephalastrum racemosum* (2 types)	289
Compactin	3β-Hydroxycompactin	53 Fungi, 8 actinomycetes	290
Compactin	6α-Hydroxyisocompactin	*Absidia coerulea*	291
Compactin	Antibiotic M_3	*Nocardia autotrophica*	292
Monacolin K	3-Hydroxymonacolin K	*Mucor hiemalis*	288
Monacolin K	6α-Hydroxyisomonacolin K	*M. hiemalis*	288
Mycophenolic acid	4-Hydroxymycophenolic acid	12 Fungi	111
Novobiocin	11-Hydroxynovobiocin	*S. benihana*	285
Novobiocin	Allylic methylhydroxylated novobiocins	*S. benihana*	283
Novobiocic Acid	Hydroxynovobiocic acid	*S. benihana*	285
Isomerization			
Novobiocin	Isomerized novobiocin	*Actinomyces sphaeroides*	18
Oxidation and Reduction			
Ascochitine	Dihydroascochitine	*Fusarium lycopersici*	230
Dihydroabikoviromycin	Abikoviromycin	*Streptomyces olivaceus*	342
Grisorixin	Grisorixin oxidized product-G_1	*Streptomyces rimosus*	44
Nanomycin D	Nanomycin A	*Streptomyces rosa*	329
Saframycin A	25-Dihydrosaframycin A	*Rhodococcus amidophilus*	321
Saframycin A	25-Dihydrosaframycin B	*R. amidophilus*	321
Oxygenation			
Fusidic Acid	3-Oxofusidic acid	*Corynebacterium simplex*	58
Fusidic Acid	6-Oxofusidic acid	*Acrocylindrium oryzae*	360
Fusidic Acid	7-Oxofusidic acid	*A. oryzae*	360
Fusidic Acid	11-Oxofusidic acid	*Fusidium coccineum*	360
Mycophenolic acid	5-Oxomycophenolic acid	12 Fungi	111
Mycophenolic acid	Mycophenolic acid lactol	*Streptomyces* sp., 5 fungi	111
Mycophenolic acid	Mycophenolic acid dihydrobenzofuran	5 Fungi and *Trichoderma koningi*	111
Mycophenolic acid	Mycophenolic acid hydroxylactone	3 Fungi	111

Table 11 (continued)
BIOTRANSFORMATIONS OF MISCELLANEOUS ANTIBIOTICS

Substrate	Transformation product	Biological agent	Ref.
Oxygenation (continued)			
Mycophenolic acid	Mycophenolic acid benzyl alcohol	*Streptomyces* sp., 5 fungi	111
Nanomycin A	Nanomycin E	*S. rosa*	137
cis-Prophenylphosphonate	Fosfomycin	20 Fungi	370
12α-Deoxytetracycline	Tetracycline	*Curvularia lunata, Curvularia pallescens, Botrytis cinerea*	97
Mixed transformations			
Antibiotic A23187	16-Hydroxy-*N*-demethyl A23187 methyl ester	*S. chartreusis*	2
4-Epicetocycline	2-Decarboxamido-2-acetyl-4-dedimethyltetracycline	*Streptomyces rimosus*	335
Novobiocin	11-Hydroxydihydronovobiocin	*S. benihana*	285
Precursor-directed biosynthesis			
5-Hydroxyindole	5-Hydroxyindolmycin	*Streptomyces griseus*	369
5-Methoxyindole	5-Methoxyindolmycin	*S. griseus*	369
Endogenous paulomycin precursor (EPP) + DL-valine	Paulomycin B	*Streptomyces paulus*	177
EPP + isobutyric acid	Paulomycin B	*S. paulus*	177
EPP + DL-isoleucine	Paulomycin A	*S. paulus*	177
EPP + 2-methylbutyric acid	Paulomycin A	*S. paulus*	177

REFERENCES

1. **Abbott, B. J. and Fukuda, D. S.,** *Antimicrob. Agents Chemother.*, 8, 282, 1975.
2. **Abbott, B. J.,** *Antimicrob. Agents Chemother.*, 16, 808, 1979.
3. **Abbott, B. J., Horton, D. R., and Whitney, J. G.,** *J. Antibiot.*, 33, 506, 1980.
4. **Abraham, E. P. and Chain, E.,** *Nature (London)*, 146, 837, 1940.
5. **Andres, W. W., McGahren, W. J., and Kunstmann, M. P.,** *Tetrahedron Lett.*, 3777, 1969.
6. **Argoudelis, A. D. and Mason, D. J.,** *J. Antibiot.*, 22, 289, 1969.
7. **Argoudelis, A. D., Coats, J. H., Mason, D. J., and Sebek, O. K.,** *J. Antibiot.*, 22, 309, 1969.
8. **Argoudelis, A. D. and Coats, J. H.,** *J. Antibiot.*, 22, 341, 1969.
9. **Argoudelis, A. D. and Coats, J. H.,** *J. Antibiot.*, 24, 206, 1971.
10. **Argoudelis, A. D. and Coats, J. H.,** *J. Am. Chem. Soc.*, 93, 534, 1971.
11. **Argoudelis, A. D., Coats, J. H., and Magerlein, B. J.,** *J. Antibiot.*, 25, 191, 1972.
12. **Argoudelis, A. D., Coats, J. H., Lemaux, P. G., and Sebek, O. K.,** *J. Antibiot.*, 25, 445, 1972.
13. **Argoudelis, A. D., Coats, J. H., Lemaux, P. G., and Sebek, O. K.,** *J. Antibiot.*, 26, 7, 1973.
14. **Argoudelis, A. D., Coats, J. H., and Mizsak, S. A.,** *J. Antibiot.*, 30, 474, 1977.
15. **Asano, N., Takeuchi, M., Ninomiya, K., Kameda, Y., and Matsui, K.,** *J. Antibiot.*, 37, 859, 1984.
16. **Aszalos, A. A., Bachur, N. R., Hamilton, B. K., Langlykke, A., Roller, P. P., Sheikh, M. Y., Sutphin, M. S., Thomas, M. C., Wareheim, D. A., and Wright, L. H.,** *J. Antibiot.*, 30, 50, 1977.
17. **Baltz, R. H. and Seno, E. T.,** *Antimicrob. Agents Chemother.*, 20, 214, 1981.
18. **Batyrova, A. S. and Egorov, N. S.,** *Antibiotiki*, 19, 1089, 1974.
19. **Benveniste, R., Yamada, F., and Davies, J.,** *Infec. Immunity*, 1, 109, 1970.
20. **Benveniste, R. and Davies, J.,** *Biochemistry*, 10, 1787, 1971.
21. **Benveniste, R., Yamada, F., and Davies, J.,** *FEBS Lett.*, 14, 293, 1971.
22. **Benveniste, R. and Davies, J.,** *Proc. Nat. Acad. Sci. U.S.A.*, 70, 2276, 1973.
23. **Biddlecome, S., Haas, M., Davies, J., Miller, G. H., Rane, D. F., and Daniels, P. J. L.,** *Antimicrob. Agents Chemother.*, 9, 951, 1976.
24. **Blumauerova, M., Kralovocova, E., Mateju, J., Jizba, J., and Vanek, Z.,** *Folia Microbiol.*, 18, 117, 1979.
25. **Blumauerova, M., Jizba, J., Stajner, K., and Vanek, Z.,** *Biotech. Lett.*, p. 471, 1979.

26. **Bod, P., Szarka, E., Gyimesi, J., Horvath, G., Vajna-Mehesfalvi, Z., and Horvath, L.,** *J. Antibiot.*, 26, 101, 1973.
27. **Boothroyd, B., Napier, E. J., and Somerfield, G. A.,** *Biochem. J.*, 80, 34, 1961.
28. **Brandl, E.,** *Z. Physiol. Chem.*, 342, 86, 1965.
29. **Brzezinska, M., Benveniste, R., Davies, J., Daniels, P. J. L., and Weinstein, J.,** *Biochemistry,* 11, 761, 1972.
30. **Brzezinska, M. and Davies, J.,** *Antimicrob. Agents Chemother.*, 3, 266, 1973.
31. **Cappelletti, L. M. and Spagnoli, R.,** *J. Antibiot.*, 36, 328, 1983.
32. **Chevereau, M., Daniels, P. J. L., Davies, J., and LeGoffic, F.,** *Biochemistry,* 13, 598, 1974.
33. **Claridge, C. A., Gourevitch, A. and Lein, J.,** *Nature (London),* 187, 237, 1960.
34. **Claridge, C. A., Luttinger, J. R., and Lein, J.,** *Proc. Soc. Exp. Biol. Med.,* 113, 1008, 1963.
35. **Claridge, C. A., Bush, J. A., Defuria, M. D., and Price, K. E.,** *Dev. Ind. Microbiol.*, 15, 101, 1974.
36. **Cleophax, J., Gero, S., Leboul, J., Akhtar, M., Barnett, J., and Pearce, C. J.,** *J. Am. Chem. Soc.*, 98, 7110, 1976.
37. **Cleophax, J., Roland, A., Colas, C., Castellanos, L., Gero, S., Sepulchre, A., and Quichlet, B.,** in *Aminocyclitol Antibiotics,* Rinehart, K. L., Jr. and Suami, T., Eds., ACS Symposium Series 125, American Chemical Society, Washington, D.C., 1980, 393.
38. **Coats, J. H. and Argoudelis, A. D.,** *J. Bacteriol.*, 108, 459, 1971.
39. **Cole, M.,** *Biochem. J.*, 115, 747, 1969.
40. **Corcoran, J. W.,** in *Methods in Enzymology,* Hash, J. H., Ed., Academic Press, New York, 1975, 487.
41. **Cornish, A., Waring, M. J., and Nolan, R. D.,** *J. Antibiot.*, 36, 1664, 1983.
42. **Courvalin, P. and Davies, J.,** *Antimicrob. Agents Chemother.*, 11, 619, 1977.
43. **Cricchio, R., Antonini, P., and Sartori, G.,** *J. Antibiot.*, 33, 842, 1980.
44. **Cuer, A., Dauphin, G., and Beloeil, J. C.,** *J. Antibiot.*, 36, 20, 1983.
45. **Daniels, P. and Rane, D.,** in *Microbiology 1979,* Schlessinger, D., Ed., American Society for Microbiology, Washington, D.C., 1979, 314.
46. **Daniels, P., Rane, D., McCombie, S., Testa, R., Wright, J., and Nagabushan, T.,** in *Aminocyclitol Antibiotics,* Rinehart, K. L., Jr. and Suami, T., Eds., ACS Symposium Series 125, American Chemical Society, Washington, D.C., 1980, 371.
47. **Daum, S. J., Rosi, D., and Goss, W. A.,** *J. Am. Chem. Soc.*, 99, 283, 1977.
48. **Daum, S. J., Rosi, D., and Goss, W. A.,** *J. Antibiot.*, 30, 98, 1977.
49. **David, L. and Kergomard, A.,** *J. Antibiot.*, 35, 1409, 1982.
50. **Delzer, J., Fiedler, H. P., Muller, H., Zahner, H., Rothmann, R., Ernst, K., and Konig, W. A.,** *J. Antibiot.*, 37, 80, 1984.
51. **Demain, A. L., Walton, R. B., Newkirk, J. F., and Miller, I. M.,** *Nature (London),* 199, 909, 1963.
52. **DeMeester, C. and Rondelet, J.,** *J. Antibiot.*, 29, 1297, 1976.
53. **DiMarco, A., Gaetani, M., and Scarpinato, B.,** *Cancer Chemother. Rep.*, 53, 33, 1969.
54. **Doi, O., Ogura, M., Tanaka, N., and Umezawa, H.,** *Appl. Microbiol.*, 16, 1276, 1968.
55. **Doi, O., Miyamoto, M., Tanaka, N., and Umezawa, H.,** *Appl. Microbiol.*, 16, 1282, 1968.
56. **Doi, O., Kondo, S., Tanaka, N., and Umezawa, H.,** *J. Antibiot.*, 22, 273, 1969.
57. **Dolak, L. A., Cialdella, J. I., Coats, J. H., Greenfield, J. C., and Marshall, V. P.,** Abstracts of the 24th Intersci. Conf. Antimicrob. Agents Chemother., Abstr. 384, Washington, D. C., October 8th to 10th, 1984.
58. **Dvonch, W., Greenspan, G., and Alburn, H. E.,** *Experientia,* 22, 517, 1966.
59. **Egami, F., Ebota, M., and Sato, R.,** *Nature (London),* 167, 118, 1951.
60. **El-Kersh, T. A. and Plourde, J. R.,** *J. Antibiot.*, 29, 292, 1976.
61. **Endo, T. and Perlman, D.,** *J. Antibiot.*, 25, 681, 1972.
62. **Endo, T. and Perlman, D.,** *J. Antibiot.*, 25, 751, 1972.
63. **Fang, J., Pearce, C. J., and Rinehart, K. L., Jr.,** *J. Antibiot.*, 37, 77, 1984.
64. **Feldman, L. J., Dill, J. K., Holmlund, C. E., Whaley, H. A., Patterson, E. L., and Bohonos, N.,** *Antimicrob. Agents Chemother.*, 54, 1963.
65. **Fildes, R. A., Potts, J. R., and Farthing, J. E.,** U.S. Patent 3,801,458, 1974.
66. **Fleming, I. D., Turner, M. K., and Napier, E. J.,** British Patent 1,473,100, 1977.
67. **Flickinger, M. C. and Perlman, D.,** *J. Antibiot.*, 28, 307, 1975.
68. **Formica, J. V. and Katz, E.,** *J. Biol. Chem.*, 248, 2066, 1973.
69. **Formica, J. V. and Apple, M. A.,** *Antimicrob. Agents Chemother.*, 9, 214, 1976.
70. **Fugono, T., Higashide, E., Suzuki, T., Yamamoto, H., Harada, S., and Kishi, T.,** *Experientia,* 26, 26, 1970.
71. **Fugono, T., Harada, S., Higashide, E., and Kishi, T.,** *J. Antibiot.*, 24, 23, 1971.
72. **Fujii, A., Takita, T., Shimada, N., and Umezawa, H.,** *J. Antibiot.*, 27, 73, 1974.
73. **Fujii, T., Matsumoto, K., and Watanabe, T.,** *Proc. Biochem.*, 11, 21, 1976.
74. **Fujisawa, Y. and Kanzaki, T.,** *Agric. Biol. Chem.*, 39, 2043, 1975.

75. **Fujiwara, T., Takehashi, Y., and Matsumoto, K.,** *J. Antibiot.*, 33, 836, 1980.
76. **Fujiwara, T. and Kondo, E.,** *J. Antibiot.*, 34, 13, 1981.
77. **Fukagawa, Y., Mutoh, Y., Ishikura, T., and Lein, J.,** *J. Antibiot.*, 37, 118, 1984.
78. **Furumai, T., Seki, Y., Takeda, K., Kinumabi, A., and Suzuki, M.,** *J. Antibiot.*, 26, 708, 1973.
79. **Furumai, T., Takeda, K., Kinumaki, A., Ito, Y., and Okuda, T.,** *J. Antibiot.*, 32, 891, 1979.
80. **Ganguly, A. K., Lee, B. K., Liu, Y. T., Lotvin, J., Sorre, O., and Vaughan, R.,** *J. Chem. Soc. Chem. Commun.*, 855, 1982.
81. **Gatenbeck, S. and Brunsberg, U.,** *Acta Chem. Scand.*, 22, 1059, 1968.
82. **Ghisalba, O., Roos, R., Schupp, T., and Nüesch, J.,** *J. Antibiot.*, 35, 74, 1982.
83. **Goi, H., Niwa, T., Nojiri, C., Miyado, S., Seki, M., and Yamada, Y.,** Japan Patent 78,094,093, 1978.
84. **Goldstein, A. W., Egan, R. S., Mueller, S. L., Martin, J. R., and Keller-Schierlein, W.,** *J. Antibiot.*, 31, 63, 1978.
85. **Goulden, S. A. and Smith, A.,** British Patent 1,474,519, 1977.
86. **Goo, Y. M.,** *Soul Taehakkyo Yakhak Nonmunjip*, 7, 47, 1982.
87. **Grafe, U., Fleck, W. F., Schade, W., Reinhardt, G., Tresselt, D., and Thrum, H.,** *J. Antibiot.*, 33, 663, 1980.
88. **Grafe, U., Reinhardt, G., Schade, W., Muhlig, P., and Thrum, H.,** *J. Antibiot.*, 33, 1083, 1980.
89. **Haas, M., Biddlecombe, S., Davies, J., Luce, C. E., and Daniels, P. J. L.,** *Antimicrob. Agents Chemother.*, 9, 945, 1976.
90. **Hamilton, B. K., Sutphin, M. S., Thomas, M. C., Wareheim, D. A., and Aszalos, A. A.,** *J. Antibiot.*, 30, 425, 1977.
91. **Hamilton-Miller, J. M. T.,** *Bacteriol. Rev.*, 30, 761, 1966.
92. **Harada, S., Yamazaki, F., Hatano, K., Tsuchiya, K., and Kishi, T.,** *J. Antibiot.*, 26, 647, 1973.
93. **Hatano, K., Higashide, E., Akiyama, S., and Yoneda, M.,** *Agric. Biol. Chem.*, 48, 1721, 1984.
94. **Hennessey, T. D. and Richmond, M. H.,** *Biochem. J.*, 109, 469, 1968.
95. **Higashide, E., Fugono, T., Hatanoe, K., and Shibata, M.,** *J. Antibiot.*, 24, 1, 1971.
96. **Hiraga, K.,** Japan Kokai Patent 76:101,694, 1976 from *Chem. Abstr.*, 85, 3876, 1976.
97. **Holmlund, C. E., Andres, W. W., and Shay, A. J.,** *J. Am. Chem. Soc.*, 81, 4750, 1959.
98. **Hoshino, T. and Fujiwara, A.,** *J. Antibiot.*, 36, 1463, 1983.
99. **Hoshino, T., Setoguchi, Y., and Fujiwara, A.,** *J. Antibiot.*, 37, 1469, 1984.
100. **Hoshino, T. and Fujiwara, A.,** *J. Antibiot.*, 37, 1473, 1984.
101. **Hou, C. T., Perlman, D., and Schallock, M. R.,** *J. Antibiot.*, 23, 35, 1970.
102. **Hou, C. T. and Perlman, D.,** *J. Biol. Chem.*, 245, 1289, 1970.
103. **Howe, R. and Moore, R. H.,** *Experientia*, 24, 904, 1968.
104. **Huang, H. T., Seto, T. A., and Shull, G. M.,** Appl. Microbiol., 11, 1, 1963.
105. **Ihn, W., Schlegel, B., Fleck, W. F., and Sedmera, P.,** *J. Antibiot.*, 33, 1457, 1980.
106. **Imanaka, H., Miyoshi, T., Konomi, T., Kubochi, Y., Hattori, S., and Kawakita, T.,** U.S. Patent 4,414,328, 1983.
107. **Ito-Kagawa, M. and Kayama, Y.,** *J. Antibiot.*, 33, 1551, 1980.
108. **Ito, M., Aida, T., and Kayama, Y.,** *Agric. Biol. Chem.*, 30, 1112, 1980.
109. **Izawa, M., Wada, Y., Kasahara, F., Asai, M., and Kishi, T.,** *J. Antibiot.*, 34, 1591, 1981.
110. **Johnson, K., Dusart, J., Campbell, J. N., and Ghuysen, J-M.,** *Antimicrob. Agents Chemother.*, 3, 289, 1973.
111. **Jones, D. F., Moore, R. H., and Crawley, G. C.,** *J. Chem. Soc. (C.)*, 1970, 1725, 1970.
112. **Kabins, S., Nathan, C., and Cohen, S.,** *Antimicrob. Agents Chemother.*, 5, 565, 1974.
113. **Kameda, Y. and Horii, S.,** *J. Chem. Soc. Chem. Commun.*, 746, 1972.
114. **Kameda, Y., Horii, S., and Yamano, T.,** *J. Antibiot.*, 28, 298, 1975.
115. **Kameda, Y., Asano, N., and Hashimoto, T.,** *J. Antibiot.*, 31, 936, 1978.
116. **Kameda, Y., Asano, N., Wakae, O., and Iwasa, T.,** *J. Antibiot.*, 33, 764, 1980.
117. **Kameda, Y., Asano, N., Teranishi, M., and Matsui, K.,** *J. Antibiot.*, 33, 1573, 1980.
118. **Kameda, Y., Asano, N., Teranishi, M., Yoshikawa, M., and Matsui, K.,** *J. Antibiot.*, 34, 1237, 1981.
119. **Karnetova, J., Mateju, J., Sedmera, P., Vokoun, J., and Vanek, Z.,** *J. Antibiot.*, 29, 1199, 1976.
119a. **Kase, H., Iida, T., Okadura, Y., Shirahata, K., and Nakayama, K.,** *J. Antibiot.*, 33, 1210, 1980.
120. **Kase, H., Odakura, Y., and Nakayama, K.,** *J. Antibiot.*, 35, 1, 1982.
121. **Kase, H., Shimura, G., Iida, T., and Nakayama, K.,** *Agric. Biol. Chem.*, 46, 515, 1982.
122. **Katz, E., Waldron, C. R., and Meloni, M. L.,** *J. Bacteriol.*, 82, 600, 1961.
123. **Katz, E., Williams, W. K., Mason, K. T., and Mauger, A. B.,** *Antimicrob. Agents Chemother.*, 11, 1056, 1977.
124. **Kaufmann, W., Bauer, K., and Offe, H. A.,** *Antimicrob. Agents Annu.*, 1960, 1, 1960.
125. **Kawabe, H., Kobayashi, F., Yamaguchi, M., Utahara, R., and Mitsuhashi, S.,** *J. Antibiot.*, 24, 651, 1971.
126. **Kawabe, H., Inoue, M., and Mitsuhashi, S.,** *Antimicrob. Agents Chemother.*, 5, 553, 1974.

127. **Kawabe, H., Naito, T., and Mitsuhashi, S.,** *Antimicrob. Agents Chemother.*, 7, 50, 1975.
128. **Kawabe, H., Kondo, S., Umezawa, H., and Mitsuhashi, S.,** *Antimicrob. Agents Chemother.*, 7, 494, 1975.
129. **Keller-Schierlein, W., Poralla, K., and Zahner, H.,** *Arch. Mikrobiol.*, 67, 339, 1969.
130. **Kida, M., Igarazi, S., Okutani, T., Asako, T., Hiraga, K., and Mitsuhashi, S.,** *Antimicrob. Agents Chemother.*, 5, 92, 1974.
131. **Kim, C. H., Otake, N., and Yonehara, H.,** *J. Antibiot.*, 27, 903, 1974.
132. **Kim, C. H., Endo, T., and Yonehara, H.,** *Agric. Biol. Chem.*, 40, 1559, 1976.
133. **Kitamura, S., Kase, H., Odakura, Y., Iida, T., Shirahata, K., and Nakayama, K.,** *J. Antibiot.*, 35, 94, 1982.
134. **Kitao, C., Ikeda, H., Hamada, H., and Omura, S.,** *J. Antibiot.*, 32, 593, 1979.
135. **Kitao, C., Hamada, H., Ikeda, H., and Omura, S.,** *J. Antibiot.*, 32, 1055, 1979.
136. **Kitao, C., Miyazawa, J., and Omura, S.,** *Agric. Biol. Chem.*, 43, 833, 1979.
137. **Kitao, C., Tanaka, H., Minami, S., and Omura, S.,** *J. Antibot.*, 33, 711, 1980.
138. **Kniep, B. and Grisebach, H.,** *J. Antibiot.*, 33, 416, 1980.
139. **Kobayashi, F., Yamaguchi, M., and Mitsuhashi, S.,** *Jpn. J. Microbiol.*, 15, 381, 1971.
140. **Kobayashi, F., Yamaguchi, M., Sato, J., and Mitsuhashi, S.,** *Jpn. J. Microbiol.*, 16, 15, 1972.
141. **Kobayashi, F., Koshi, T., Eda, J., Yoshimura, Y., and Mitsuhashi, S.,** *Antimicrob. Agents Chemother.*, 4, 1, 1973.
142. **Kojima, M., Inouye, S., and Niida, T.,** *J. Antibiot.*, 26, 246, 1973.
143. **Kojima, M. and Satoh, A.,** *J. Antibiot.*, 26, 784, 1973.
144. **Kojima, M., Ezaki, N., Amano, S., Inouye, S., and Niida, T.,** *J. Antibiot.*, 28, 42, 1975.
145. **Komiyama, T., Oki, T., and Inui, T. A.,** *J. Antibiot.*, 32, 1219, 1979.
146. **Komori, T., Kunugita, K., Nakahara, K., Aoki, H., and Imanaka, H.,** *Agric. Biol. Chem.*, 42, 1439, 1978.
147. **Kondo, S., Okanishi, M., Utahara, R., Maeda, K., and Umezawa, H.,** *J. Antibiot.*, 21, 22, 1968.
148. **Konecny, J. and Sieber, M.,** *Biotechnol. Bioeng.*, 22, 2013, 1980.
149. **Kuwabara, S. and Abraham, E. P.,** *Biochem. J.*, 103, 27C, 1967.
150. **Lancini, G., Thiemann, J. E., Sartori, G., and Sensi, P.,** *Experientia*, 33, 899, 1967.
151. **Lancini, G., Gallo, G. G., Sartori, G., and Sensi, P.,** *J. Antibiot.*, 22, 369, 1969.
152. **Lancini, G. and Hengeller, C.,** *J. Antibiot.*, 22, 637, 1969.
153. **Lancini, G. and White, R. J.,** *Process Biochem.*, 8, 14, 1973.
154. **Lancini, G. and Sartori, G.,** *J. Antibiot.*, 29, 466, 1976.
155. **Lancini, G. and Parenti, F.,** in *Antibiotics and Other Secondary Metabolites: Biosynthesis and Production*, Hutter, R., Leisinger, T., Nuesch, J., and Wehrli, W., Eds., Academic Press, New York, 1978, 129.
156. **Lee, B. K., Bailey, J. V., Condon, R. G., Marquez, J. A., and Wagman, G. H.,** *Antimicrob. Agents Chemother.*, 12, 335, 1977.
157. **LeGoffic, F., Martel, A., and Witchitz, J.,** *Antimicrob. Agents Chemother.*, 6, 680, 1974.
158. **LeGoffic, F. and Martel, A.,** *Biochimie*, 56, 893, 1974.
159. **LeGoffic, F., Martel, A., Capmau, M. L., Boca, B., Goebel, P., Charden, H., Soussy, C. J., Duval, J., and Bouanchaud, D. H.,** *Antimicrob. Agents Chemother.*, 10, 258, 1976.
160. **LeGoffic, F., Martel, A., Moreau, N., Capmau, M. L., Soussy, C. J., and Duval, J.,** *Antimicrob. Agents Chemother.*, 12, 26, 1977.
161. **LeMahieu, R. A., Ax, H. A., Blount, J. F., Carson, M., Despraux, C. W., Pruess, D. L., Scannell, J. D., Weiss, T., and Kierstead, R. W.,** *J. Antibiot.*, 29, 728, 1976.
162. **Lotvin, J., Puar, M. S., Patel, M., Lee, B. K., Schumacher, D., and Waitz, J. G.,** *J. Antibiot.*, 35, 1407, 1982.
163. **Lowe, D. A., Romancik, G., and Elander, R. P.,** *Dev. Ind. Microbiol.*, 22, 163, 1981.
164. **Lowe, D. A.,** *Dev. Ind. Microbiol.*, 26, 143, 1985.
165. **McCarville, M. and Marshall, V. P.,** *Biochem. Biophys. Res. Commun.*, 74, 331, 1977.
166. **Maeda, K., Kondo, S., Okanishi, M., Utahara, R., and Umezawa, H.,** *J. Antibiot.*, 21, 458, 1968.
167. **Maezawa, I., Hori, T., Kinumaki, A., and Suzuki, M.,** *J. Antibiot.*, 26, 771, 1973.
168. **Maezawa, I., Kinumaki, A., and Suzuki, M.,** *J. Antibiot.*, 29, 1203, 1976.
169. **Majer, J., Martin, J. R., Egan, R. S., and Corcoran, J. W.,** *J. Am. Chem. Soc.*, 99, 1620, 1977.
170. **Malik, V. S. and Vining, L. C.,** *Can. J. Microbiol.*, 16, 173, 1970.
171. **Marconi, W., Bartoli, F., Cecere, F., Galli, G., and Morisi, F.,** *Agric. Biol. Chem.*, 39, 277, 1975.
172. **Marshall, V. P., Reisender, E. A., Reineke, L. M., Johnson, J. H., and Wiley, P. F.,** *Biochemistry*, 15, 4139, 1976.
173. **Marshall, V. P., Reisender, E. A., and Wiley, P. F.,** *J. Antibiot.*, 29, 966, 1976.
174. **Marshall, V. P., McGovren, J. P., Richard, F. A., Richard, R. E., and Wiley, P. F.,** *J. Antibiot.*, 31, 336, 1978.

175. **Marshall, V. P.**, in *Experiences in Biochemical Perception,* Ornsten, L. N. and Sligar, S. G., Eds., Academic Press, New York, 1982, 75.
176. **Marshall, V. P. and Wiley, P. F.**, in *Microbial Transformations of Bioactive Compounds,* Rosazza, J. P., Ed., CRC Press, Boca Raton, Fla., 1982, 45.
177. **Marshall, V. P., Cialdella, J., Fox, J., and Laborde, A. L.**, *J. Antibiot.*, 37, 923, 1984.
178. **Matsuda, T., Yamaguchi, T., Fujii, T., Matsumoto, K., Morishita, M., Fukushima, M., and Shibuya, Y.**, U.S. patent 3,960,662, 1976.
179. **Matsuhashi, Y., Yagisawa, M., Kondo, S., Takeuchi, T., and Umezawa, H.**, *J. Antibiot.*, 28, 442, 1975.
180. **Matsuhashi, Y., Sawa, T., Kondo, S., and Takeuchi, T.**, *J. Antibiot.*, 30, 435, 1977.
181. **Matsuhashi, Y., Ogawa, H., and Nagaoka, K.**, *J. Antibiot.*, 32, 777, 1979.
182. **Matsumoto, K., Yamamoto, S., and Fujii, T.**, Japan Patent 77,125,696, 1977.
183. **Matsuzawa, Y., Yoshimoto, A., Oki, T., Naganawa, H., Takeuchi, T., and Umezawa, H.**, *J. Antibiot.*, 33, 1341, 1980.
184. **Matsuzawa, Y., Yoshimoto, A., Shibamoto, N., Naganawa, H., Takeuchi, T., Oki, T., and Umezawa, H.**, *J. Antibiot.*, 34, 959, 1981.
185. **Miller, A. L. and Walker, J. B.**, *J. Bacteriol.*, 99, 401, 1969.
186. **Miller, A. L. and Walker, J. B.**, *J. Bacteriol.*, 104, 8, 1970.
187. **Minshew, B., Holmes, R. K., Sanford, J. B., and Baxter, C. R.**, *Antimicrob. Agents Chemother.*, 6, 492, 1974.
188. **Mitsuhashi, S., Kobayashi, F., and Yamaguchi, M.**, *J. Antibiot.*, 24, 400, 1971.
189. **Mitsuhashi, S., Tanaka, T., Kawabe, H., and Umezawa, H.**, *Microbiol. Immun.*, 21, 325, 1977.
190. **Morohoshi, T., Toriya, M., Yokoiyama, S., Fujimoto, K., Hayano, K., Goto, S., and Tsuji, A.**, *J. Antibiot.*, 37, 1687, 1984.
191. **Naganawa, H., Kondo, S., Maeda, K., and Umezawa, H.**, *J. Antibiot.*, 24, 823, 1971.
192. **Naganawa, H., Yagisawa, M., Kondo, S., Takeuchi, T., and Umezawa, H.**, *J. Antibiot.*, 24, 913, 1971.
193. **Nagaoka, K. and Demain, A. L.**, *J. Antibiot.*, 28, 627, 1975.
194. **Nakahama, K., Izawa, M., Muroi, M., Kishi, M., Uchida, M., and Igarasi, S.**, *J. Antibiot.*, 27, 425, 1974.
195. **Nakahama, K., Kishi, T., and Igarasi, S.**, *J. Antibiot.*, 27, 433, 1974.
196. **Nakahama, K., Kishi, T., and Igarasi, S.**, *J. Antibiot.*, 27, 487, 1974.
197. **Nakahama, K. and Igarasi, S.**, *J. Antibiot.*, 27, 605, 1974.
198. **Nakahama, K., Harada, S., and Igarasi, S.**, *J. Antibiot.*, 28, 390, 1975.
199. **Nakahama, K., Izawa, M., Asai, M., Kida, M., and Kishi, T.**, *J. Antibiot.*, 34, 1581, 1981.
200. **Nara, T., Okachi, R., and Misawa, M.**, *J. Antibiot.*, 24, 321, 1971.
201. **Nara, T., Misawa, M., Okachi, R., and Yamamoto, M.**, *Agric. Biol. Chem.*, 35, 1676, 1971.
202. **Narimatsu, Y., Uyeda, M., and Watanabe, K.**, Japan Kokai, Patent 77:108,085, 1977.
203. **Niida, F., Inoue, T., Shomura, T., Omoto, S., Watanabe, H., Nomiya, B., and Koeda, T.**, Japan Kokai, Patent 73:72,389,1973.
204. **Nimi, O., Ito, G., Sueda, S., and Nomi, R.**, *Agric. Biol. Chem.*, 35, 848, 1971.
205. **Nimi, O., Ito, G., Ohata, Y., Furayama, S., and Nomi, R.**, *Agric. Biol. Chem.*, 35, 856, 1971.
206. **Ninet, L., Florent, J., Lunel, J., Renaut, J., Abraham, A., Lombardi, B., and Tissier, R.**, 5th Int. Fermentation Symp., Abstr. 17.06, Berlin, July 1976.
207. **Nishiura, T., Kawada, Y., Shiomi, Y., O'Hara, K., and Kono, M.**, *Antimicrob. Agents Chemother.*, 13, 1036, 1978.
208. **Niwa, T., Nojiri, C., Goi, H., Miyado, S., Kai, F., Seki, S., Yamada, Y., and Niida, T.**, U.S. Patent 4,141,790, 1979.
209. **Nokano, H., Matsuhoshi, Y., Takeuchi, T., and Umezawa, H.**, *J. Antibiot.*, 30, 76, 1977.
210. **Nomi, R., Nimi, O., and Kado, T.**, *Agric. Biol. Chem.*, 32, 1256, 1968.
211. **Nomi, R., Nimi, O., and Kado, T.**, *Agric. Biol. Chem.*, 32, 1454, 1968.
212. **Nozaki, Y., Kitano, K., and Imada, A.**, *Agric. Biol. Chem.*, 48, 37, 1984.
213. **Nys, P. S., Satarova, D. E., Podshibyakina, L., Korchagin, V. B., and Savitskaya, E. M.**, *Antibiotiki*, 25, 914, 1980.
214. **Ochi, K., Yoshima, S., and Eguchi, Y.**, *J. Antibiot.*, 28, 965, 1975.
215. **Odakura, Y., Kase, H., and Nakayama, K.**, *J. Antibiot.*, 36, 125, 1983.
216. **Ogawa, Y., Mizukoshi, S., and Mori, H.**, *J. Antibiot.*, 36, 1561, 1983.
217. **O'Hara, K., Kono, M., and Mitsuhashi, S.**, *J. Antibiot.*, 27, 349, 1974.
218. **O'Hara, K., Kono, M., and Matsuhashi, S.**, *Antimicrob. Agents Chemother.*, 5, 558, 1974.
219. **Oka, Y., Ishida, H., Morioka, M., Numasaki, Y., Yamafuji, T., Osono, T., and Umezawa, H.**, *J. Antibiot.*, 34, 777, 1981.
220. **Okachi, R., Kato, F., Miyamura, Y., and Nara, T.**, *Agric. Biol. Chem.*, 31, 1953, 1973.

221. **Okamoto, R., Fukumoto, T., Takamatsu, A., and Takeuchi, T.,** Japan Kokai, 77:41,294, 1977.
222. **Okamoto, R., Fukumoto, T., Imakuku, K., Okufo, T., Kiyoshima, K., Takamatsu, A., and Takeuchi, T.,** *J. Ferm. Technol.*, 57, 519, 1979.
223. **Okaoto, R., Fukumoto, T., Inui, T., and Takeuchi, T.,** Japan Kokai, Patent 79:28,892, 1979.
224. **Okamoto, R., Fukumoto, T., Nomura, H., Kiyoshima, K., Nakamura, K., and Takamatsu, A.,** *J. Antibiot.*, 33, 1300, 1980.
225. **Okanishi, M., Kondo, S., Utahara, R., and Umezawa, H.,** *J. Antibiot.*, 21, 13, 1968.
226. **Oki, T., Yoshimoto, A., Matsuzawa, Y., Takeuchi, T., and Umezawa, H.,** *J. Antibiot.*, 33, 1331, 1980.
227. **Oki, T., Matsuzawa, Y., Kiyoshima, K., Yoshimoto, A., Naganawa, H., Takeuchi, T., and Umezawa, H.,** *J. Antibiot.*, 34, 783, 1981.
228. **Oki, T., Takatsuki, Y., Tobe, H., Yoshimoto, A., Takeuchi, T., and Umezawa, H.,** *J. Antibiot.*, 34, 1229, 1981.
229. **Oki, T.,** in *Anthracycline Antibiotics,* El Khadem, H. S., Ed., Academic Press, New York, 1982, 75.
230. **Oku, H. and Nakanishi, T.,** Naturwissenschaften, 51, 538, 1964.
231. **Okubo, T., Inoue, M., Nagashima, M., Okii, M., Iida, T., and Mitsuhashi, S.,** *J. Antibiot.*, 38, 122, 1985.
232. **Okuda, S., Isaka, H., Iida, M., Minemura, Y., Iizuka, H., and Isuda, K.,** *J. Pharm. Soc. Jpn.*, 87, 1003, 1967.
233. **Okuda, T. and Ito, Y.,** Hand. Exp., Pharmacol., 62, 111, 1982.
234. **Okumura, Y., Takei, T., Sakamoto, M., Ishikura, T., and Fukagawa, Y.,** *J. Antibiot.*, 32, 584, 1979.
235. **Okumura, Y., Takei, T., Sakamoto, M., Ishikura, T., and Fukagawa, Y.,** *J. Antibiot.*, 32, 1602, 1979.
236. **Okutani, T., Asako, T., Yoshioka, K., Hirago, K., and Kida, M.,** *J. Am. Chem. Soc.*, 99, 1278, 1977.
237. **Omanne, B., Benveniste, R., Tipper, D., and Davies, J.,** *J. Bacteriol.*, 100, 1144, 1969.
238. **Omura, S., Miyazawa, J., Takeshima, H., Kitao, C., Atsumi, K., and Aizawa, M.,** *J. Antibiot.*, 29, 1131, 1976.
239. **Omura, S., Miyazawa, J., Takeshima, H., Kitao, C., and Aizawa, M.,** *J. Antibiot.*, 30, 192, 1977.
240. **Omura, S., Ikeda, H., and Kitao, C.,** *J. Biochem.*, 86, 1753, 1979.
241. **Omura, S., Kitao, C., Hamada, H., and Ikeda, H.,** *Chem. Pharm. Bull.*, 27, 176, 1979.
242. **Omura, S., Kitao, C., and Sadakane, N.,** *J. Antibiot.*, 33, 911, 1980.
243. **Omura, S., Sadakane, N., Kitao, C., Matsubara, H., and Nakagawa, A.,** *J. Antibiot.*, 33, 913, 1980.
244. **Omura, S., Ikeda, H., Matsubara, K., and Sadakane, N.,** *J. Antibiot.*, 33, 1570, 1980.
245. **Omura, S., Kitao, C., and Matsubara, H.,** *Chem. Pharm. Bull.*, 28, 1963, 1980.
246. **Omura, S., Sadakane, N., and Matsubara, H.,** *Chem. Pharm. Bull.*, 30, 223, 1982.
247. **Omura, S., Sadakane, N., Tanaka, Y., and Matsubara, H.,** *J. Antibiot.*, 36, 927, 1983.
248. **Ozaki, M., Kariya, J., Kato, H., and Kimura, T.,** *Agric. Biol. Chem.*, 36, 451, 1972.
249. **Pandy, R. C., Toussaint, M. W., McGuire, J. C., and Thomas, M. C.,** 21st Interrsci. Conf. Antimicrob. Agents Chemother., Abstr. 182, Chicago, 1981.
250. **Pearce, C. J., Barnett, J., Anthony, C., Akhtar, M., and Gero, S.,** *Biochem. J.*, 159, 601, 1976.
251. **Perlman, D. and Langlykke, A. F.,** *J. Am. Chem. Soc.*, 70, 3968, 1948.
252. **Perlman, D.,** *Process Biochem.*, 13, 13, 1971.
253. **Perlman, D. and Sebek, O. K.,** *Pure Appl. Chem.*, 28, 367, 1971.
254. **Perlman, D.,** *Cancer Chemother. Rep.*, 58, 93, 1974.
255. **Pollock, M. R.,** *J. Gen. Microbiol.*, 26, 239, 1961.
256. **Pollock, M. R.,** *Biochem. J.*, 94, 666, 1965.
257. **Prince, A. S. and Jacoby, G. A.,** *Antimicrob. Agents Chemother.*, 22, 525, 1982.
258. **Prokopenko, V. M., Belyaeva, I. D., and Nefelova, M. V.,** *Antibiotiki,* 23, 99, 1978.
259. **Reuckert, P. W., Wiley, P. F., McGovren, J. P., and Marshall, V. P.,** *J. Antibiot.*, 32, 141, 1979.
260. **Richmond, M. H.,** *Biochem. J.*, 94, 584, 1965.
261. **Richmond, M. H.,** *Methods Enzymol.*, 43, 672, 1975.
262. **Rinehart, K. L., Jr.,** in *Neomycin and Related Antibiotics,* John Wiley & Sons, New York, 1964, 93.
263. **Rinehart, K. L., Jr.,** *Pure Appl. Chem.*, 49, 1361, 1977.
264. **Rinehart, K. L., Jr.,** Jpn. J. Antibiot., 32 (Suppl.), S32, 1979.
265. **Rinehart, K. L., Jr.,** in *Microbiology 1979,* Schlessinger, D., Ed., American Society for Microbiology, Washington, D.C., 1979, 308.
266. **Rinehart, K. L., Jr.,** in *Aminocyclitol Antibiotics,* Rinehart, K. L., Jr. and Suami, T., Eds., ACS Symposium Series 125, American Chemical Society, Washington, D.C., 1980, 335.
267. **Rinehart, K. L., Jr., Fang, J. R., Jin, W. Z., Pearce, C. J., Tadano, K. I., and Toyokuni, T.,** *Dev. Ind. Microbiol.*, 26, 117, 1985.
268. **Rolinson, G. N., Batchelor, F. R., Butterworth, D., Cameron-Wood, J., Cole, M., Eustace, G. C., Hart, M. V., Richards, M., and Chain, E. B.,** *Nature (London),* 187, 236, 1960.

269. **Rosi, D., Goss, W. A., and Daum, S. J.,** *J. Antibiot.*, 30, 88, 1977.
270. **Rottman, F. and Guarino, A. J.,** *Biochim. Biophys. Acta,* 80, 632, 1964.
271. **Sadakane, N., Tanaka, Y., and Omura, S.,** *J. Antibiot.*, 35, 680, 1982.
272. **Sakaguchi, K. and Murao, S. J.,** *J. Agric. Chem. Soc. (Jpn),* 23, 411, 1950.
273. **Sakaguchi, Y., Takaishi, N., and Hachimori, A.,** *J. Antibiot.*, 27, 248, 1974.
274. **Sato, S., Iida, T., Okachi, R., Shirahata, K., and Nara, T.,** *J. Antibiot.*, 30, 1025, 1977.
275. **Sawa, T., Fukugawa, Y., Homma, I., Wakashiro, T., Takeuchi, T., and Hori, M.,** *J. Antibiot.*, 21, 334, 1968.
276. **Schmidt-Kastner, G.,** *Ann. N.Y. Acad. Sci.*, 89, 299, 1960.
277. **Schmitz, H. and Claridge, C. A.,** *J. Antibiot.*, 30, 635, 1977.
278. **Schupp, T., Traxler, P., and Auden, J. A. L.,** *J. Antibiot.*, 34, 965, 1981.
279. **Sebek, O. K. and Perlman, D.,** *Adv. Appl. Microbiol.*, 14, 123, 1971.
280. **Sebek, O. K. and Hoeksema, H.,** *J. Antibiot.*, 25, 434, 1972.
281. **Sebek, O. K.,** *Lloydia,* 37, 115, 1974.
282. **Sebek, O. K.,** *Acta Microbiol. Acad. Sci. Hung.*, 22, 381, 1975.
283. **Sebek, O. K. and Dolak, L. A.,** U.S. Patent 4,128,563, 1978.
284. **Sebek, O. K.,** *Econ. Microbiol.*, 5, 575, 1980.
285. **Sebek, O. K. and Dolak, L. A.,** *J. Antibiot.*, 37, 136, 1984.
286. **Sebek, O. K.,** in *Biotechnology,* Rehm, H-J. and Reed, G., Eds., Verlag Chemie, Weinheim, 1984, 240.
287. **Seong, B. L., Son, H. J., Mheen, T-I., and Moon, H. H.,** *J. Antibiot.*, 36, 1402, 1983.
288. **Serizawa, N., Nakagawa, K., Hamano, K., Tsujita, Y., Terahara, A., and Kuwano, H.,** *J. Antibiot.*, 36, 604, 1983.
289. **Serizawa, N., Nakagawa, K., Hamano, K., Tsujita, Y., Terahara, A., and Kuwano, H.,** *J. Antibiot.*, 36, 608, 1983.
290. **Serizawa, N., Serizawa, S., Nakagawa, K., Furuga, K., Okazaki, T., and Terahara, A.,** *J. Antibiot.*, 36, 887, 1983.
291. **Serizawa, N., Nakagawa, K., Hamano, K., Tsujita, Y., Terahara, A., Kuwano, H., and Tanaka, M.,** *J. Antibiot.*, 36, 918, 1983.
292. **Serizawa, N., Nakagawa, K., Tsujita, Y., and Terahara, A.,** *Agric. Biol. Chem.*, 48, 2581, 1984.
293. **Shaw, W. V.,** *J. Biol. Chem.*, 242, 687, 1967.
294. **Shaw, W. V. and Brodsky, R. F.,** *J. Bacteriol.*, 95, 28, 1968.
295. **Shibata, M., Uyeda, M., and Mori, S.,** *J. Antibiot.*, 28, 434, 1975.
296. **Shibata, M., Uyeda, M., and Mori, S.,** *J. Antibiot.*, 29, 824, 1976.
297. **Shibata, M. and Uyeda, M.,** *Annu. Rep. Ferm. Proc.*, 2, 267, 1978.
298. **Shibuya, Y., Matsumoto, K., and Fujii, T.,** *Agric. Biol. Chem.*, 45, 1561, 1981.
299. **Shier, W. T., Rinehart, K. L., Jr., and Gottlieb, D.,** *Proc. Natl. Acad. Sci. U.S.A.*, 63, 198, 1969.
300. **Shier, W. T., Rinehart, K. L., Jr., and Gottlieb, D.,** U.S. Patent 3,669,838; as cited in *Chem. Abstr.*, 77, 154584x, 1972.
301. **Shier, W. T., Ogawa, S., Hichens, P., and Rinehart, K. L., Jr.,** *J. Antibiot.*, 26, 551, 1973.
302. **Shier, W. T., Schaefer, P. C., Gottlieb, D., and Rinehart, K. L., Jr.,** *Biochemistry,* 13, 5073, 1974.
302a. **Shimizu, M., Masuike, T., Fujita, H., Kimura, K., Okachi, R., and Nara, T.,** *Agric. Biol. Chem.*, 39, 1225, 1975.
303. **Shirafuji, H., Kida, M., Asako, T., and Yoneda, M.,** *Agric. Biol. Chem.*, 43, 2579, 1979.
304. **Shirato, S., Hoshida, K., and Miyazaki, Y.,** *J. Ferm. Technol.*, 46, 233, 1968.
305. **Shoji, J., Kato, T., and Hinoo, H.,** *J. Antibiot.*, 30, 1035, 1977.
306. **Shomura, T., Inolue, S., Ezaki, N., Tsurioka, T., Amano, T., Watanabe, H., and Niida, T.,** Japan Kokai, Patent 73: 29, 148, 1973.
307. **Singh, K. and Rakhit, S.,** *J. Antibiot.*, 24, 704, 1971.
308. **Singh, K. and Rakhit, S.,** *J. Antibiot.*, 32, 78, 1979.
309. **Smith, A. and Smith, D. H.,** *J. Infect. Dis.*, 129, 391, 1974.
310. **Smith, A. and Larner, R. W.,** U.S. Patent 3,976,546, 1976.
311. **Smith, G. N., Worrel, C. S., and Lilligren, B. S.,** *Science,* 110, 297, 1949.
312. **Smith, G. N. and Worrel, C. S.,** *Arch. Biochem.*, 28, 232, 1950.
313. **Spagnoli, R., Cappelletti, L., and Toscano, L.,** *J. Antibiot.*, 36, 365, 1983.
314. **Spagnoli, R. and Toscano, L.,** *J. Antibiot.*, 36, 435, 1983.
315. **Suhadolnik, R. J.,** *Nucleoside Antibiotics,* Wiley-Interscience, New York, 1970, 327.
316. **Sussmuth, R., Haag, R., and Lingins, F.,** *J. Antibiot.*, 32, 1293, 1979.
317. **Suzuki, Y. and Okamoto, S.,** *J. Biol. Chem.*, 242, 4722, 1967.
318. **Suzuki, M., Takamaki, T., Miyagawa, K., Ono, H., Higashide, E., and Ochida, M.,** *Agric. Biol. Chem.*, 41, 419, 1977.
319. **Sykes, R. B. and Richmond, M. H.,** *Nature (London),* 226, 952, 1970.
320. **Takahashi, T., Yamazaki, Y., Kato, K., and Isono, M.,** *J. Am. Chem. Soc.*, 94, 4035, 1972.

321. **Takahashi, K., Yazawa, K., Kishi, K., Mikami, Y., Aral, T., and Kuto, A.,** *J. Antibiot.*, 35, 196, 1983.
322. **Takasawa, S., Utahara, R., Okanishi, M., Maeda, K., and Umezawa, H.,** *J. Antibiot.*, 21, 477, 1968.
323. **Takeda, K., Kinumaki, A., Furumai, T., Yamaguchi, T., Oshima, S., and Ito, Y.,** *J. Antibiot.*, 31, 247, 1978.
324. **Takeda, K., Aihara, K., Furumai, T., and Ito, Y.,** *J. Antibiot.*, 31, 250, 1978.
325. **Takeda, K., Okuno, S., Ohashi, Y., and Furumai, T.,** *J. Antibiot.*, 31, 1023, 1978.
326. **Takeda, K., Kinumaki, A., Hayasaka, H., Yamaguchi, T., and Ito, Y.,** *J. Antibiot.*, 31, 1031, 1978.
327. **Takeda, K., Kinumaki, A., Okuno, S., Matsushita, T., and Ito, Y.,** *J. Antibiot.*, 31, 1039, 1978.
328. **Takeda, K., Aihara, K., Furumai, T., and Ito, Y.,** *J. Antibiot.*, 32, 18, 1979.
329. **Tanaka, H., Minami-Kakinuma, S., and Omura, S.,** *J. Antibiot.*, 35, 1565, 1982.
330. **Testa, R. L., Wagman, G. H., Daniels, P. J. L., and Weinstein, M. J.,** *J. Antibiot.*, 27, 917, 1974.
331. **Testa, R. L. and Tilley, B. C.,** *J. Antibiot.*, 28, 573, 1975.
332. **Testa, R. L. and Tilley, B. C.,** *J. Antibiot.*, 29, 140, 1976.
333. **Taylor, H. D. and Schmitz, H.,** *J. Antibiot.*, 29, 532, 1976.
334. **Theriault, R. J.,** U.S. Patent 3,784,447, 1974.
335. **Theriault, R. J.,** *J. Antibiot.*, 35, 364, 1982.
336. **Tobe, H., Yoshimoto, A., Ishikura, T., Naganawa, H., Takeuchi, T., and Umezawa, H.,** *J. Antibiot.*, 35, 1641, 1982.
337. **Torii, H., Asano, T., Matsumoto, N., Kato, K., Tsushima, S., and Kakinuma, A.,** *Agric. Biol. Chem.*, 44, 1431, 1980.
338. **Toscano, L., Fioriello, G., Spagnoli, R., Cappelletti, L., and Zanuso, G.,** *J. Antibiot.*, 36, 1439, 1983.
339. **Toscano, L., Fioriello, G., Spagnoli, R., and Cappelletti, L.,** *J. Antibiot.*, 36, 1585, 1983.
340. Toyo Jozo Co., Ltd., Japan Kokai, Patent 81:154,998, 1981.
341. **Traxler, P., Schupp, T., and Wehrli, W.,** *J. Antibiot.*, 35, 594, 1982.
342. **Tsuruoka, T., Shomura, T., Ogawa, Y., Ezeki, N., Watanabe, H., Amano, S., Inouye, S., and Niida, T.,** *J. Antibiot.*, 26, 168, 1973.
343. **Uematsu, T. and Suhadolnik, R. J.,** *Arch. Biochem. Biophys.*, 162, 614, 1974.
344. **Umezawa, H., Sawa, T., Fukagawa, Y., Koyama, G., Murase, M., Hamada, M., and Takeuchi, T.,** *J. Antibiot.*, 18, 178, 1965.
345. **Umezawa, H., Okanishi, M., Utahara, R., Maeda, K., and Kondo, S.,** *J. Antibiot.*, 20, 136, 1967.
346. **Umezawa, H., Sawa, T., Fukagawa, Y., Homma, I., Ishizuka, M., and Takeuchi, T.,** *J. Antibiot.*, 20, 308, 1967.
347. **Umezawa, H., Okanishi, M., Kondo, S., Hamana, K., Utahara, R., and Maeda, K.,** *Science*, 157, 1559, 1967.
348. **Umezawa, H., Tapasawa, S., Okanishi, M., and Utahara, R.,** *J. Antibiot.*, 21, 81, 1968.
349. **Umezawa, H., Doi, O., Ogura, M., Kondo, S., and Tanaka, N.,** *J. Antibiot.*, 21, 154, 1968.
350. **Umezawa, H.,** *Pure Appl. Chem.*, 28, 665, 1971.
351. **Umezawa, H. Takahashi, Y., Fujii, A., Saino, T., Shirai, T., and Takita, T.,** *J. Antibiot.*, 26, 117, 1973.
352. **Umezawa, H., Yagisawa, M., Matsuhashi, Y., Naganawa, H., Yamamoto, H., Kondo, S., and Takeuchi, T.,** *J. Antibiot.*, 26, 612, 1973.
353. **Umezawa, H.,** *Adv. Carbohyd. Chem. Biochem.*, 30, 183, 1974.
354. **Umezawa, Y., Yagisawa, M., Sawa, T., Takeuchi, T., and Umezawa, H.,** *J. Antibiot.*, 28, 845, 1975.
355. **Uyeda, M., Mori, S., Morita, M., Ogata, T., Mori, M., and Shibata, M.,** *J. Antibiot.*, 30, 1130, 1977.
356. **Uyeda, M., Nakamichi, K., Shigemi, K., and Shibata, M.,** *Agric. Biol. Chem.*, 44, 1399, 1980.
357. **Uyeda, M., Kirbata, K., Miyamura, S., and Shibata, M.,** *J. Antibiot.*, 36, 1722, 1983.
358. **Vandamme, E. J.,** in *Economic Microbiology*, Rose, A. H., Ed., Academic Press, New York, 1980, 467.
359. **Vanek, Z., Tax, J., Komersova, I., and Eckardt, K.,** *Folia Microbiol.*, 18, 524, 1973.
360. **VonDaehne, W., Lorch, H., and Godtfredsen, W. O.,** *Tetrahedron Lett.*, 47, 4843, 1968.
361. **Wagner, C., Eckardt, K., Schumann, G., Ihn, W., and Tresselt, D.,** *J. Antibiot.*, 37, 691, 1984.
362. **Waitz, J., Miller, G., Moss, E., Jr., and Chiu, P.,** *Antimicrob. Agents Chemother.*, 13, 41, 1978.
363. **Walker, M. S. and Walker, J. B.,** *J. Biol. Chem.*, 245, 6683, 1970.
364. **Walker, J. B. and Skorvaga, M.,** *J. Biol. Chem.*, 248, 2435, 1973.
365. **Walton, R. B.,** *Dev. Ind. Microbiol.*, 5, 349, 1964.
366. **Warren, R. A. J. and Neilands, J. B.,** *J. Gen. Microbiol.*, 35, 459, 1964.
367. **Wayczikowska, B., Pass, L., Girdwoyn, M., and Raczynska-Bojanowska, K.,** *Acta Biochem. Pol.*, 20, 285, 1973.
368. **Weber, J. M. and Perlman, D.,** *J. Antibiot.*, 31, 373, 1978.
369. **Werner, R. G. and Demain, A. L.,** *J. Antibiot.*, 34, 551, 1981.

370. **White, R. F., Bernbaum, J., Meyer, R. T., tenBroeke, J., Chemerda, J. M., and Demain, A. L.,** *Appl. Microbiol.*, 22, 55, 1971.
371. **White, R. J., Martinelli, E., and Lancini, G.,** *Proc. Natl. Acad. Sci. U.S.A.*, 71, 3260, 1974.
372. **Wiley, P. F. and Marshall, V. P.,** *J. Antibiot.*, 28, 838, 1975.
373. **Wiley, P. F., Koert, J. M., Elrod, D. W., Reisender, E. A., and Marshall, V. P.,** *J. Antibiot.*, 30, 649, 1977.
374. **Wiley, P. F., Elrod, D. W., Slavicek, J. M., and Marshall, V. P.,** *J. Antibiot.*, 33, 819, 1980.
375. **Williams, J. W. and Northrop, D. B.,** *Biochemistry*, 15, 125, 1976.
376. **Withander, L. and Heding, H.,** *J. Antibiot.*, 29, 774, 1976.
377. **Yagisawa, M., Naganawa, H., Kondo, S., Hamada, M., Takeuchi, T., and Umezawa H.,** *J. Antibiot.*, 24, 911, 1971.
378. **Yagisawa, M., Naganawa, H., Kondo, S., Takeuchi, T., and Umezawa, H.,** *J. Antibiot.*, 25, 492, 1972.
379. **Yagisawa, M., Naganawa, H., Kondo, S., Takeuchi, T., and Umezawa, H.,** *J. Antibiot.*, 25, 495, 1972.
380. **Yagisawa, M., Yamamoto, H., Naganawa, H., Kondo, S., Takeuchi, T., and Umezawa, H.,** *J. Antibiot.*, 25, 748, 1972.
381. **Yagisawa, M., Kondo, S., Takeuchi, T., and Umezawa, H.,** *J. Antibiot.*, 28, 486, 1975.
382. **Yamada, T., Tipper, D., and Davies, J.,** *Nature (London)*, 219, 288, 1968.
383. **Yamaguchi, M., Mitsuhashi, S., Kobayashi, F., and Zenda, H.,** *J. Antibiot.*, 27, 507, 1974.
384. **Yamaguchi, J., Shibata, H., Seto, H., and Misato, T.,** *J. Antibiot.*, 28, 7, 1975.
385. **Yamamoto, H., Yagisawa, M., Naganawa, H., Kondo, S., Takeuchi, T., and Umezawa, H.,** *J. Antibiot.*, 25, 746, 1972.
386. **Yamamoto, H., Hotta, K., Okomi, Y., and Umezawa, H.,** *J. Antibiot.*, 35, 1020, 1982.
387. **Yoshida, T., Mauger, A., Weissbach, H., and Katz, E.,** *J. Bacteriol.*, 95, 952, 1968.
388. **Yoshida, M., Konomi, T., Kohsaka, M., Baldwin, J. E., Herchen, S., Singh, P., Hunt, N. A., and Demain, A. L.,** *Proc. Natl. Acad. Sci. U.S.A.*, 75, 6253, 1978.
389. **Yoshimoto, A., Ogasawara, T., Kitamura, I., Oki, T., and Inui, T.,** *J. Antibiot.*, 32, 472, 1979.
390. **Yoshimoto, A., Matsuzawa, Y., Oki, T., Naganawa, H., Takeuchi, T., and Umezawa, H.,** *J. Antibiot.*, 33, 1150, 1980.
391. **Yoshimoto, A., Oki, T., Takeuchi, T., and Umezawa, H.,** *J. Antibiot.*, 33, 1158, 1980.
392. **Yoshimoto, A., Oki, T., and Umezawa, H.,** *J. Antibiot.*, 33, 1199, 1980.
393. **Yoshimoto, A., Matsuzawa, Y., Oki, T., Takeuchi, T., and Umezawa, H.,** *J. Antibiot.*, 34, 951, 1981.
394. **Yoshimoto, A., Matsuzawa, Y., Matsushita, Y., Oki, T., Takeuchi, T., and Umezawa, H.,** *J. Antibiot.*, 34, 1492, 1981.
395. **Yoshimoto, A., Johdo, O., Takatsuki, Y., Ishikura, T., Sawa, T., Takeuchi, T., and Umezawa, H.,** *J. Antibiot.*, 37, 935, 1984.
396. **Yukioka, M., Saito, Y., and Otani, S.,** *J. Biochem.*, 60, 295, 1966.
397. **Zanca, D. M. and Martin, J. F.,** *J. Antibiot.*, 36, 700, 1983.

Chapter 3

MICROBIAL TRANSFORMATIONS: MODELS OF MAMMALIAN METABOLISM AND CATALYSTS IN SYNTHETIC MEDICINAL CHEMISTRY

Patrick J. Davis

TABLE OF CONTENTS

I. INTRODUCTION

Microbial transformations (biotransformations) include the use of microbiological systems to catalyze the introduction of new functional groups or the modification of existing functional groups on chemically defined substrates. From the earliest developments involving poorly understood and uncontrolled processes, microbial transformations have continued to evolve to more well-defined, highly predictable biocatalysts, and have found new applications in biotechnology. The focus of this discussion will be on two areas in which microbial transformations contribute to medicinal and natural products chemistry. First, the concept of "microbial models of mammalian metabolism" will be discussed, a discipline which represents a novel application of microbial transformations to facilitate the study of mammalian biotransformations of drugs and other xenobiotics. The second area to be discussed is the expanding role of microbial systems as biocatalysts for conducting highly stereo- and regioselective transformations of drugs, intermediates, and natural products.

II. MICROBIAL MODELS OF MAMMALIAN METABOLISM

A. Background

A common approach in the physical and biological sciences is to utilize simpler model systems to facilitate the study of more complex systems. One of the obvious advantages of such an approach is that it allows for the study of particular aspects of the parent system while avoiding complexities and interrelationships of that parent system, which would make such studies difficult or impossible. Thus, while the model will surely differ from the parent system (if it were identical in all attributes, it obviously would not be a model), the real value of a model is determined by how well it facilitates the study of and predicts responses in the parent system.

In biomedical research, animal systems continue to serve an important role as models for biological responses in man. It has been estimated that at least 70 million animals are used annually in the U.S. alone for toxicity testing as well as for biomedical and behavioral research.[1] Worldwide, this figure may be as high as 250 million animals.[2] In recent years, interest has turned to the development of alternative systems for decreasing the use of (and, where appropriate, the replacement of) animals in laboratory studies.[3-5] This is in part due to the strong political involvement of animal welfare groups; the enormous expense of animal studies (including purchase or breeding, maintenance, and the actual expense of experimentation); and the recognized difficulties of dealing with such complex systems experimentally. Established alternative systems and newer systems under current investigation in toxicology and metabolism studies include: in vivo systems (animals rendered nonsentient; invertebrates); in vitro systems derived from animals (isolated organs and tissues, tissue and cell cultures, and microsomal systems); simulated tissues and body fluids; chemical models; isolated enzymes; mathematical models and other computer simulations; plants, and microbial systems.[4-6] The focus of this discussion will be the use of microbial cultures to facilitate the study of mammalian metabolism of drugs and other xenobiotics.

The concept of "microbial models of mammalian metabolism" can be defined as the use of microorganisms (bacteria, yeasts, and fungi) to facilitate the study of xenobiotic biotransformations in mammals, including man. This discipline was first formalized by Rosazza and Smith[7] as a result of a systematic study of microbiological hydroxylations on a number of aromatic substrates. Direct parallels were observed with patterns of hydroxylations described earlier for in vitro and in vivo mammalian systems. This concept was formalized in an initial review,[8] which, along with subsequent reviews,[9-16] provided an overview of (1) the methodologies involved, (2) parallels in the types of reactions observed, (3) the enzymatic (biochemical) basis for observed parallels, and (4) numerous specific examples supporting the basis and utility of the microbial models concept.

Table 1
EXAMPLES OF PHASE-ONE BIOTRANSFORMATIONS OBSERVED IN MICROORGANISMS

- Oxidations
  - Hydroxylations (aliphatic and aromatic)
  - Dealkylations (e.g., $-SCH_3$, $-NCH_3$, $-OCH_3$)
  - Heteroatom oxidations (e.g., N to N-oxide, S to S-oxide)
  - Oxidative deaminations
- Dehydrogenations
- Hydrolyses
  - Esters
  - Amides
  - Carbamates
  - Epoxides
- Reductions
  - Ketones
  - Olefins
  - N-oxides
  - S-oxides

The fundamental premise upon which the microbial models concept is based is that the same Phase 1 routes of metabolism of xenobiotics observed in mammals are paralleled in microorganisms. Table 1 lists examples of specific type-reactions which have been observed with a variety of xenobiotics in both types of systems to illustrate this point. Thus, the broad metabolic diversity of mammalian (principally liver) metabolism is exhibited by microbial systems as well. In terms of Phase 2 (conjugation reactions), it should be pointed out that microbes, like mammals, do conjugate polar function groups. It appears, however, that microbes typically use different sugars in this process, e.g., to form glucosides, rather than glucuronides, as occurs in mammals. There are examples, however, in which fungi do metabolize xenobiotics by Phase 1 routes of metabolism, followed by conjugation to form glucuronides.[17,18]

A number of standard "animal models of human metabolism" are in common usage, including metabolism in whole animals, isolated organs and tissues, cell cultures, microsomes, and purified enzymes, each with advantages and disadvantages in comparison to one another.[19] Microbial systems show distinct advantages in certain areas which make them attractive to drug metabolism studies as follows:

1. Ease of experimental design — The first advantage is the ease of manipulation of experimental parameters. For example, a reductive reaction (such as ketone reduction to the corresponding alcohol) can be enhanced by rendering the culture more reductive by lowering aeration or by the addition of glucose.
2. Large-scale bioconversions — Once a culture conducting the desired transformation is discovered, milligram or even gram quantities of metabolites can be produced by routine fermentation scale-up techniques. At this level, the full-structure elucidation of the metabolite is markedly facilitated. This facet is particularly important if the same metabolite is suspected to occur in mammals, but the structure is unknown or equivocal because metabolite levels are limiting. Further, the metabolite can be evaluated for biological activity, including toxicity, yielding information which is exceedingly important in the early phases of drug development. In addition, the metabolite may serve as a standard reference material for conducting mammalian metabolism studies to determine if that metabolite is formed in higher organisms.
3. Single or multiple pathway modeling — A single microbial model (one culture) may exhibit the entire complex of divergent routes of mammalian metabolism seen with a

particular drug in numerous mammalian species. For example, *Cunninghamella bainieri* was shown to produce nine known oxidative mammalian metabolites of *N-n*-propylamphetamine, plus a tenth metabolite as well.[20] Alternatively, one culture may exhibit a single route of metabolism in lieu of all others. This is particularly useful if the pathway is minor and/or quantitatively masked by other routes of metabolism in mammals, and thus hard to study. In this case, the ultimate "microbial model" may be a composite of cultures each exhibiting the individual reactions.

4. Cost — The methodologies employed involve a substantially reduced cost over mammalian studies.
5. Lessened animal demand — The use of microbial systems in metabolism studies should lessen the demand for animals which would be beneficial from both humanitarian and economic standpoints.

While such advantages may be cited, the author would not suggest that microbial models will replace all animal metabolism studies, any more than the animal models have "replaced" human metabolism studies. But such systems should be particularly useful in predicting routes of metabolism resulting in bioactivation, detoxification, or generation of toxic metabolites, and to uncover aspects of mammalian metabolism difficult to approach by other methods.

Several approaches are utilized in the development of microbial models of mammalian metabolism. Studies may be "retrospective", i.e., conducted after the mammalian routes of metabolism have been elucidated. Such studies can be used to confirm tentative assignments of metabolites in mammals, and support the (now well-established) premise that the same routes of metabolism are observed in both systems. Alternatively, studies may be "prospective", in which the routes of metabolism are first elucidated in microbial systems and then used to predict (and the metabolites facilitate the confirmation of) routes of metabolism in mammals. "Parallel" studies involve the simultaneous study of biotransformations in mammalian and microbial species. Most studies to date have been retrospective, but the trend is to make such studies more prospective by using the results to explore unanswered questions in mammalian metabolism. For example, while the mammalian routes of metabolism may have been generally elucidated, microbial studies may reveal previously undetected mammalian metabolites. The search for this metabolite in mammalian systems is then facilitated by having the microbiologically generated metabolite on hand.[21,22] Alternatively, the microbial system may be utilized to elucidate the previously unexplored stereochemical disposition of the drug in mammalian systems (vide infra) or used to study mechanistic questions concerning the formation of the metabolite.

A schematic is depicted in Figure 1 of a general strategy for developing a microbial model of the mammalian metabolism of a drug or other xenobiotic. Where one starts in this scheme depends on whether the model will be retrospective, prospective, or parallel in relation to the mammalian studies. The study usually begins with "defining the desired biotransformations". If the mammalian studies have already been conducted, these routes are already defined; i.e., one is looking for specific known, suspected, or tentative metabolites (reported or suggested) to occur in mammals. If that information is not available (as in a prospective or parallel study), one considers the structure of the substrate and known Phase 1 routes of metabolism to predict logical biotransformation reactions (e.g., carbonyl reduction, *N*- and *O*-demethylation, etc.).

The next step is analytical development as a prelude to screening studies. Standard methods of thin-layer chromatography (TLC), high-performance liquid chromatography (HPLC), gas chromatography (GC), and GC/MS are usually employed. If mammalian studies were conducted previously, it is probable that some type of analytical development has already been undertaken which can then be modified to analyze for the drug and metabolites in microbial

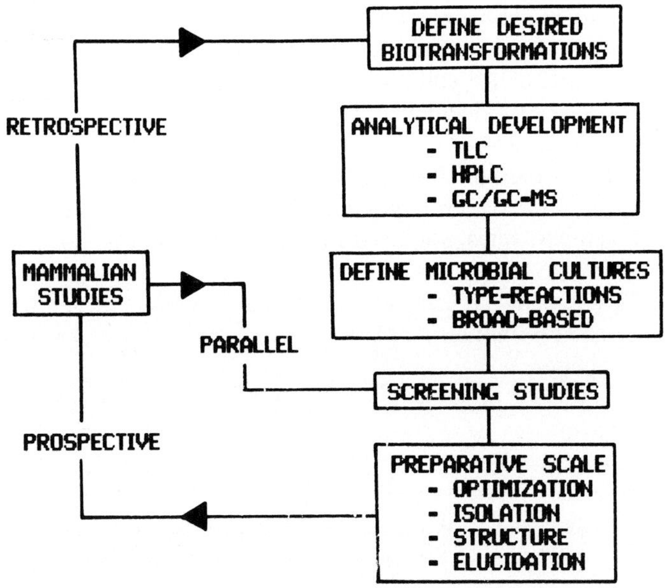

FIGURE 1. Schematic of a strategy for the development of microbial models of mammalian metabolism.

culture (rather than blood, urine, etc.). If that information is not available, analytical development must be based on predicted routes of metabolism from the previous step.

The crucial step in microbial models development is defining which microbial cultures will be examined for the ability to metabolize the compound under investigation. As with other studies in microbial transformations, this choice may involve a rather empirical approach using a broad "random screen" in which many cultures, chosen more or less at random from the culture collection, are screened for their ability to biotransform the compound in any fashion. More directed approaches based on literature precedent typically yield more predictable results with considerably fewer cultures. One may use a "type-reaction screen", in which cultures are chosen based on their reported ability to catalyze specific Phase 1 biotransformations (i.e., the specific type-reactions indicated in Table 1), e.g., ketone reduction, aromatic hydroxylation, etc. In this type of study, the ultimate "model" will represent a "composite" of several cultures, each exhibiting the individual routes of metabolism. Alternatively, one may choose (again, based on literature precedent) cultures known to exhibit the broad range of reactions observed in mammalian systems. Fungi of the genus *Cunninghamella* fall into this category.

"Screening" is the preliminary evaluation of the metabolic capabilities of the selected cultures, involving uniform microbiological culturing techniques and the analytical system(s) developed (above) to monitor biotransformations. Parallel mammalian metabolism studies would also be conducted at this point as well. "Hits" (cultures exhibiting metabolism) are re-run to confirm metabolic capability and then examined in more detail as a prelude to preparative scale production of metabolites. Optimization of the biotransformations (if necessary) involves the systematic alteration of fermentation parameters (medium composition, aeration, pH, the influence of growth on metabolism, etc.) and other parameters (time-course studies to determine optimal time of substrate addition and harvesting, mode of substrate addition, etc.) to enhance yields. Larger-scale fermentations are then conducted for the isolation, purification, and structural elucidation of the metabolite(s) produced.

In a prospective study, the investigator is now ready to proceed with mammalian studies

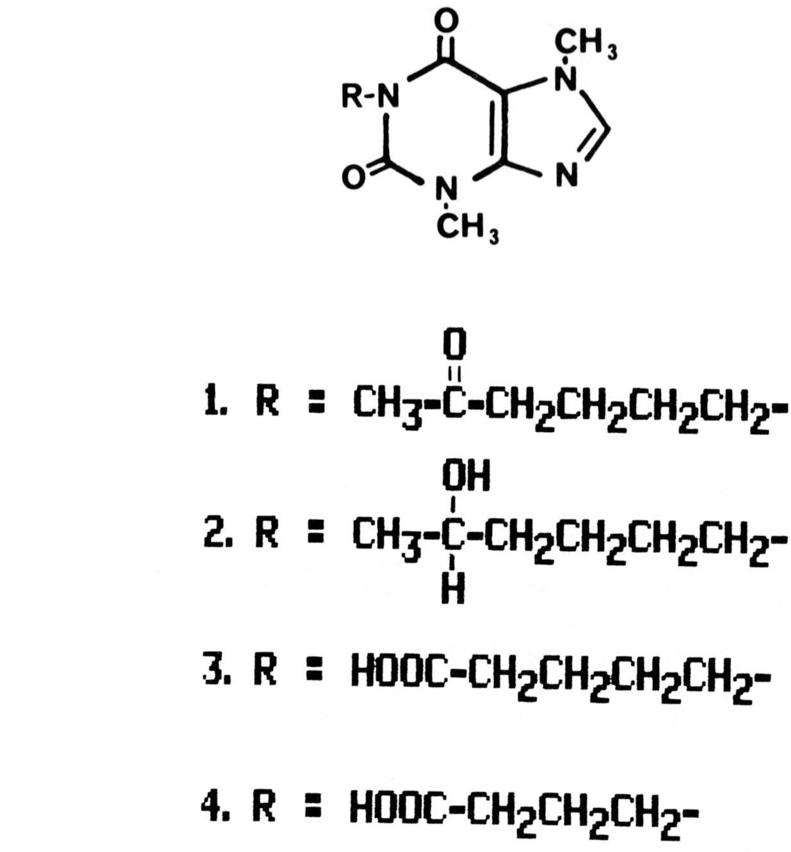

FIGURE 2. Structures of pentoxifylline (**1**) and metabolites **2** to **4**.

to confirm routes of metabolism predicted by the microbial system, a process that would be facilitated by the metabolites and analytical systems developed using the model. Even in a "retrospective" study (i.e., the mammalian routes of metabolism have already been explored), the microbial model, isolated metabolites, or analytical systems developed may be used in a prospective fashion to confirm tentative metabolites or explore unanswered questions in mammalian metabolism, such as the stereochemical course or mechanism of biotransformation.

B. Microbial Models — Pentoxifylline

In order to illustrate some of the previous points, the development of microbial models of pentoxifylline metabolism will be discussed.[23-25] Pentoxifylline (**1**; see Figure 2; pentoxifylline is marketed by Hoechst-Roussel as Trental®) is a xanthine agent recently approved in the U.S. for the treatment of intermittent claudication, a disorder suffered by over one million Americans. This disorder is a type of chronic, occlusive arterial disease with a loss of elasticity of red blood cell (RBC) membranes and poor penetration into tissues, resulting in poor oxygenation and severe, intermittent pain. Pentoxifylline functions as a "hemorheologic" agent, improving the elasticity of the RBC membranes, and thus penetration into and oxygenation of tissues. The major routes of metabolism of pentoxifylline in mammals (including man) focus on the ketonic alkyl side-chain (see Figure 2). One important biotransformation is ketone reduction to form pentoxifylline alcohol (**2**). In addition, oxidative cleavage of the side-chain on either side of the carbonyl yields the carboxybutyl metabolite

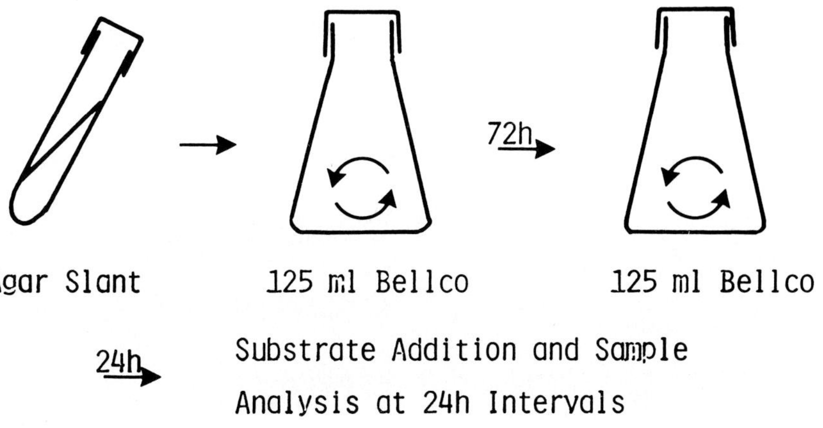

FIGURE 3. A general screening procedure used in microbial transformations.

(**3**) and the carboxypropyl metabolite (**4**). Our intent was to develop microbial models for these various routes of metabolism with the intent of preparing metabolites for biological activity and using the models to address unexplored aspects of pentoxifylline metabolism in mammals (e.g., the stereochemical course of ketone reduction). The overall goal of this research is to develop a composite of microbes exhibiting the broad range of Phase 1 biotransformations seen in mammals for use in drug development and related studies involving drug metabolism.

Since the desired biotransformations were already defined by the mammalian studies, the next step in models development (see Figure 1) was the development of appropriate analytical methodology. Routine screening of microbial cultures is usually monitored by TLC, and three such systems were developed which adequately separate the parent compound (**1**) from the three metabolites (**2** to **4**).[24] For confirmation of TLC results and for conducting more quantitative studies, an extraction procedure and a HPLC system was then developed to analyze and quantitate (**1** to **4**) in culture samples, utilizing an *N*-*n*-propyl-pentoxifylline analogue, as the internal standard.[24]

The next step was to define microbial cultures with a high probability of exhibiting the desired biotransformations. The initial studies were to focus on ketone reduction to prepare metabolite (**2**). The choice of appropriate cultures is markedly facilitated by compilations, such as that by Kieslich,[26] which tabulates cultures according to the specific type-reactions they catalyze. The author's laboratory also maintains a computerized database of more recent literature which facilitates choosing appropriate cultures. Based on literature precedent, 14 cultures were chosen for initial exploration in ketone reduction.

The general screening procedure use is outlined in Figure 3. Cultures are maintained at 5°C on slants of various types of agar, depending on the particular requirements of the class of microorganisms (e.g., Sabaroud-maltose agar, sporulation agar, etc.). To initiate "Stage 1" cultures, the surface growth of bacteria and yeasts or spores of fungal cultures are aseptically transferred to 125-mℓ Bellco-DeLong flasks containing one fifth their volume (25 mℓ) of a rich medium (dextrose 25 g, soybean meal [20 mesh] 5 g, sodium chloride 5 g, yeast extract 5 g, dipotassium hydrogen phosphate 5 g, double-distilled deionized water, 1ℓ; pH adjusted to 7.0; all medium is autoclaved in individual flasks). After incubation in an environmentally controlled shaker at 28°C and at 250 rpm, a 2-mℓ portion of each culture was used to inoculate a second-stage flask of the same dimensions and medium composition. After incubation for an additional 24 hr, all cultures are typically fully grown, and the appropriate substrate (in this case, pentoxifylline [**1**]) is added to a level of 0.5 mg/mℓ in

Table 2
SCREENING RESULTS — PENTOXIFYLLINE METABOLISM

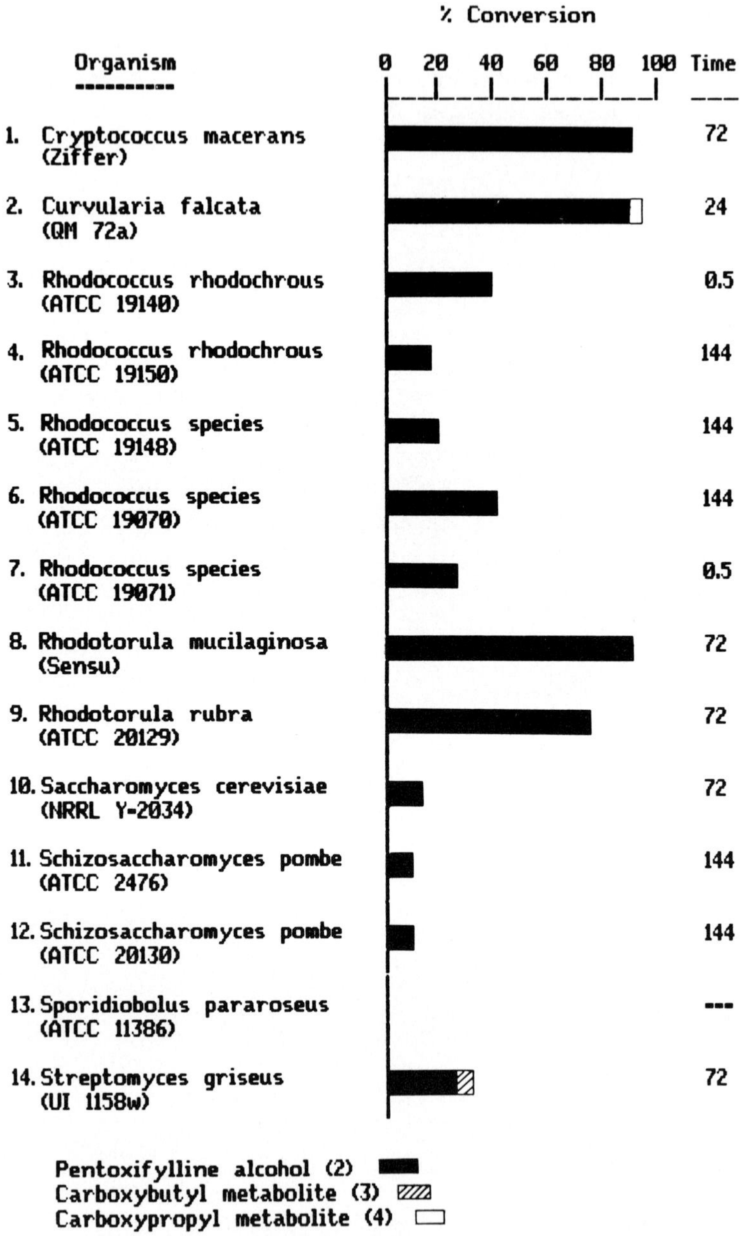

Time = Hr to maximal production of alcohol metabolite 2.

the culture (12.5 mg in 50 μℓ dimethylformamide (DMF) per flask). Incubation is allowed to continue, and samples of 1.0 mℓ are aseptically harvested at 24-hr intervals, extracted, and analyzed by TLC. Cultures exhibiting metabolism are further analyzed by HPLC.

The results of the screening studies (Table 2) show that 13 of the 14 cultures examined were capable of reducing pentoxifylline to the corresponding alcohol (**2**) in yields ranging from 6% to a nearly quantitative conversion.[23] All positive cultures were reexamined to confirm the production of this metabolite and to generate "time-course" curves (percentage conversion vs. time of incubation). The time to maximum conversion of each culture (also indicated in Table 1) ranges from a very rapid metabolism (0.5 hr) to a rather slow or delayed reduction (144 hr). Time-course studies are valuable as a prelude to larger-scale fermentations for predicting the optimal time for harvesting cultures.

In addition to the expected production of the alcohol metabolite (**2**) was the unexpected observation that two cultures produced the carboxylic acid metabolites resulting from oxidative cleavage of the ketonic side-chain. Specifically, *Curvularia falcata* produced the carboxypropyl metabolite (**4**), and *Streptomyces griseus* produced the carboxybutyl metabolite (**3**) albeit in low, overall yields.[23]

Rhodotorula rubra was chosen for preparative-scale production, isolation, and characterization of the alcohol metabolite. For this study, five 125-mℓ Stage 1 flasks were used to inoculate five 1-ℓ Stage 2 flasks, and the substrate pentoxifylline (500 mg) in DMF was distributed among the flasks at 24 hr in Stage 2. After the biotransformation ceased at 144 hr, the culture was exhaustively extracted and the product purified by TLC and semipreparative HPLC to yield 100 mg (**2**). This product was identical chromatographically (TLC, HPLC) and spectrally (IR, MS, PMR) to standard pentoxifylline alcohol. However, its melting point (107°C) was substantially lower than the standard (124 to 125°C).[23] One obvious explanation would be that the reduction of pentoxifylline by this organism (which generates a chiral center) could be highly stereoselective, producing predominantly (or exclusively) the *S*- or the *R*-pentoxifylline alcohol, which would be expected to have a different melting point from the standard alcohol, which is a racemate (*RS*-pentoxifylline alcohol). Since the metabolite was optically active, we were convinced that this was the case. Our efforts then focused on the development of appropriate stereochemical methodology to determine the stereochemical course of reduction with this organism and to apply this methodology to the other 12 organisms which reduce pentoxifylline. Further, it was hoped that such methods could be applied to mammalian systems, including humans, where the stereochemical course of reduction has not been deduced.

A classical approach to the separation of enantiomers involves derivatization with a chiral derivatizing agent (CDA) to yield diastereomers which are much easier to resolve due to differences in physical properties. The chiral acid known as Mosher's reagent (α-methoxy-α-[trifluoromethyl]phenylacetic acid; MTPA), developed by Dale and Mosher in 1967[27] has been shown to be of general utility as a CDA in determining the stereochemistry of a wide range of analytes and also yields diastereomers amenable to chromatographic separation. Thus, derivatization of standard *RS*-pentoxifylline alcohol with *R* — (+) — MTPA, using dicyclohexylcarbodiimide as the coupling agent and imidazole as the catalyst, yielded the diastereomeric *S*-alcohol-*R*-MTPA ester (**5**) and the *R*-alcohol-*R*-MTPA ester (**6**) (see Figure 4).[25] Proton NMR was then utilized to deduce the stereochemistry of the alcohol(s) present by using the chiral NMR model developed by Dale and Mosher for related compounds. The perspective drawings shown in Figure 4 are based on that model. The fairly rigid conformations shown are based on strong dipole-dipole interactions between the ester ketone, the trifluoromethyl group, and the carbinyl (C*H*-O-) proton.[27] Note that the methyl group of the pentoxifylline alcohol is influenced by the phenyl ring of the MTPA as follows: in the *R*-alcohol-*R*-acid ester (**6**), the phenyl ring shields the methyl group (which is observed to resonate at 1.26 ppm); while in the *S*-alcohol-*R*-acid ester (**5**), the phenyl ring has little influence on the methyl group, which would thus be expected to resonate at a more deshielded position (1.34 ppm). The PMR spectrum of **5/6** (Figure 5A) yields the expected results, i.e., separation of the methyl doublets, based on differential anisotropic shielding, and equal peak integrals for the two esters (since they were derived from the racemate). The *R*-MTPA ester of the alcohol from *R. rubra* (Figure 5B) exhibits only the downfield doublet (corresponding to the *S*-alcohol) with virtually no *R*-alcohol-*R*-ester present (98:2 ratio of **5:6**).[25] Thus, reduction appears to be highly stereoselective for the *S*-alcohol with this organism.

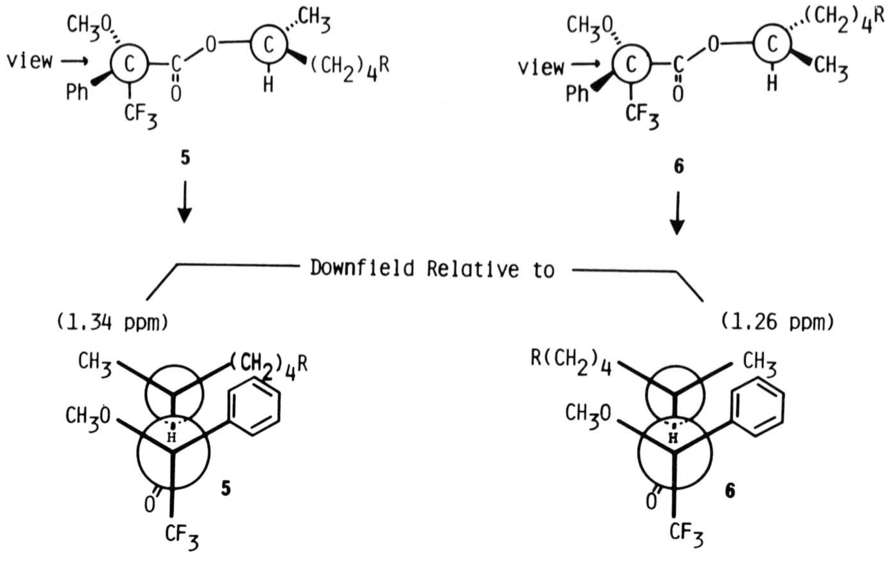

FIGURE 4. Perspective drawings for the diastereomeric S-pentoxifylline alcohol-R-MTPA ester (**5**) and R-pentoxifylline alcohol-R-MTPA ester (**6**) and for the PMR configuration correlation model to assign absolute configuration.

These results were confirmed using optical rotatory dispersion (ORD), where aliphatic secondary alcohols of the *S*-configuration exhibit a positive-plane curve, and *R*-alcohols exhibit a negative-plane curve. While these methods allowed us to assign the stereoselectivity in reduction with this one organism, preparative scale production, derivatization, and PMR- and ORD-spectral analyses would be prohibitive for all 12 other cultures, as well as with **2** produced in mammalian systems.

Fortunately, we were able to develop a reverse-phase HPLC system which allowed for the separation of these same diastereomeric esters (**5** and **6**), as shown in (Figures 6A and 6B).[25] The esters derived from *RS*-pentoxifylline yielded two peaks corresponding to the two esters, with equal peak areas, as expected. In contrast, the ester(s) resulting from derivatization of the *R. rubra* alcohol metabolite yielded one major peak corresponding to the ester with the shorter retention time and a minor peak for the later-eluting ester. Since the metabolite had been assigned the S-configuration (by the Mosher's procedure and ORD), we were able to assign the elution order as the *S*-alcohol-*R*-MTPA ester (**5**) at 41 min, and the *R*-alcohol-*R*-MTPA ester (**6**) at 45 min. Thus, from this chromatogram, he could confirm that the reduction of pentoxifylline by *R. rubra* is highly stereoselective for the *S*-alcohol with minor amounts of the *R*-alcohol produced (in a ratio of 98:2). Further, with this procedure, we could examine the stereochemical course of pentoxifylline reduction with the other organisms on an analytical scale, simply by derivatization of the product with Mosher's reagent, followed by HPLC. The results of these stereochemical time-course studies are shown in Table 3, which includes the percent of *S*- and/or *R*-alcohol produced, and the time required for maximal production. It is obvious from these studies that most organisms produce predominantly the *S*-alcohol, which is the usual stereochemical course of ketone reduction in mammals.[28] A few organisms produce low levels of the *R*-alcohol as well, and two are nonstereoselective in that they produce racemic pentoxifylline alcohol.

To this point, the microbial models developed for ketone reduction (as well as oxidative side-chain cleavage) would be considered "restrospective", i.e., the mammalian routes of biotransformations had already been elucidated. Efforts then focused on utilizing these models in a more prospective fashion to approach an unexplored aspect of mammalian

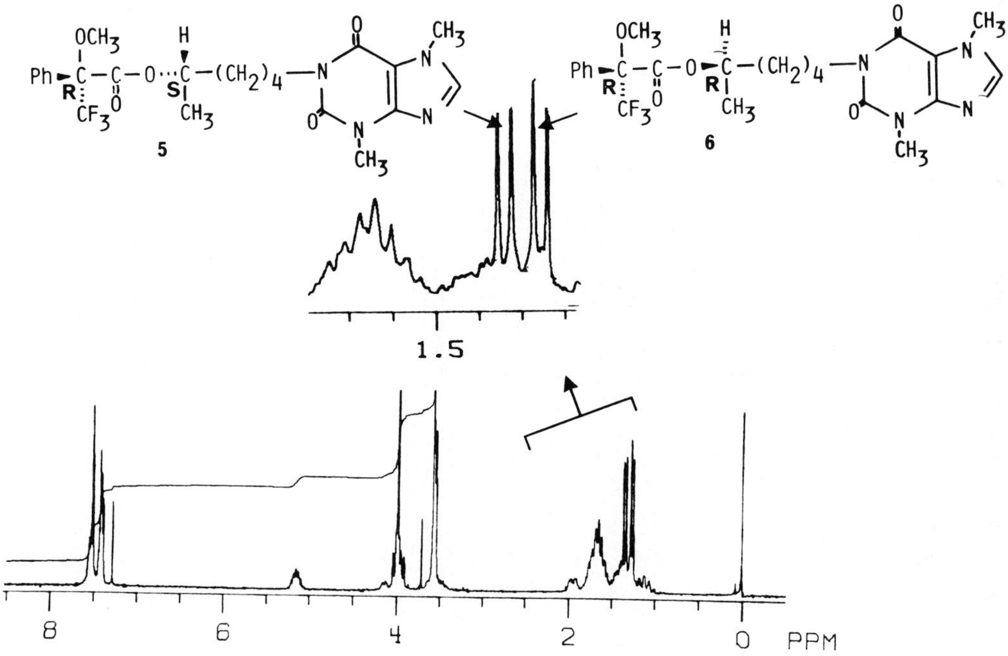

A

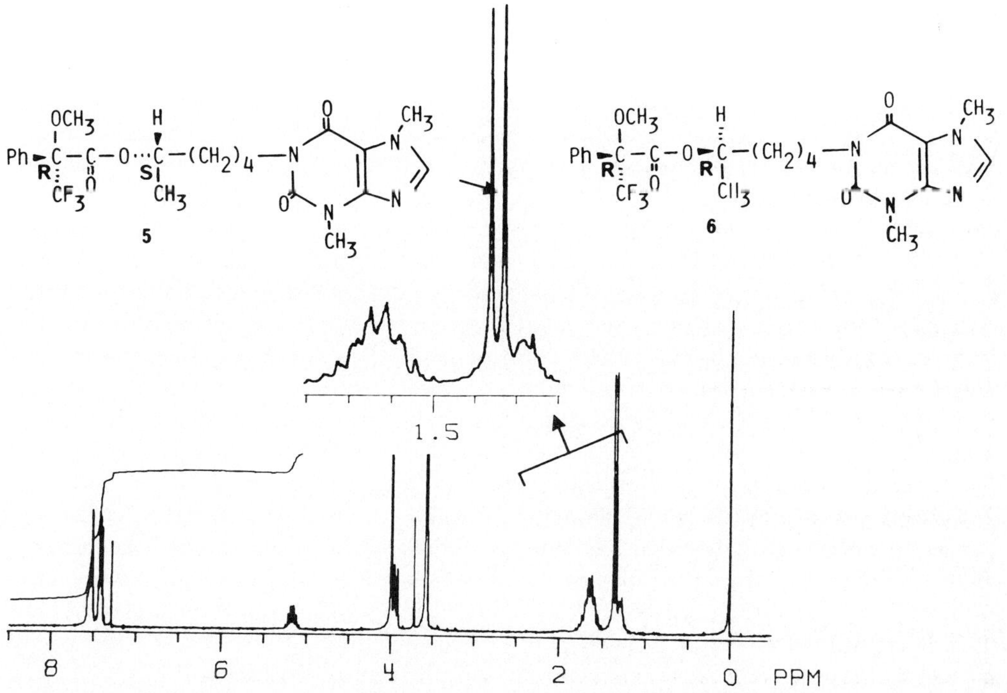

B

FIGURE 5. 200 MHz PMR spectra of the diastereomeric MTPA esters **5** and **6**. Figure 5A is the spectrum of **5**/**6** derived from standard racemic pentoxifylline alcohol (*RS*-**2**) with the region from 1.0 to 2.0 ppm expanded. Figure 5B is the spectrum of the pentoxifylline alcohol(s) obtained from the microbial reduction of pentoxifylline by *Rhodotorula rubra* with the region from 1.0 to 2.0 ppm expanded.

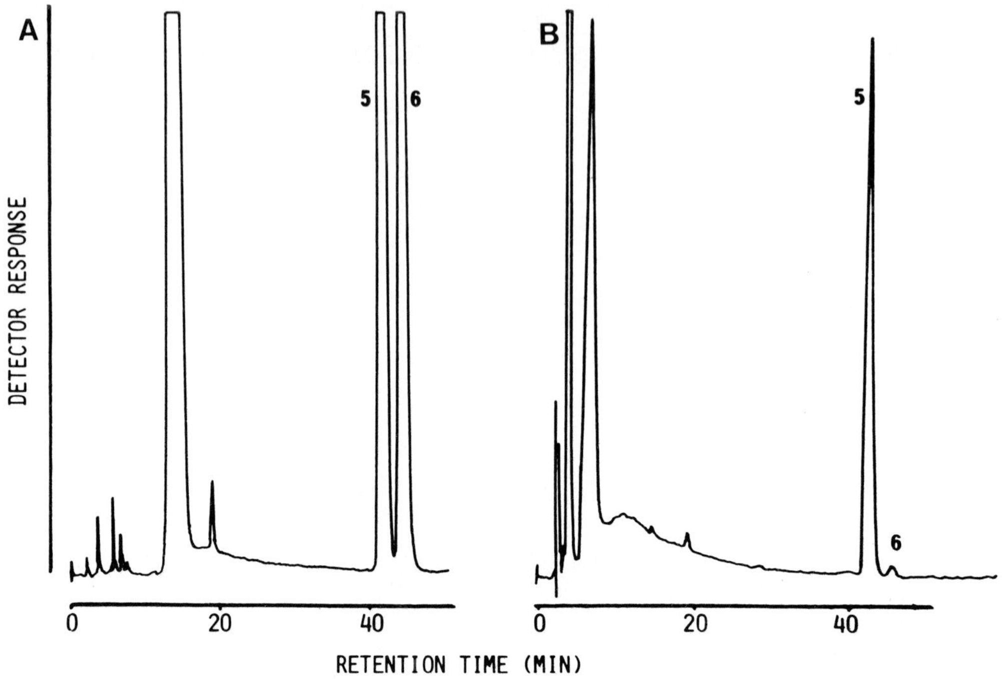

FIGURE 6. HPLC analysis of the diastereomeric MTPA esters **5** and **6**. Figure 6A is the chromatogram of **5/6** derived from the MTPA derivatization of standard racemic pentoxifylline alcohol (*RS*-**2**). Figure 6B is the chromatogram of **5/6** derived from the derivatization of pentoxifylline alcohol obtained from the microbial reduction of pentoxifylline by *Rhodotorula rubra*.

metabolism: specifically the stereochemical course of pentoxifylline reduction in humans. To do so, the author applied the analytical methodology conveniently developed with the model systems to selected human samples from a larger pharmacokinetic study on pentoxifylline metabolism in humans.[29] Six subject-plasma samples were selected which were shown in the previous study to contain blood levels of 1.9 to 2.8 μg/mℓ. Following derivatization with Mosher's reagent and HPLC analyses, the results (Table 4) clearly indicate a preferential reduction to the *S*-alcohol, with all subjects exhibiting lesser amounts of the *R*-alcohol.[66] Thus, the analytical methodology developed allows us to explore the stereochemical course of pentoxifylline alcohol formation in humans and indicates a similar stereoselectivity in the microbial models and human samples analyzed. This methodology should be useful in following the entire stereochemical time course of pentoxifylline metabolism in man.

C. Other Examples of Microbial Models

A number of other microbial models of the mammalian metabolism of drugs are currently being explored in the author's laboratory, two of which will be briefly discussed (see Figure 7). Sulindac (**7**) is a potent, broad-spectrum, antiinflammatory prodrug of the arylacetic acid class. In mammals, sulindac is metabolized by a reversible reduction at the sulfoxide moiety to yield the pharmacologically active sulfide (**8**), followed by reoxidation to the sulfoxide and further oxidation to the sulfone (**9**). We examined 24 fungal, yeast, and bacterial cultures for sulindac biotransformations, based on their reported abilities to catalyze the desired reduction/oxidations on sulfur-containing xenobiotics.[30] *Arthrobacter* species (ATCC 19140) and *Sporobolomyces pararoseus* (ATCC 11386) catalyzed only reduction to the sulfide, while *Aspergillus alliaceus* (NRRL 315) was shown to oxidize sulindac to the corresponding

Table 3
MICROBIAL STEREOSPECIFICITY IN PENTOXIFYLLINE REDUCTION

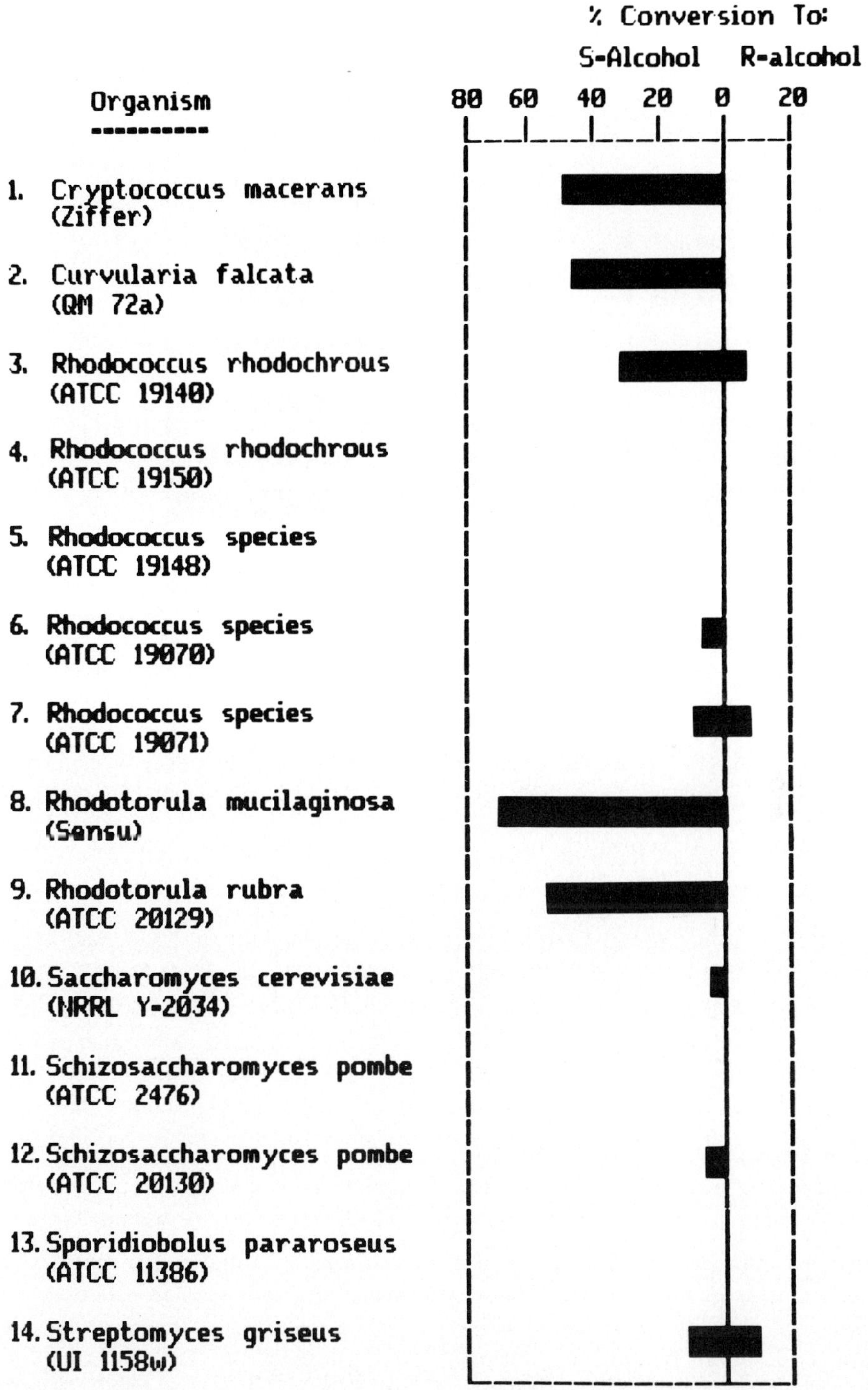

Table 4
HUMAN STEREOSPECIFICITY IN PENTOXIFYLLINE REDUCTION

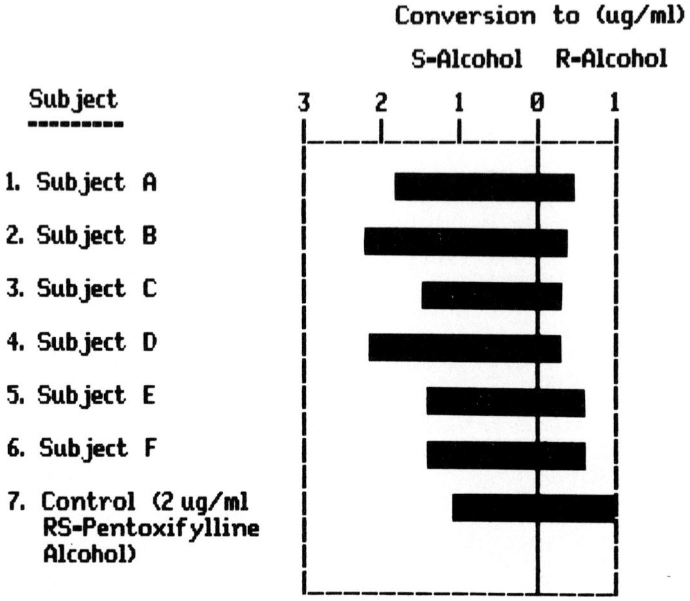

sulfone. Of particular interest was the observation that *Nocardia corallina* (ATCC 19070) produced both the sulfide and sulfone metabolites. More thorough, analytical studies demonstrated that this organism exhibited an aeration-dependent, reversible reduction of sulindac to the sulfide during the early phase of metabolism, followed by oxidation back to sulindac, and then to the sulfone. Thus, this organism parallels the total composite of major Phase 1 reduction/oxidation biotransformations observed for this drug in mammals. Current efforts are focusing on the development of appropriate analytical methodology for determining the stereochemical course of these reactions (the sulfoxide moiety is chiral). It is anticipated that this methodology will also be applicable to examining the stereochemical course of sulindac metabolism in mammals (including man), as described above for pentoxifylline.

The 4-hydroxycoumarin, warfarin (**10**), is a vitamin K antagonist which has been used for decades as an anticoagulant. More recently, interest in warfarin has been based on its use as a ''metabolic probe'' of mammalian enzyme systems, since one can deduce the presence of specific enzymes by the pattern of metabolites produced, rather than by isolating the individual enzymes.[31] The major routes of warfarin metabolism include complex patterns of aromatic hydroxylations at the 6-, 7-, 8-, 9- (benzylic), and 4′-positions. The substrate stereoselectivity and regioselectivity (i.e., which stereoisomer of warfarin is hydroxylated to which phenol(s)) is controlled by individual cytochrome P-450 isozymes, and serves as a basis for discriminating the individual isozymes in vivo.[31] In addition, ketone reduction to the corresponding alcohol(s) is a highly stereoselective process in mammals. The authors explored the microbiological metabolism of warfarin as an initial entree into developing microbial models of the individual Phase 1 biotransformations seen in mammals, and to determine the chirality of microbial reductase systems for application in stereoselective synthesis (vide infra). Nineteen cultures were examined based on their reported abilities to reduce various ketonic substrates. Of these, *N. corallina* (ATCC 19070) exhibited complete substrate and product stereoselectivity, as it reduced only *S*-warfarin to the corresponding *S*-alcohol (**11**). *Arthrobacter* species (ATCC 19140) exhibited marked substrate and complete product stereoselectivity in that it reduced principally *S*-warfarin, and to a lesser extent *R*-warfarin, to the corresponding *S*-alcohols (9**S-11** and 9**R-11**), respectively).[32] The product

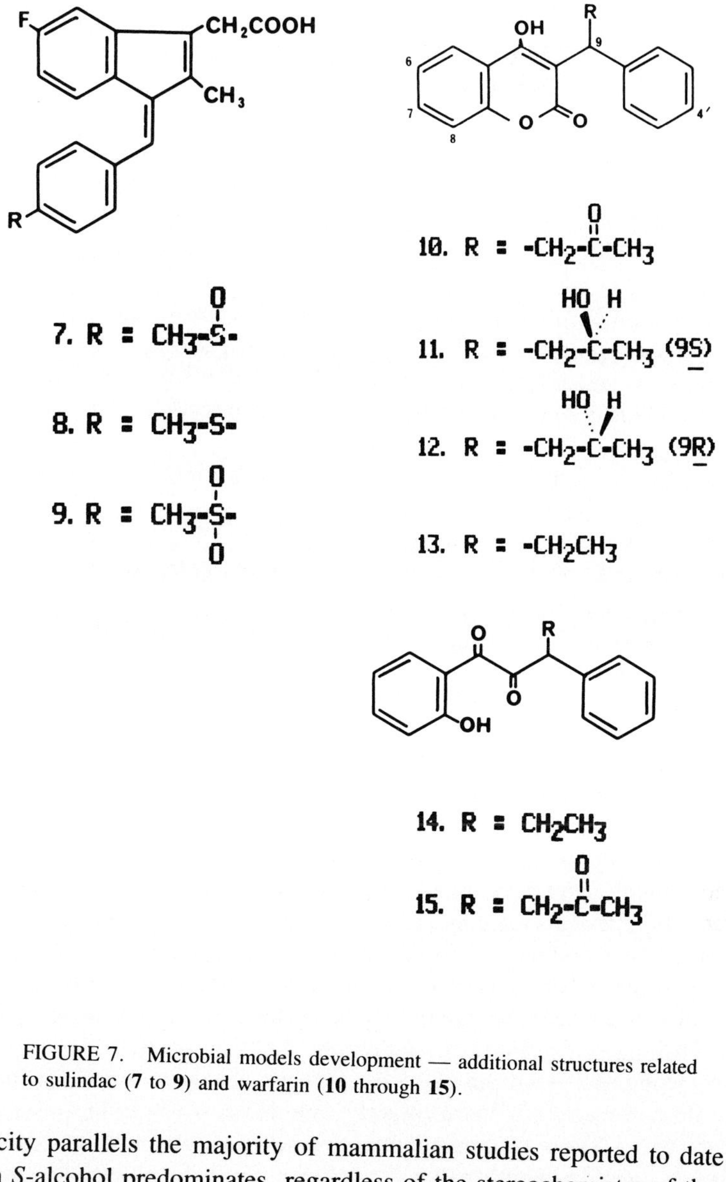

FIGURE 7. Microbial models development — additional structures related to sulindac (**7** to **9**) and warfarin (**10** through **15**).

stereospecificity parallels the majority of mammalian studies reported to date, in that reduction to an *S*-alcohol predominates, regardless of the stereochemistry of the substrate.

Warfarin and related compounds are also reported to undergo apparent hydrolytic cleavage of the lactone ring followed by oxidative decarboxylation in mammals to yield the corresponding alpha-diketone metabolite.[33] The authors found that *Aspergillus niger* (X-172) catalyzes this same reaction, first demonstrated on the related compound phenprocoumon (**13**) to yield **14** (chosen to simplify structure of the metabolite), and then shown with warfarin to yield **15**.[64]

III. MICROBIAL SYSTEMS AS BIOCATALYSTS IN MEDICINAL CHEMISTRY

A. Background

One of the most widely exploited uses of microbial transformations is for biocatalysis in synthetic medicinal and organic chemistry. Such methods are complimentary to classical synthetic methods, and are used when clear advantages exist. Often cited advantages of microbial systems over more classic reagents include the following:[34]

1. Mild reaction conditions — Microbial transformations are typically conducted under mild, physiological-reaction conditions (ambient temperature, physiological pH, simple buffers). Consequently, instability of substrates or products due to extremes of pH or temperature typically does not present a problem. Occasionally, the high aeration rates used in oxidative reactions will yield difficulties in the spontaneous air oxidation of labile products. This can often be overcome by the use of biocompatible antioxidants such as ascorbic acid, bisulfite, dithiothreitol, etc.
2. Catalysis at ''nonactivated'' positions — Synthetic reagents usually require ''activating'' functional groups at or near the site of modification. In contrast, microbial transformations may involve catalysis and functionalization at completely ''nonactivated'' positions. A classic example is the report by Peterson and Murray[35] at Upjohn that *Rhizopus arrhizus* catalyzes the 11α-hydroxylation of the steroid progesterone in high yields (a key step in the manufacture of corticosteroids). Up until that time, the reaction required a series of low-yielding chemical steps — low-yielding because of the nonactivated (essentially aliphatic) nature of the position to be hydroxylated. Many investigators consider this discovery the birth of modern microbial transformations. The utility of this reaction catalyzed an intense exploration for other organisms catalyzing other reactions, with the result that nearly every available position on the steroidal nucleus can be hydroxylated by the proper choice of microorganism.[36]
3. Regioselectivity — Microbial systems often exhibit unexcelled regioselectivity with polyfunctional molecules. That is, with multiple, similar functional groups, one may be selectively altered in lieu of the others by the proper choice of microorganism. By contrast, chemical reagents may be indiscriminate, yielding complex mixtures of products. For example, Rosazza and co-workers[37] demonstrated the regioselective O-demethylation of the 10,11-dimethyl ether of the catecholic aporphine, apomorphine. Specifically, *Streptomyces rimosus* O-demethylated at the C-11 position to yield apocodeine, while *Cunninghamella elegans* attached the C-10 methoxyl to produce isoapocodeine. While apocodeine is easily prepared chemically, this represents the most practical synthesis of isoapocodeine to date.
4. Stereoselectivity — One of the most widely exploited attributes of microbial transformations is the high degree of stereoselectivity often observed in terms of substrate metabolism (i.e., one stereoisomer is preferentially metabolized), or in product formation (i.e., one stereoisomeric product predominates). In a few cases, microbial cultures exhibiting reverse (hence, complimentary) stereoselectivity have been described. This property will be discussed in the examples to follow.
5. Predictability — Not in contrast to, but in common with synthetic reagents, microbial transformations can exhibit a high degree of type-reaction predictability and specificity. By reaction specificity, it is meant that conversion at one type of functional group may occur (e.g., amide hydrolysis), while sparing related functional groups (e.g., esters, carbamates). Type-reaction predictability means that one can choose cultures for a particular biotransformation with a high probability of success if that choice is based upon literature precedent that such organisms can catalyze the desired type-reaction (e.g., ketone reduction) rather than using random screening of a whole culture collection. The discussion (above) of cultures initially examined for pentoxifylline reduction is a case in point. Complications of type-reactions catalyzed by microorganisms,[26] or by transformations according to structural or drug class[38,39] have rapidly advanced and facilitated this more systematic approach to finding cultures catalyzing the desired reaction.

In a few cases, sufficient numbers of substrate analogues have been examined with a particular microorganism catalyzing a particular reaction, to have allowed the formulation of an ''active-site map'' or ''set of rules'' useful in predicting substrate requirements for

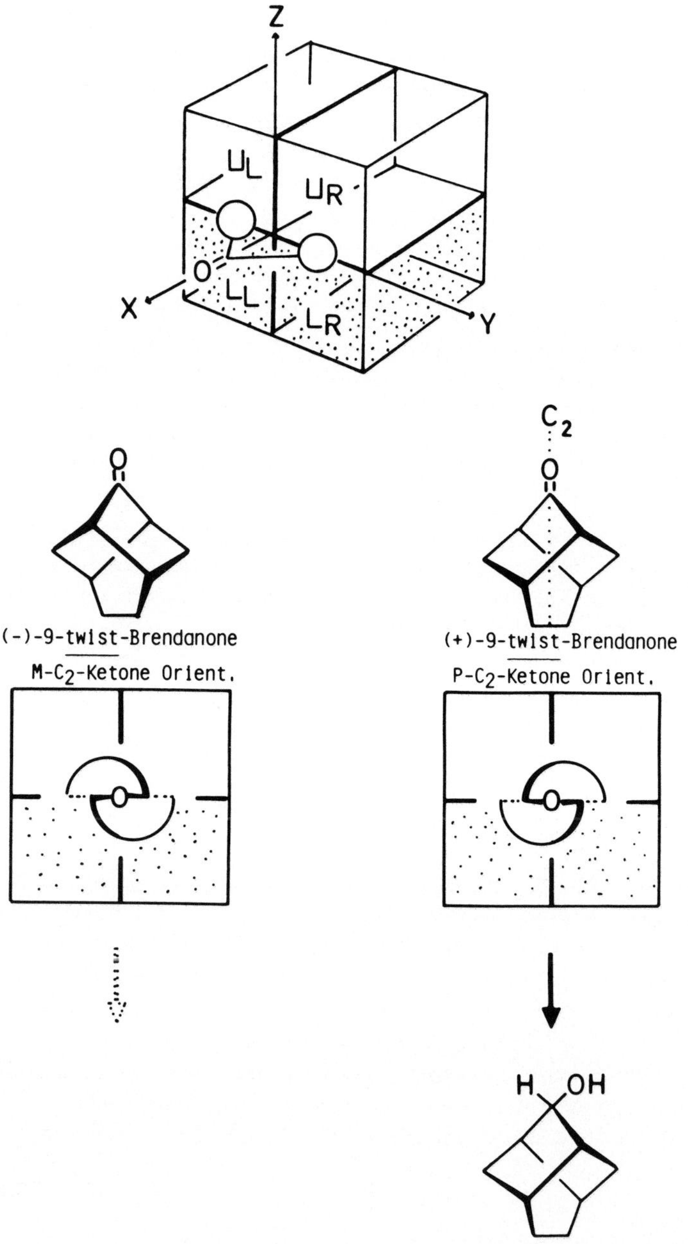

FIGURE 8. Quadrant projection formula for microbial reductases from *Curvularia falcata* and *Rhodotorula rubra* (from Reference 41).

catalysis and stereochemical outcome.[40] A recent example is that of Nakazaki and co-workers on ketone reductions by *Curvularia falcata* and *Rhodotorula rubra*.[41] Quadrant projection formulas (e.g., Figure 8) depicting the active sites of the enzymes responsible for reduction were used to predict preferred modes of substrate binding; to specify the approach of the hydride equivalent involved in reduction (i.e., which face of the ketone is attacked); and, as a result, to preduct structural requirements and the stereochemical outcome of the reduction. Other examples of active site maps or rules are given in Table 5.[34,41-55] Clearly, it

Table 5
STRUCTURAL SPECIFICITY MAPS AND OTHER PREDICTIVE TOOLS OF MICROBIAL (AND RELATED) REACTIONS

Correlation	Organism(s) and description	Ref.
Bertrand-Hudson rule	*Acetobacter* sp.: only polyols with *cis*-secondary hydroxyl-groups adjacent to a primary alcohol are oxidized to the corresponding ketones.	42
Prelog's rule	Yeasts and fungi: reduction of unsymmetrical ketones yields principally S-alcohols, and is generally controlled by steric bulk of functional groups flanking the ketone.	43, 44
Sih-Rosazza keto-reductase proposal	General: keto-enol tautomerism of a ketonic substrate controls the degree of microbial reduction; ketonic character favors reduction.	45
Diamond lattice section	Horse liver alcohol dehydrogenase: keto-reductase reduces mono- to polycyclic substrates consisting of cyclohexane rings, the composite of which forms a lattice describing allowed and forbidden substitution.	43, 46, 47
Diamond lattice map	*Pseudomonas testosteroni:* 3-hydroxysteroid dehydrogenase (as keto-reductase); 3-keto steroidal substrates superimposed define a lattice describing allowed and forbidden substitution.	48, 49
Triangulation map	*Calonectria decora:* steroidal hydroxylase catalyzes hydroxylation on steroid skeleton 4 Å from one and 6 Å from a second polar binding site (ketone or hydroxyl) to give a 4 Å × 6 Å × 6 Å triangular site; reciprocal binding is allowed (Å = angstrom).	50, 51
Reciprocal binding plot	*Rhizopus nigricans:* reciprocal binding on oxygenated steroids controls site(s) of hydroxylation.	52
Reciprocal binding plot	*Aspergillus tamarii:* reciprocal and multiple binding sites on oxygenated steroids control site(s) of hydroxylation.	53
Quadrant section rule	*Sporotrichum (Beauveria) sulfurescens:* aliphatic hydroxylation of amide substrates occurs 5.5 Å from the ketonic (amide) binding site; quadrant rule allows prediction of orientation (stereochemistry) of hydroxyl group.	54
Quadrant rules	*Curvularia falcata* and *Rhodotorula rubra:* quadrant rules predict stereospecificity of ketone reduction of open, cyclic, and caged gyrochiral ketones of C_1, C_s, or C_2 symmetry.	41
Heteroatom triangulation map for dealkylation	*Streptomyces griseus:* N- and O-dealkylation (presumably via C(α)-hydroxylation) of alkaloidal substrates occurs 6.7 to 7 Å from the electrophilic site and 6 Å from a pi-binding site to give a 6.7 to 7 Å × 6 Å × 5 Å triangular site.	55

is this type of approach which results in a more predictable biocatalyst and the attainment of the status of "reagent" for use in synthesis and modification of structurally-related substrates.

B. Biocatalysis — Aporphines

In order to discuss some of the foregoing points with specific examples, microbial transformation studies on a class of alkaloids, known as the aporphines will be discussed. S-(+)-Glaucine ([**16**], Figure 9) is a potent antitussive and is structurally related to a number of alkaloidal antitumor agents. As with most natural products, only one stereoisomer is produced in nature, and the biological evaluation of "unnatural stereoisomers" is limited by their lack of availability. For a number of reasons, including a desire to prepare the unnatural R-(−)-glaucine (**17**), microbial transformations of **16** were undertaken. Initial studies in this regard included the observation that the fungus *Fusarium solani* (ATCC

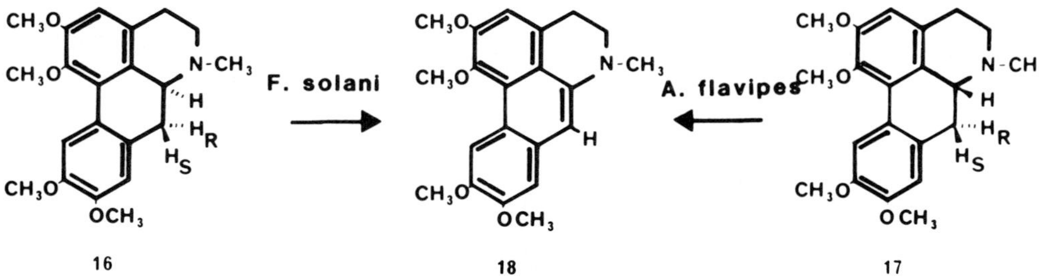

FIGURE 9. Stereoselective fungal metabolism of S-(+)-glaucine (**16**) by *Fusarium solani* and R-(−)-glaucine (**17**) by *Aspergillus flavipes* to yield dehydroglaucine (**18**).

12823) oxidized S-(+)-glaucine to the achiral product, dehydroglaucine (**18**).[56] Following the development of appropriate analytical methodology,[57] we were able to discern that the oxidation is quantitative and completely stereospecific, since R-(−)- glaucine (**17**) is not metabolized by this organism. This observation suggested the possible utility of *F. solani* in the destructive resolution of racemic glaucine (**16/17**) by substrate-stereoselective metabolism. After subjecting a total of 1.5 g of the racemic substrate to this fungus, R-(−)-glaucine was obtained in 90% theoretical yield and 96% enantiomeric excess.[58] Since this system is capable of the stereoselective metabolism of other aporphines (see below), it could be of very practical significance in other resolutions, since most naturally occurring aporphines are of the S-configuration, or are produced as racemates to total synthesis.

With R-(−)-glaucine on hand, the authors were then able to address the possibility that other organisms might exhibit the reverse stereospecificity of that seen with *F. solani*. The occurrence of such complimentary stereoselective biocatalysts has been described previously for other classes of compounds and type reactions.[59,60] Indeed, *Aspergillus flavipes* was shown to oxidize **17** to dehydroglaucine quantitatively and with complete stereospecificity, since the S-(+)-enantiomer (**16**) was not metabolized to any extent.[58] Such a system could be used in a destructive resolution (as described above) when the S-(+)-enantiomer is desired.

The stereoselectivity of these biocatalysts makes them attractive for application to other aporphines and related systems as well. In addition, the "catalyst" nature of *F. solani* is further enchanced by the observation that (like many microbes) it is catalytically active in "cell suspension", i.e., harvested cells simply resuspended in buffer (a growing culture is not necessary).[61] The reaction is also much faster; complete conversion of 500 μg/mℓ of S-(+)-glaucine occurs in 8 hr of cell suspension, as opposed to 3 days in growing culture.

An elucidation of the stereochemical course of these reactions is important not only mechanistically, but for predicting what other aporphines and related compounds will be accepted by the enzyme systems involved. While the stereospecificity at the sole chiral center (C-6a) was clear, that of the C-7-methylene (a diastereotopic center) was not. That is, while we knew that with *F. solani* the 6a*S*-chirality is necessary, the reaction could involve the C-7-*pro*-S hydrogen in an overall *trans*-elimination, or the C-7-*pro*-R hydrogen in an overall *cis*-elimination (see Figure 9). This question was addressed by the synthesis and metabolism of 7-methyl blocked analogues of glaucine ([**19**] to [**22**]; Figure 10) which, depending on the stereochemistry of the C-7-methyl group, would allow the reaction to proceed only in a *cis*- or only in a *trans*-fashion.[61] Transformation experiments with these substrates and analysis of residual substrates by chiral lanthanide shift reagent NMR and ORD demonstrated that both cultures catalyze an overall "*cis*"-elimination of hydrogen at the 6a,7-center,[61] and, as expected (based on the required stereochemistry at C6a), *F. solani* stereoselectively metabolizes the *6aS,7S*-stereoisomer (**19**), while *A. flavipes* is stereoselective for the enantiomeric *6aR,7R*-stereoisomer (**20**).[62]

In an effort to explore the scope and specificity of these reactions (i.e., what structural

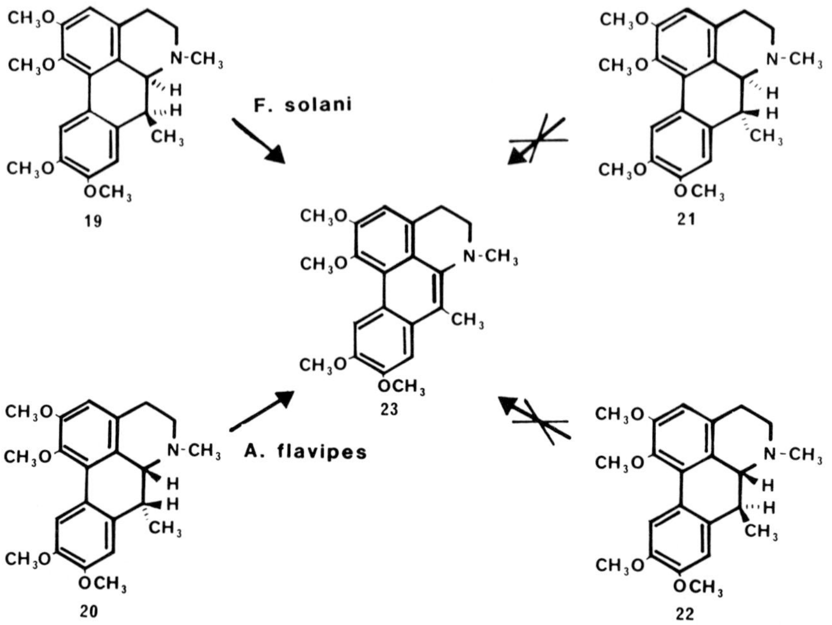

FIGURE 10. Stereoselective fungal metabolism of 7-methyl-glaucine substrates **19** through **22** to 7-methyldehydroglaucine (**23**).

features are required, allowed, or disallowed), it was our desire to examine a number of glaucine analogues as substrates with *F. solani*. Prior to such studies, however, it was important to determine if the enzyme(s) responsible for catalysis were constitutive (i.e., the enzyme is normally present irrespective of the presence of the substrate) or inducible (i.e., the presence of the substrates initiates *de novo* enzyme synthesis). It is possible that certain analogues may not undergo metabolism, simply because they are not inducers of the requisite enzyme(s), rather than being poor substrates. Cell suspension experiments with antibiotic inhibitors (to block *de novo* protein synthesis) show that the enzyme system of *F. solani* is nonconstitutive and stereospecifically induced by S-(+)-glaucine (**16**).[65] The author now can preinduce cells with **16** to insure adequate enzyme levels prior to examining the metabolism of substrate analogues in cell suspension.

Figure 11 represents a structural specificity map for *F. solani,* constructed from the information discussed above, and from examining various ''desmethoxyl-analogues'' of glaucine as substrates. The intent of the latter study was to delineate which of the alkoxy-groups were required rather than simply tolerated in the substrate. The stereochemical requirements at positions 6a and 7 depict the necessity for a 6a*S*-chirality as well as a *cis*-orientation of hydrogen to be complimentary to a ''hydrogen acceptor'' (A) on the enzyme. An R-group at position 7 other than hydrogen is tolerated (e.g., 7-methylglaucine), provided it replaces the 7-*pro*-S hydrogen. Replacement of the 7-*pro*-R hydrogen would obviously preclude a *cis*-elimination of hydrogen. Analogue studies have shown that the methoxyl-groups at positions 1, 9, and/or 10 are acceptable (as in glaucine), but are not required for catalysis.[63] An oxygenated functionality is required at position 2 (perhaps as a binding site), but this does not have to be a methoxyl-group; for example, an O-acetyl function is acceptable at this position. Clearly, the enzyme system of *F. solani* has a broader scope and specificity than only for glaucine. It is anticipated that further refinement of this structural specificity map will be useful in predicting stereoselective biotransformations of other aporphines and related compounds using this microbial biocatalyst.

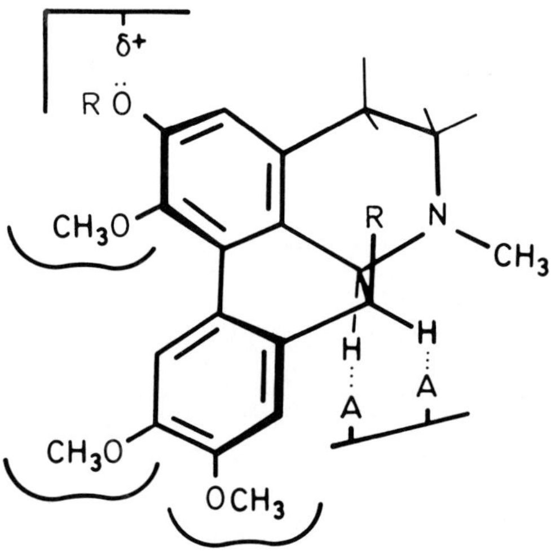

CH_3O- ACCEPTED, NOT REQUIRED

RO- REQUIRED

R= H, CH_3 ACCEPTABLE

FIGURE 11. Structural specificity map for the aporphine-oxidizing system of *Fusarium solani* (ATCC 12823).

IV. CONCLUSIONS

Microorganisms have a long history of providing mankind with valuable products of primary and secondary metabolism (alcoholic beverages, vitamins, amino acids, organic acids, solvents, antibiotics, etc.). In addition, microbial systems enjoy an expanding role in catalyzing specific biotransformations of interest on both the laboratory and industrial scale. The foregoing discussion focused on two areas in which microbial transformations facilitate studies in medicinal chemistry and related areas. Microbial models of mammalian Phase 1 biotransformations have excellent potential for facilitating drug metabolism studies by predicting routes and other aspects of metabolism, providing preparative quantities of metabolites, lowering costs, and lessening animal demand. Microbes also serve an ever-expanding role as synthetic biocatalysts, due principally to the high degree of stereo- and/or regioselectivity often attained. The continued development of structural specificity maps and related predictive tools will continue to facilitate their use as reagents in synthetic, organic, medicinal, and natural products chemistry.

ACKNOWLEDGMENTS

The pentoxifylline work was supported in part by Hoechst-Roussel Pharmaceuticals, Inc. (Somerville, N.J.). The aporphine (glaucine) studies were supported by Grant F-780 from the Robert A. Welch Foundation. Studies concerning sulindac and warfarin were supported by Biomedical Research Support Grants from the National Institutes of Health administered through the University Research Institute, at the University of Texas.

REFERENCES

1. **Dagani, R.,** Alternative methods could cut animal use in toxicity testing, *Chem. Eng. News,* p. 7, Oct. 31, 1983.
2. **Rollin, B. E.,** *Animal Rights and Human Mortality,* Prometheus, Buffalo, 1981, p. 182.
3. **Dagani, R.,** In vitro methods may offer alternatives to animal testing, *Chem. Eng. News,* p. 25, Nov. 12, 1984.
4. **Dawson, M.,** Three R's of animal experimentation: reduction, replacement, and refinement, *Pharm. J.,* p. 607, May 19, 1984.
5. **Holden, C.,** New focus on replacing animals in the lab, *Science,* 215, 35, 1982.
6. **Balls, M., Riddell, R. J., and Worden, A. N., Eds.,** *Animals and Alternatives in Toxicity Testing,* Academic Press, London, 1983.
7. **Rosazza, J. P. and Smith, R. V.,** Microbial models for mammalian metabolism, *Arch. Biochem. Biophys.,* 161, 551, 1974.
8. **Smith, R. V. and Rosazza, J. P.,** Microbial models for mammalian metabolism, *J. Pharm. Sci.,* 64, 1737, 1975.
9. **Rosazza, J. P. and Smith, R. V.,** Microbial models for drug metabolism, *Adv. Appl. Microbiol.,* 25, 169, 1979.
10. **Rosazza, J. P. and Smith, R. V.,** Microbial models of mammalian metabolism, in *Applied Microbiology,* Vol. 25, Perlman, D., Ed., Academic Press, New York, 1979, 169.
11. **Smith, R. V. and Rosazza, J. P.,** Microbial transformations as a means of preparing mammalian drug metabolites, in *Microbial Transformations of Bioactive Compounds,* Vol. 2, Rosazza, J. P., Ed., CRC Press, Boca Raton, Fla., 1982, 1.
12. **Smith, R. V. and Rosazza, J. P.,** Microbial models of mammalian metabolism, *J. Nat. Prod.,* 46, 79, 1983.
13. **Smith, R. V. and Davis, J. P.,** Induction of xenobiotic monooxygenases, in *Adv. Biochem. Eng.,* Vol. 14, Ghasi, T. K., Fiechter, A., and Blakebrough, N., Eds., Springer-Verlag, Berlin, 1980, 61.
14. **Smith, R. V., Acosta, D., and Rosazza, J. P.,** Microbial and cellular models for mammalian metabolism, *Adv. Biochem. Eng.,* 5, 70, 1977.
15. **Rosazza, J. P.,** Antitumor antibiotic bioactivation, biotransformation, and derivatization by microbial systems, *Recent Results Cancer Res.,* 63, 58, 1978.
16. **Clark, A. M., McChesney, J. D., and Hufford, C. D.,** The use of microorganisms for the study of drug metabolism, *Medicinal Res. Rev.,* 5, 231, 1985.
17. **Wackett, L. P. and Gibson, D. T.,** Metabolism of xenobiotic compounds by enzymes in cell extracts of the fungus *Cunninghamella elegans, Biochem. J.,* 205, 117, 1982.
18. **Cerniglia, C. E., Freeman, J. P., and Mitchum, R. K.,** Glucuronide and sulfate conjugates in the fungal metabolism of aromatic hydrocarbons, *Appl. Environ. Microbiol.,* 43, 1070, 1982.
19. **Bridges, J. W., Chasseaud, L. F., Cohen, G. M., and Walker, C. H.,** Applications of pharmacokinetics, in *Animals and Alternatives in Toxicity Testing,* Balls, M., Riddell, R. J., and Worden, A. N., Eds., Academic Press, New York, 1983, 31.
20. **Coutts, R. T., Foster, B. C., Grahm, R. J., and Myers, G. E.,** Metabolism of (+/−) — N — (n-propyl)amphetamine by *Cunninghamella echinulata, Appl. Environ. Microbiol.,* 37, 429, 1979.
21. **Baker, J. K., Clark, A. M., and Hufford, C. D.,** High performance liquid chromatographic analysis of the metabolism of primaquine and the identification of a new mammalian metabolite, *J. Chromatogr.,* 230, 69, 1982.
22. **Iishi, U., Hayashi, M. H., Niwaguchi, T., and Nakahara, Y.,** Studies on lysergic acid diethylamide and related compounds. VII. Microbiological transformations of lysergic acid diethylamide and related compounds, *Chem. Pharm. Bull.,* 27, 1570, 1979.
23. **Davis, P. J., Yang, S. K., and Smith, R. V.,** Microbial models of mammalian metabolism: microbial reduction and oxidation of pentoxifylline, *Appl. Environ. Microbiol.,* 48, 327, 1984.
24. **Smith, R. V., Yang, S. K., Davis, P. J., and Bauza, M. T.,** Determination of pentoxifylline and its major metabolites in microbial extracts by thin-layer and high-performance liquid chromatography, *J. Chromatogr.,* 281, 281, 1983.
25. **Davis, P. J., Yang, S. K., and Smith, R. V.,** Microbial models of mammalian metabolism, stereospecificity in ketone reduction with pentoxifylline, *Xenobiotica,* 15, 1001, 1985.
26. **Kieslich, K.,** *Microbial Transformations of Non-Steroidal Cyclic Compounds,* Georg Thieme Verlag, Stuttgart, 1976.
27. **Dale, J. A. and Mosher, H. S.,** Nuclear magnetic resonance enantiomer reagents. Configurational correlations via nuclear magnetic resonance chemical shifts of diastereomeric mandelate, O-methylmandelate, and *alpha*-methoxy-*alpha*-trifluoromethyl phenylacetate (MTPA) esters, *J. Am. Chem. Soc.,* 95, 512, 1973.
28. **You, K., Arnold, L. J., Jr., Allison, W. S., and Kaplan, N. O.,** Enzyme stereospecificity of nicotinamide nucleotides, *Trends Biochem. Sci.,* 3, 265, 1978.

29. **Smith, R. V., Waller, E. S., Doluisio, J. T., Bauza, M. T., Puri, S. K., and Lassman, H. B.,** Pharmacokinetics of orally administered pentoxifylline in men, *J. Pharm. Sci.,* 75, 47, 1986.
30. **Davis, P. J. and Guenthner, L. E.,** Sulindac oxidation/reduction by microbial cultures; microbial models of mammalian metabolism, *Xenobiotica,* 15, 845, 1985.
31. **Porter, W. R., Whesler, C., and Trager, W. F.,** Changes in the Metabolic Profiles of R- and S-Warfarin and R- and S-Phenprocoumon as a Probe to Categorize the Effect of Inducing Agents on Microsomal Hydroxylases, *Biochem. Pharmacol.,* 30, 3099, 1981.
32. **Davis, P. J. and Rizzo, J. D.,** Stereoselective reduction of warfarin by selected bacterial cultures, *Appl. Environ. Microbiol.,* 43, 884, 1982.
33. **Thonnart, N.,** In vitro metabolism of clocoumarol, a new 4-hydroxycoumarin anticoagulant; identification of two additional metabolites, *Drug Metab. Dispos.,* 7, 450, 1979.
34. **Sih, C. J. and Rosazza, J. P.,** Microbial transformations in organic synthesis, in *Applications of Biochemical Systems in Organic Chemistry,* Jones, J. B., Sih, C. J., and Perlman, D., Eds., John Wiley & Sons, New York, 1976, 69.
35. **Peterson, D. H. and Murray, H. C.,** Microbiological oxygenation of steroids at carbon 11, *J. Am. Chem. Soc.,* 74, 1871, 1952.
36. **Charney, W. and Herzog, H. L.,** *Microbial Transformations of Steroids,* Academic Press, New York, 1967.
37. **Rosazza, J. P., Stocklinsii, A. W., Gustafson, M. A., Adrian, J., and Smith, R. V.,** Microbial models of mammalian metabolism. O-Dealkylation of 10,11-dimethoxyaporphine, *J. Med. Chem.,* 18, 791, 1975.
38. **Rosazza, J. P., Ed.,** *Microbial Transformations of Physiologically Active Substances,* CRC Press, Boca Raton, Fla., 1982.
39. **Rhem, H. J. and Reed, G., Eds.,** *Microbial Transformations,* Vol. VI-B, B in *Biotechnology: a Comprehensive Treatise in Eight Volumes,* Kieslich, K., Ed., Verlag-Chemie, Weinheim, 1984.
40. **Abbott, B. J.,** Some new approaches to biotransformations, *Dev. Indust. Microbiol.,* 20, 345, 1979.
41. **Nakazaki, M., Chikamatsu, H., Naemura, K., and Asao, M.,** Microbial stereodifferentiating reduction of carbonyl compounds; proposed quadrant rule, *J. Org. Chem.,* 45, 4432, 1980.
42. **Hann, R. M., Tilden, E. B., and Hudson, C. S.,** The oxidation of sugar alcohols by *Acetobacter suboxydans, J. Am. Chem. Soc.,* 60, 1201, 1938.
43. **Prelog, V.,** Specification of the stereospecificity of some oxido-reductases by diamond lattice sections, *Pure Appl. Chem.,* 9, 119, 1964.
44. **MacLeod, R., Prosser, H., Fikentsher, L., Lanzi, J., and Mosher, M. S.,** Asymmetric reductions. XII. Stereoselective ketone reduction by fermenting yeasts, *Biochemistry,* 3, 838, 1964.
45. **Sih, C. J. and Rosazza, J. P.,** Microbial transformations in organic synthesis, in *Applications of Biochemical Systems in Organic Chemistry,* Jones, J. B., Sih, C. J., and Perlman, D., Eds., John Wiley & Sons, New York, 1976, 69.
46. **Prelog, V.,** Product and substrate stereoselectivity in the enzymic reduction of carbonyl compounds, *Colloq. Ges. Physiol. Chem.,* 14, 288, 1963.
47. **Graves, J. M. H., Clark, A., and Ringold, H. J.,** Stereochemical aspects of the substrate specificity of horse liver alcohol dehydrogenase, *Biochemistry,* 4, 2655, 1965.
48. **Ringold, H. J., Bellas, T., and Clark, A.,** Adamantanone as a probe for the dimensions and characteristics of the substrate binding pocket of certain alcohol dehydrogenases, *Biochem. Biophys. Res. Commun.,* 27, 361, 1967.
49. **Ringold, H. J., Graves, J. M. H., Clark, A., and Bellas, T.,** Experimental strategy for the delineation of the enzyme region involved in substrate recognition, *Proc. Natl. Acad. Sci. U.S.A.,* 56, 255, 1966.
50. **Bell, A. M., Cherry, P. C., Clark, I. M., Denny, W. A., Jones, E. R. H., Meakins, G. D., and Woodgate, P. D.,** Microbiological hydroxylation of steroids. IV. The pattern of dihydroxylations of monooxygenated 5-*alpha*-androstanes with cultures of the fungus *Calonectria decora, J. Chem. Soc. Perkin Trans., 1,* 2081, 1972.
51. **Jones, E. R. H., Meakins, G. D., Miners, J. O., Mirrington, R. N., and Wilkins, A. L.,** microbiological hydroxylation of steroids. XII. Hydroxylation of 3,20-, 7,20-, and 11,20-dioxygenated 5-*alpha*-pregnanes, *J. Chem. Soc. Perkin Trans., 1,* 1842, 1976.
52. **Browne, J. W., Denny, W. A., Jones, E. R. H., Meakins, C. R., Morisawa, Y., Pendlebury, A., and Prognell, J.,** Microbiological hydroxylation of steroids. VII. The pattern of dihydroxylation of monooxo-5-*alpha*-androstanes and -5-*alpha*-estranes with the fungus *Rhizopus nigricans, J. Chem. Soc. Perkin Trans., 1,* 1493, 1973; Microbiological hydroxylations of steroids. VIII. The pattern of monohydroxylation of diketones and keto-alcohols derived from 5-*alpha*-androstanes with cultures of the fungus *Rhizopus nigricans,* 1500, 1973.
53. **Brannon, D. R., Parrish, F. W., Wiley, B. J., and Long, L.,** Microbial transformations of a series of androgens with *Aspergillus tamarii, J. Org. Chem.,* 32, 1521, 1967.

54. **Herr, M. E., Johnson, R. A., Murray, H. C., Reinede, L. M., and Fonken, G. S.,** The microbiological oxygenation of I-acylated 1-adamantanamines. Stereochemistry and structural determinations, *J. Org. Chem.*, 33, 3201, 1968; Stereochemistry of microbiological hydroxylations, 33, 3217, 1968.
55. **Sariaslani, F. S. and Rosazza, J. P.,** Biocatalysts in natural products chemistry, *Enzyme Microb. Technol.*, 6, 242, 1984.
56. **Davis, P. J. and Rosazza, J. P.,** Microbiological systems in organic synthesis. I. The stereospecific oxidation of glaucine by *Fusarium solani, Bioorg, Chem.*, 10, 97, 1981.
57. **Davis, P. J.,** High-performance liquid chromatographic determination of glaucine and its metabolite, dehydroglaucine in microbial culture, *J. Chromatogr.*, 193, 170, 1980.
58. **Davis, P. J. and Talaat, R. E.,** Microbial transformations in organic synthesis. II. Preparative-scale resolution of RS-glaucine by *Fusarium solani,* and stereospecific oxidation of R-glaucine by *Aspergillus flavipes, Appl. Environ. Microbiol.*, 41, 1243, 1981.
59. **Sih, C. J.,** Microbes in Synthesis, *Am. J. Pharm. Educ.*, 41, 432, 1977.
60. **Ohta, H. and Tetsukawa, H.,** Preparation of chiral secondary alcohols via enantioselective hydrolysis of the corresponding esters with a microorganism, *Agric. Biol. Chem.*, 48, 863, 1980.
61. **Kerr, K. M. and Davis, P. J.,** Microbiological systems in organic synthesis. III. Synthesis, characterization, and microbial transformation of *cis-* and *trans-*7-methylglaucine, *J. Org. Chem.*, 48, 928, 1983.
62. **Kerr, K. M. and Davis, P. J.,** Microbial transformations in organic synthesis. IV. Stereoselective fungal metabolism of 7-Methylglaucine, *J. Org. Chem.*, 51, 1741, 1986.
63. **Talaat, R. E. and Davis, P. J.,** Structural specificity in the oxidation of aporphines using *Fusarium solani, Helv. Chim. Acta,* submitted.
64. **Rizzo, J. D. and Davis, P. J.,** Microbial metabolism of the coumarin anticoagulants warfarin and phenprocoumon, Abstr. 1st Nat. Mtg. Am. Assoc. Pharm. Sci. (Medicinal and Natural Products Chemistry Div.), Washington, D.C., November 1986.
65. **Talaat, R. E. and Davis, P. J.,** Induction of glaucine oxidation in *Fusarium solani,* Abstr. 37th Southwest Regional Mtg. American Chemical Society (Biological Chemistry Division), San Antonio, December 1981.
66. **Davis, P. J.,** unpublished results.

Chapter 4

MICROBIAL TRANSFORMATIONS OF CARDENOLIDES AND OTHER DRUGS: CONTRIBUTION TO THE UNDERSTANDING OF MAMMALIAN DRUG METABOLISM

Larry W. Robertson

TABLE OF CONTENTS

I. GENERAL INTRODUCTION TO MICROBIAL TRANSFORMATIONS

The ability of microorganisms to catalyze a wide variety of chemical reactions, often referred to as "microbial transformations" or "biotransformations", on drugs and other chemicals, is now well recognized. Outstanding successes on an industrial scale with the 11α-hydroxylation of progesterone by fungal cultures led to the greatly enhanced production of corticosteroids and many other steroid derivatives.[1-3] Knowledge gained in the steroid field has been extended to additional classes of compounds, and the literature now documents a large number of successful microbial transformations of alcohols, alicyclics, alkaloids, alkanes, antibiotics, antitumor agents, aromatics, cannabinoids, olefins, pesticides, polyfunctional drug molecules, and terpenoids.[4-7] Transformation reactions performed by microorganisms and their derived enzymes are usually characterized by high degrees of stereospecificity and regiospecificity, as well as by often remarkable substrate specificity. It is often possible to make a specific change in one functional group or moiety in a complex drug molecule, for example, without altering the remainder of the molecule. In addition, microbial transformation reactions are generally performed under mild reaction conditions, in aqueous culture media at 25 or 37°C, so that heat-labile or pH-sensitive substrates are less likely to be destroyed than in standard synthetic chemical reactions.[6]

Two specific applications of microbial transformations have made and will continue to make significant contributions to the understanding of drug metabolism in humans and animals. First, the resident microorganisms of the gastrointestinal tract are now recognized to play a key role in the metabolism of drugs and other xenobiotics.[8,9] Second, microbially catalyzed reactions may be convenient models of mammalian metabolism, allowing the production of useful quantities of known or suspected drug metabolites.[10] The cardenolide metabolism study is making use of both types of applications.

II. TRANSFORMATIONS OF DRUGS AND OTHER XENOBIOTICS BY MICROFLORA OF THE GASTROINTESTINAL TRACT

Although the study of drug metabolism in man and animals has largely focused on biotransformations in the liver and other mammalian organs and tissues, the contribution of the microflora of the GI tract is receiving increasing recognition as a major factor in drug transformations.[8,9] The numbers and variety of microorganisms differ greatly in the specific segments of the GI tract, esophagus, stomach, small intestine, cecum, and large intestine. The quantity and taxonomic types of microbial cells per gram of contents generally increase greatly in successively lower regions, with human feces containing an estimated 400 species and a composition which is approximately 40% microorganisms by weight. While many prokaryotic and some eukaryotic microbes readily colonize certain regions in the GI tract and are considered resident flora, the qualitative and quantitative composition of the population is subject to alteration by disease states, administration of antibiotics, changes in GI motility, aging, and diet.[11]

The study of the GI flora and its contribution to the understanding of the metabolism of drugs and other chemicals is an extremely complicated undertaking. Because early attempts to enumerate the microflora of the feces relied on aerobic or imperfect anaerobic cultivation techniques, *Escherichia coli* (cultivated at concentrations of about 10^8/g from feces) was thought to be the major species present. More careful investigations using direct microscopic examination and improved anaerobic culture techniques have shown that the majority of the approximately 10^{11} bacteria per gram are strict anaerobes such as *Eubacterium* and *Bacteroides* species.[8,9,11] The assignment of all or part of the transformation of a drug to the GI flora also requires careful investigation. Human feces has traditionally served as a ready source of intestinal microorganisms for study as drug transforming agents. While incubation of drugs with suspensions or dilutions of feces under appropriate (probably anaerobic or microaerobic) conditions gives an indication of the innate metabolic capabilities of the microorganisms present, such studies cannot confirm the ability of the microorganisms to catalyze the same biotransformations in the actual microenvironments present within the GI tract. An indirect indication of GI microbial metabolism is a decrease or cessation of the metabolic conversion following orally administered antibiotics. While sampling of various segments of the intact GI tract contents following oral drug administration would be very informative, the difficulty of collecting samples in areas other than the extreme proximal and distal regions makes this approach impractical. Comparison of biotransformation and metabolite accumulation in normal and germfree animals (usually rats) is a more certain approach to attributing a metabolic reaction to the GI microflora. Observations of lack of metabolite formation in germfree animals can be followed by colonizing the animals with cultures of one or more microorganisms which are known or suspected to perform the transformation. Formation of the specific metabolite in the resulting gnotobiotic animals is good evidence for the involvement of the microbial flora.[8,9]

Biotransformation of drugs by GI microorganisms may be either presystemic or postsystemic. Obviously, the portion of a drug which is not absorbed in the stomach or small intestine is available directly to the GI flora. Absorbed drugs and their metabolites may be excreted in the bile and thus be subject to further transformation in the lower GI tract.[8] Conjugated metabolites, such as glucuronic acid derivatives, may be deconjugated by intestinal bacteria, and the unconjugated drug or metabolite reabsorbed. Like transformations by liver enzymes, drug modifications catalyzed by microbial enzymes may result in less toxic or more toxic, and in some cases carcinogenic metabolites.[8]

A. Examples of Types of Transformation Reactions Catalyzed by the Intestinal Flora

The GI microflora exists in an environment which is free of molecular oxygen. Members of the flora are, therefore, obligate or facultative anaerobes, and they generally catalyze the reduction of xenobiotic compounds.[8] This is in sharp contrast to the cytochrome P-450 system of mammalian liver, which requires molecular oxygen and tends to oxidize drugs and other chemicals. Another major set of reactions catalyzed by the flora is deconjugation (hydrolytic) reactions.[8] This is again the opposite of the common liver reaction of conjugation by the addition of glucuronic acid, sulfate, etc., to the drug or Phase 1 metabolite. Enterohepatic circulation of some drugs and metabolites results in the deconjugation by the flora of metabolites formed by the liver, often followed by reabsorption of the deconjugated drug or metabolite.

The following reactions are specific examples of types of transformation reactions which are known to be catalyzed by members of the intestinal microflora. The reactions were selected to represent the range of substrates and reaction types rather than to provide an exhaustive review of the literature.

1. Reductive Chemical Reactions

The substrates for reductive (''biohydrogenation'') reactions vary from naturally occurring intermediates in the human body, such as bile acids and steroid hormones, to drugs, food constituents, and other xenobiotics.

a. Carbon-Carbon Double Bond Reduction

Double bond reduction is typified by the reduction of 16-dehydroprogesterone to 17-isoprogesterone by *Eubacterium* species strain 144, and the same organism can reduce progesterone to 5β-pregnadione (major product) and 5α-pregnadione (minor product).[12]

b. Carbonyl Group Reduction

The biohydrogenation of the sterol cholest-4-ene-3-one by *Eubacterium* species ATCC

21408 for form 5β-cholestan-3β-ol is an example of carbonyl reduction which probably occurs quite commonly in the oxygen-free atmosphere of the lower GI tract.[13]

c. Dehydroxylation (Reductive)

A bacterium similar to *Eubacterium lentum* has been reported to reductively dehydroxylate deoxycorticosterone at the C-21 position, forming high yields of progesterone.[14] Intestinal bacteria have also been shown to dehydroxylate caffeic acid at the 4-position of the 3,4 catechol.[15,16] Similar 4-dehydroxylation of L-dopa and dopamine by anaerobic microorganisms in the human GI tract leads to appearance of 3-hydroxyphenylacetic acid and 3-hydroxyphenethylamine in the urine of Parkinsonian patients treated with L-dopa.[17] Neither L-dopa nor dopamine is converted to 4-dehydroxylated metabolites in the germfree rat.[18]

d. Azo Bond Reduction

Azo bond reduction by the intestinal flora results in the cleavage of the azo linkage to release free sulfanilamide from either prontosil or neoprontosil.[19] Although some liver preparations can also catalyze azo bond reduction, it has been shown that the level of urinary sulfanilamide is greatly reduced when rats dosed with either prontosil or neoprontosil are also given an antibiotic to suppress the GI flora.[19] Sulfasalazine is similarly cleaved to yield sulfapyridine and 5-aminosalicylic acid in humans and normal rats, while germfree rats cannot perform the reductive cleavage and excrete only the intact drug in their urine.[20]

e. Nitro Group Reductions

Chloramphenicol metabolites in which the nitro group has been reduced to an amino group are excreted in the urine of both humans and rats and are detected in the feces of rats and dogs. The reduction of the chloramphenicol nitro group has been attributed to the GI microflora.[21,22] Reduction of the nitro group of metronidazole, which appears to be catalyzed by the flora, is necessary for the antibacterial, antitrichomonal, tumorigenic, and mutagenic activities of the drug.[23,24] Reduction of the nitro group of *p*-nitrobenzenesulfonamide occurs at a much lower than normal rate in germfree rats.[25] Reduction of the nitro group of nitrobenzene appears to be a key step in the development of nitrobenzene-induced methemoglobinemia in rats; and germfree rats or rats pretreated with antibiotics to inhibit the flora do not suffer from the toxic effect.[26]

f. Reduction of N-Hydroxy Compounds

The potential carcinogens, *N*-hydroxyfluorenylacetamide and *N*-hydroxyacetylaminobiphenyl, are transformed via N dehydroxylation by intestinal microflora.[8,27] This reductive reaction is a protective mechanism which converts the compounds to less carcinogenic metabolites.[27]

2. Oxidative (Dehydrogenation) Reactions

While oxidative reactions in the highly anaerobic environment of the lower GI tract are rarer than reductions, there are examples of such reactions. *Eubacterium lentum* has been reported to catalyze the dehydrogenation of 3α-hydroxy-5β-pregnan-20-one to 5β-pregnane-3,20-dione, as well a similar oxidation of the 3α-hydroxy group of 3α,21-dihydroxy-5β-pregnan-20-one to form the corresponding 3,20-dione.[14]

3. Hydrolytic (Deconjugation) Reactions

a. Glucuronides

Deconjugation of glucuronides is a common fate for such metabolites which enter the lower intestinal tract, and the β-glucuronidase activity which catalyzes the reaction appears to reside in the GI flora and not the intestinal wall.[8,28] For example, the glucuronide of diethylstilbestrol is readily deconjugated by the intestinal flora, thus allowing absorption of the free diethylstilbestrol.[29] There are reports of the influence of diet on the amount or activity of fecal bacterial β-glucuronidase, with meat diets causing an increase in the reaction in humans.[30]

b. Other Glycosides

Cycasin (methylazoxymethanol-β-D-glucoside) is a potent carcinogen in normal but not germfree rats, the carcinogenic aglycone being released in the intestinal tract by bacterial β-glycosidases.[31] The plant glycoside amygdalin (laetrile) is likewise hydrolyzed by the intestinal flora to release the aglycone and cyanide. Germfree rats suffer no cyanide toxicity when administered doses which would be fatal to normal rats.[32] Rutin, a common flavonoid glycoside component of foods, is hydrolyzed by *Streptococcus faecium* and an additional *Streptococcus* species to release free quercetin, the mutagenic aglycone.[33] The β-glucosidase activity seems to be inducible in the isolated bacteria and in the bacterial population in the intestinal tract.[33]

c. Sulfate Bonds

The hydrolysis of the sulfamate bond of the artificial sweetener cyclamate to form cyclohexylamine is apparently due entirely to the GI flora, and the hydrolytic activity is inducible in the fecal bacteria of some lab animals and human subjects.[34,35] Hydrolysis of the sulfate ester of the laxative 4,4′-dihydroxydiphenyl-(pyridyl-2)-methane is necessary for the compound to be active, and the hydrolysis appears to be entirely due to the action of intestinal bacteria.[36]

4. Epimerization

A *Eubacterium aerofaciens* strain transformed chenodeoxycholic acid to ursodeoxycholic acid when cocultured with 7α-hydroxysteroid dehydrogenase-elaborating organisms.[37]

5. Deamination (Transamination)

The deamination of the antifungal agent 5-fluorocytosine by human intestinal microflora to form 5-fluorouracil has recently been reported. The conversion of the less toxic agent to the more toxic antitumor compound in certain individuals may explain the occurrence of serious toxicity in some patients receiving the oral antifungal agent.[38]

6. Formation of Nitrosamines and Other Carcinogens

The anaerobic flora of the rat cecum has been implicated in the formation of the carcinogen dimethylnitrosamine from dimethylamine and nitrite.[39] The metabolism of bile acids and other steroids by colon bacteria as a factor in the development of colon cancer has also been examined.[40]

7. Methylation and Demethylation

The methylation of inorganic mercury to form methylmercury, which more readily enters the central nervous system is catalyzed by suspensions of cecal contents under either anaerobic or aerobic conditions. Formation of methylmercury is decreased by filtering, autoclaving, or treating the cecal suspension with antibiotics. Animals with blind jejunal loops, which increase microbial flora and contact time with mercury, form increased amounts of methylmercury and exhibit greater toxicity from similar doses of mercury than normal animals.[8,41,42] Rats can also apparently demethylate methylmercury to some extent, as inorganic mercury appears in the feces of animals fed high levels of methylmercury and low levels of inorganic mercury.[8] The GI flora are also capable of biomethylating selenium.[43]

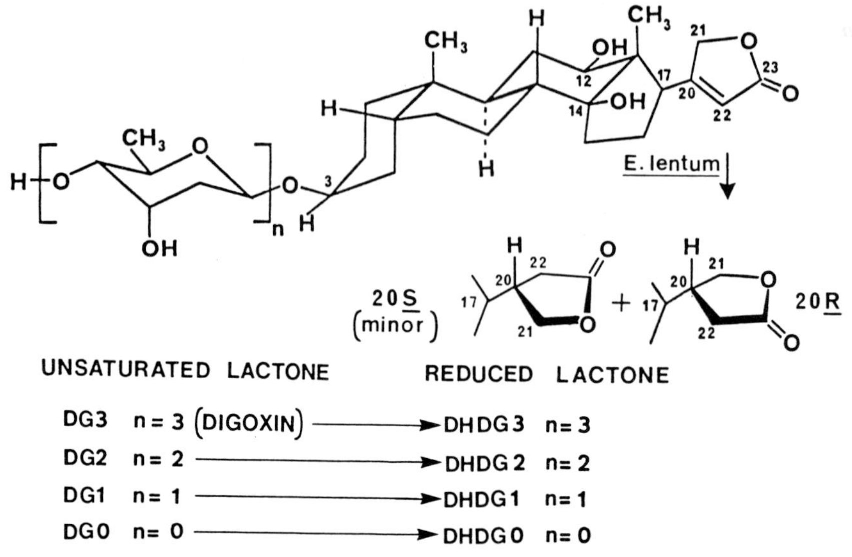

FIGURE 1. Metabolism of digoxin by sugar hydrolysis and *Eubacterium lentum*-catalyzed lactone ring reduction.

B. Reduction of Cardiac Glycosides by *Eubacterium lentum*

Preparations of digitalis, which is the dried leaf of *Digitalis purpurea* L., have been used widely in medicine since the late 18th century.[44] The characteristic glycosides present in *Digitalis* species and certain other plant species act on the myocardium by inhibiting Na^+,K^+-ATPase.[45] Digoxin, digitoxin, gitoxin, and related cardiac glycosides are used clinically in the therapy of congestive heart failure and certain types of cardiac arrhythmias. The effective therapeutic dose for all of the drugs is approximately 50 to 60% that of the toxic dose, and the therapeutic plasma levels for digoxin are in the 0.5 to 2.0 ng/mℓ range.[45] There is, therefore, a need for careful dosage adjustment and monitoring of plasma levels in order to insure an adequate therapeutic level but avoid toxic side effects.

The structure of the cardioactive glycosides is typified by that of digoxin (DG3) as shown in Figure 1. The minimum structural requirements for biological activity appear to include the following: a steroid nucleus with the A/B and C/D ring junctions in the *cis*-configuration; an α,β-unsaturated lactone (cardenolide) ring at the 17β-position of the steroid nucleus; a 14β-hydroxy group; and at least one sugar attached to the 3β-hydroxy group.[44,45] Additional hydroxy groups on the steroid nucleus generally speed the onset of action and the elimination from the body.

1. Studies with Digoxin

Digoxin is currently the most widely used drug in the U.S. for treatment of congestive heart failure and certain cardiac arrhythmias. Because of the potency and potential for adverse side effects of the drug, it is desirable to learn as much as possible concerning the absorption, plasma levels, metabolism, and elimination of digoxin in a variety of patients. Conditions which may significantly alter these parameters, and hence the plasma levels of the drug, include achlorhydria, bacterial overgrowth in the upper GI tract, jejunal diverticula, blind loops, scleroderma, malabsorption, and general effects of aging on the motility and function of the GI tract. The current active investigations in the laboratory of the author are primarily concerned with the effects of achlorhydria and bacterial overgrowth on digoxin metabolism and disposition in aged patients.

a. Digoxin Metabolism and Disposition in Humans

It was long believed that digoxin was not extensively degraded or altered by metabolic enzymes, but was excreted as the intact glycoside in the urine and feces.[46] This belief may have largely been due to the small concentrations of digoxin and metabolites present and the lack of specificity of the available assay methods. More recently, it has become evident that a significant portion of the drug is converted to inactive metabolites in approximately 10% of patients.[47] The two widely recognized paths of digoxin transformation are the hydrolytic and reductive routes.[48] The hydrolytic pathway results in the formation of active metabolites having fewer sugars attached at C-3 (DG2 and DG1 in Figure 1) and a less cardioactive aglycone, or genin (DG0).[49,50] Although it has been shown that gastric pHs in persons with normal stomach acid are sufficiently low to allow a rapid hydrolysis of the sugar side chain, a contribution to hydrolysis by bacterial or mammalian enzymes cannot be ruled out.[51,52] The reductive route involves the reduction of the C-20—C-22 double bond of the unsaturated-lactone ring, and this results in a cardioinactive metabolite.[50,53,54] While the formation of the reduced metabolite in significant amounts in some patients, evidenced by urinary excretion of dihydrodigoxin, was recognized for some time, it has only recently been recognized that the transformation occurs in the GI tract.[47] Early evidence for digoxin inactivation by bacterial flora was the elimination of dihydrodigoxin (DHDG3, Figure 1) formation and excretion in urine by treatment of subjects with oral antibiotics. *E. lentum,* a Gram-positive, obligate anaerobe and normal member of the GI flora, was subsequently identified as a single bacterial species which can perform the reduction.[55,56] Workers proceeded to examine additional *Eubacterium* species for the ability to form digoxin reduction products which they measured using a radioimmune assay (RIA). The RIA antibodies could distinguish between the unsaturated and reduced lactone ring, but gave no information concerning the rest of the molecule, such as whether the sugars at C-3 were present or had been hydrolyzed. *E. lentum* was the only one of 13 *Eubacterium* species which could convert digoxin to its reduction products, and 18 out of 28 *E. lentum* strains examined could perform the reaction.[56] The successful organisms included the type strain *E. lentum* ATCC 25559.[56]

Reduction of the unsaturated ring of digoxin forms a new chiral center at C-20.[57,58] Chemical catalytic hydrogenation using standard methods results in the formation of two dihydrodigoxin epimers, 20*R*-DHDG3 and 20*S*-DHDG3 being formed in a ratio of about 3:1. Specific analysis in an HPLC system which clearly separates the two epimers shows that the DHDG3 epimer found in human urine is the 20*R*.[59,60]

Although there is evidence for additional metabolism, perhaps by conjugation, of the aglycone (DG0) and glycosidic intermediates, the products are incompletely characterized.[49,61] These further modifications might be catalyzed by enzymes of microbial or human origin, or they might be due to purely chemical reactions or combinations of microbial and mammalian reactions.

***b. Conversion of Digoxin to 20*R*-Dihydrodigoxin by* E. lentum**

Although it has been shown that *E. lentum* cultures could convert digoxin to reduction products, the RIA detection method could not determine the stereochemistry of the reduction products or whether the sugar side chain was hydrolyzed or still intact.[55,56] Likewise, it was not known whether *E. lentum* could only reduce digoxin, or whether it could also reduce the products of the hydrolytic pathway (DG2, DG1, and DG0), which would normally be available to the intestinal microflora. To investigate the stereospecificity and substrate specificity of the reduction, digoxin and each of its sugar-hydrolyzed metabolites were incubated with pure cultures of *E. lentum* ATCC 25559, and with diluted fecal samples obtained from a volunteer who is known to be a dihydrodigoxin excretor.[62]

E. lentum cultures and fecal samples were grown and incubated in thioglycollate medium (TG) without glucose or indicator, and brain heart infusion broth with cysteine (BHIC).[63]

The cardenolide substrates were added to the incubation media (freshly autoclaved) to give an overall concentration of 10 μg/mℓ of digoxin, with equimolar concentrations of the hydrolyzed compounds being used. The tubes were then inoculated with 1 mℓ of a 2-day old *E. lentum* culture, capped tightly and incubated at 37°C for 7 days, a time which was consistent with that used by the original investigators.[55,56] Diluted suspensions of fecal samples were used to inoculate additional sets of culture tubes containing digoxin and its hydrolysis products, and the tubes were incubated as detailed above. Appropriate controls, including autoclaved cultures and fecal samples, were run to insure that any reduction measured would be the result of biologically mediated reactions.

When the incubations were finished, a 1-mℓ aliquot of the supernatant of each tube was extracted with 2 mℓ of methylene chloride containing digitoxin (100 ng) as an internal standard. After the aqueous layer was removed, the methylene chloride extract was washed with aqueous sodium bicarbonate, evaporated to dryness, and derivatized with 1-naphthoyl chloride to obtain a fluorescent derivative for HPLC assay.[58,59] Liquid chromatography was performed on a normal phase column (Lichrosorb® Si 60, Jones Chromatography, 5-μm particle size, 25 cm × 4.6 mm i.d.), with a mobile phase of hexane/methylene chloride/acetonitrile (5:1:1), at a flow rate of 2.2 mℓ/min. The HPLC was equipped with a model 110A pump (Beckman), a model FS970 (Kratos) fluorometer, a 200-μℓ injection loop, and a dual-channel recorder.[58,59] Figure 2 shows the separation of internal standard (DTO), digoxin (DG3), and reference metabolites.

Incubation of digoxin (10 μg/mℓ) with growing cultures of *E. lentum* for 7 days at 37°C results in a nearly quantitative stereospecific reduction of the glycoside to its dihydro metabolite, 20*R*-DHDG3, as shown in chromatogram II of Figure 2. Comparison with the reference clearly shows that the main product is the 20*R* epimer, which is the stereochemistry of the reduction product detected in human urine.[59,60] In addition, the same 20*R*-dihydrodigoxin product is formed in the fecal-suspension incubations, which were inoculated with feces from a volunteer who was known to excrete the reduction product (20*R*-DHDG3) in his urine. No conversion was seen in the autoclaved control tubes, and there were no interfering substances from either the microorganisms, feces, or culture media which had retention times near those of the dihydro metabolites.

Examination of chromatogram II in Figure 2 shows no evidence of hydrolysis of the sugar moieties from the C-3 position of the steroid nucleus. However, *E. lentum* is capable of converting (reducing) each of the sugar-hydrolyzed metabolites (DG2, DG1, and DG0) to its dihydro derivative. Both the pure cultures and the fecal incubates resulted in the formation of pairs of reduction products, which were identified by cochromatography with reference compounds. In each case the 20*R* epimer constituted greater than 99% of the products formed. Table 1 shows the representative retention times for each of the possible sugar-hydrolyzed and reduced metabolites.

A time-course study on a series of pure *E. lentum* cultures inoculated with digoxin involved the analysis of one culture per hour for 30 hr. Dihydrodigoxin begins to appear in the culture after approximately 16 hours of growth and incubation, with conversion to the dihydro metabolite being complete after 24 hr. HPLC analysis detected a disappearance of the digoxin starting material corresponding to the appearance of the reduction product.

In order to prepare useful quantities of pure 20*R* epimers of dihydrodigoxin and the corresponding reduction products of the sugar-hydrolyzed metabolites, the reduction reaction was scaled up to produce a larger quantity of 20*R*-dihydrodigoxin. For the scale-up, larger-volume cultures containing both *E. lentum* ATCC 25559 and *Escherichia coli* ATCC 10536 were tested for their ability to reduce larger quantities of digoxin. The use of a facultative anaerobe such as *E. coli* to provide a more anaerobic environment for an obligate anaerobe such as *E. lentum* is a common practice.[14] Results of a preliminary concentration study are shown in Table 2. Essentially quantitative reductions were obtained with concentrations up

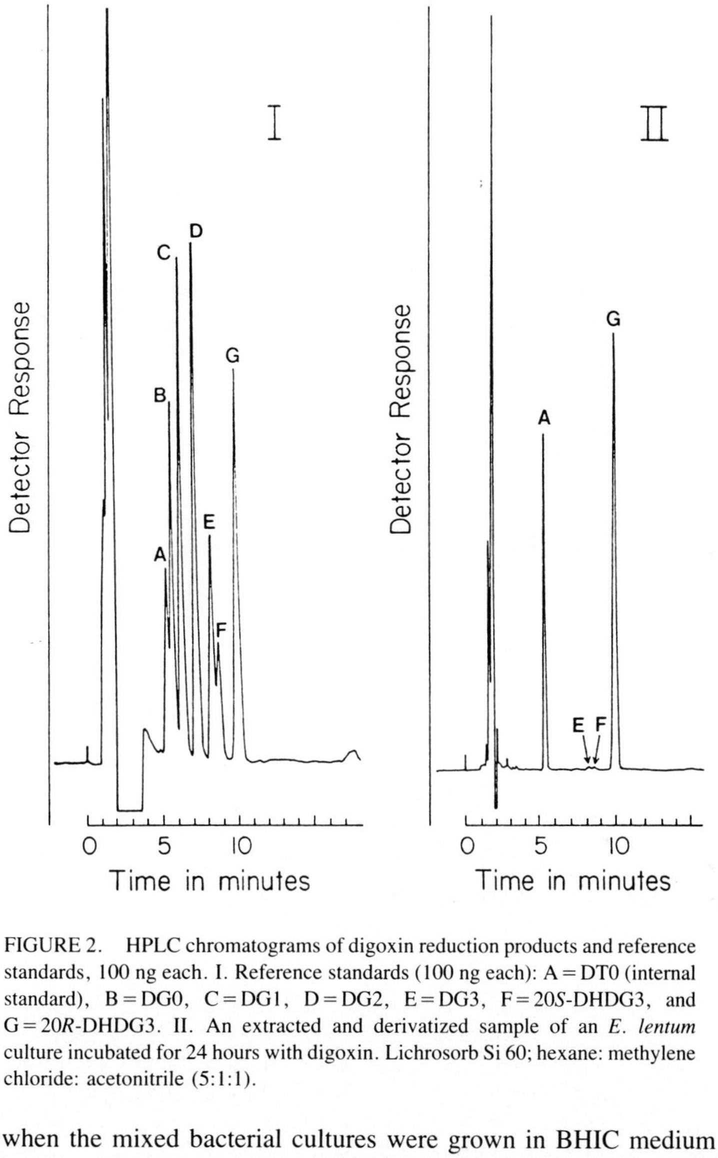

FIGURE 2. HPLC chromatograms of digoxin reduction products and reference standards, 100 ng each. I. Reference standards (100 ng each): A = DT0 (internal standard), B = DG0, C = DG1, D = DG2, E = DG3, F = 20*S*-DHDG3, and G = 20*R*-DHDG3. II. An extracted and derivatized sample of an *E. lentum* culture incubated for 24 hours with digoxin. Lichrosorb Si 60; hexane: methylene chloride: acetonitrile (5:1:1).

to 80 μg/mℓ when the mixed bacterial cultures were grown in BHIC medium at 37°C for 48 hr. In every case, the digoxin must be added at the time of culture inoculation, as the reduction appears to occur only in cultures in the active growth phase. Older *E. lentum* cultures or resting-cell suspensions do not seem to be capable of reducing digoxin, at least under the currently used culture conditions.

For the actual scale-up, the substrate, digoxin (20 mg in 0.4 mℓ of ethanol), was added to each of 25 Erlenmeyer flasks (500 mℓ, containing 250 mℓ of BHIC medium). Each flask was then inoculated with 1 mℓ of a 24-hr old *E. lentum* broth culture, and 0.1 mℓ of a 24-hr old *E. coli* broth culture, and the flasks were incubated at 37°C for 48 hr. After incubation, each culture was extracted twice with methylene chloride, and the combined solvent extracts were evaporated to give a brown colored solid. The solid was subjected to chromatography on a silica gel column using a methylene chloride and methanol (10:1) solvent system. Pure crystalline dihydrodigoxin was obtained from the column. Similar scale-up procedures but with smaller amounts of starting material, were used to prepare the dihydro metabolites (20*R*) of the sugar-hydrolyzed metabolites, DG2, DG1, and DG0.

The availability of pure 20*R*-dihydro metabolites of digoxin and its hydrolytic products

Table 1
CHROMATOGRAPHIC (HPLC) DETECTION OF REDUCED PRODUCTS FROM DIGOXIN AND SUGAR-HYDROLYZED METABOLITES[a]

Compound derivatized	Abbrev.	Retention time (min)
Digitoxigenin	DT0	5.2
Digoxigenin	DG0	5.6
20*S*-Dihydrodigoxigenin	20*S*-DHDG0	6.1
20*R*-Dihydrodigoxigenin	20*R*-DHDG0	6.5
Digoxigenin mono-digitoxoside	DG1	6.3
20*S*-Dihydrodigoxigenin monodigitoxoside	20*S*-DHDG1	6.5
20*R*-Dihydrodigoxigenin monodigitoxoside	20*R*-DHDG1	7.2
Digoxigenin bis-digitoxoside	DG2	7.4
20*S*-Dihydrodigoxigenin bisdigitoxoside	20*S*-DHDG2	7.6
20*R*-Dihydrodigoxigenin bisdigitoxoside	20*R*-DHDG2	7.9
Digoxin	DG3	8.2
20*S*-Dihydrodigoxin	20*S*-DHDG3	8.6
20*R*-Dihydrodigoxin	20*R*-DHDG3	10.2

[a] Lichrosorb Si 60; hexane: methylene chloride: acetonitrile (5:1:1).

Table 2
CONVERSION OF DIGOXIN TO 20*R*-DIHYDRODIGOXIN BY *EUBACTERIUM LENTUM* ATCC 25559 AND *ESCHERICHIA COLI* ATCC 10536 MIXED CULTURES: A PRELIMINARY CONCENTRATION STUDY[a]

Digoxin conc. (μg/mℓ)	Conversion (reduction) (%)
10	100
20	100
40	100
80	100
160	<100
320	< 50

[a] Cultures were grown in stationary culture in BHIC medium at 37°C for 48 hr.

served to furnish a very clean set of standards for the analytical laboratory examining fecal and urine samples of patients and subjects in the clinical study.[64] In addition, it was possible to obtain mass spectra and other chemical and physical data on the reduction products. Figure 3 shows the comparison of two sophisticated, mass-spectral acquisition techniques as applied to the reduction product of major interest, 20*R*-dihydrodigoxin. The fast atom bombardment (FAB) spectrum shows a pseudomolecular ion at m/z 805 $(M + Na)^+$, which corresponds to a molecular ion of 782, indicating the addition of two mass units to digoxin (mol wt 780). The laser desorption/fourier transform/ion cyclotron resonance (LD/FT/ICR)

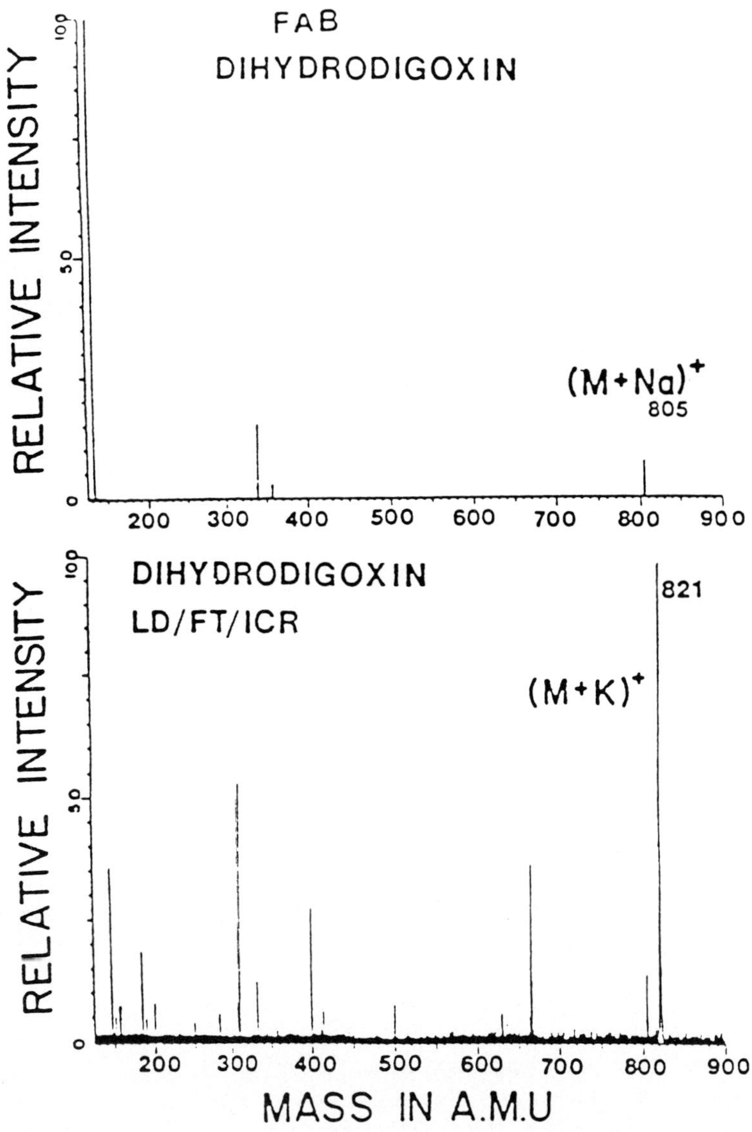

FIGURE 3. Comparison of mass spectra of dihydrodigoxin in fast atom bombardment (FAB) and laser desorption/fourier transform/ion cyclotron resonance (LD/FT/ICR) mass spectrometry.

spectrum shows a pseudomolecular ion at m/z 821 $(M + K)^+$, corresponding again to a molecular ion of 782 for the dihydrodigoxin. Table 3 compares the pseudomolecular ions and corresponding molecular weights for each of the reduced and sugar-hydrolyzed metabolites.[65]

c. Clinical Significance of Bacterial Reduction of Digoxin

A minority (about 10%) of patients taking digoxin excrete significant quantities of the dihydro metabolite in the urine.[47] However, in some patients within this minority group, dihydrodigoxin may account for 50% or more of the total cardiac glycosides and metabolites in the urine and feces.[47,58] In such persons, reduction represents a major route of inactivation. In a recent clinical study, one subject excreted 97% of the single dose administered within 3 days, based on percentage molar excretion: 45% was as digoxin (45% in urine, a trace in

Table 3
MASS SPECTROMETRIC FRAGMENTATION OF DIGOXIN AND ITS REDUCED AND SUGAR-HYDROLYZED METABOLITES: COMPARISON OF FAB AND LD/FT/ICR MASS SPECTRA

	FAB mass spectrum		LD/FT/ICR mass spectrum	
Compound (mol wt)	**Pseudomolecular ion (R.I.%)**	**Key fragments**	**Pseudomolecular ion (R.I.%)**	**Key fragments**
DG3 (780)	781 $(M + H)^+$ (0.2)	651, 521, 391, 373, 355, 337	819 $(M + K)^+$ (100)	651, 521, 391
DHDG3 (782)	805 $(M + Na)^+$ (0.6)	357, 339	821 $(M + K)^+$ (100)	357, 339
DG2 (650)	651 $(M + H)^+$ (1.8)	503, 391	689 $(M + K)^+$ (100)	355, 337
DHDG2 (652)	675 $(M + Na)^+$ (1.9)	339	691 $(M + K)^+$ (100)	561, 431, 339
DG1 (520)	521 $(M + H)^+$ (1.4)	373, 355, 337	559 $(M + K)^+$ (100)	373, 355, 337
DHDG1 (522)	545 $(M + Na)^+$ (1.3)	357, 339	561 $(M + K)^+$ (100)	—
DG0 (390)	391 $(M + H)^+$ (14.1)	373, 355, 337	429 $(M + K)^+$ (100)	373, 355, 337
DHDG0 (392)	—	357, 339	431 $(M + K)^+$ (100)	357, 339

feces) and 50% was as 20*R*-dihydrodigoxin (24% in urine, 26% in feces).[64] In this individual, reduction appears to be the major route of inactivation. Patients who consistently show this level of inactivation should, perhaps, have their dose adjusted upward. Another point to consider, however, is what happens when such a patient receives oral antibiotics. Inhibition of the intestinal flora, which is responsible for the reductive inactivation, could rapidly produce significant increases in the plasma level of the active digoxin. In some patients of this type, oral antibiotics and digoxin might constitute a harmful drug interaction.

One point which is not well understood is why there is such a large variation among individuals in the ability to form dihydrodigoxin. *E. lentum* is now recognized as a very common resident of the human intestinal tract. As part of the ongoing study in the author's laboratory, 11 subjects were dosed with digoxin, and a total urine and feces collection was done for 3 days. Only one patient formed very large amounts of the dihydrodigoxin, with 50% recovered in urine and feces as the reduced product. Two additional subjects formed moderate amounts — 10% (7% in urine, 3% feces) and 13% (2% in urine, 11% in feces); but in the remaining 8 subjects, total dihydrodigoxin excretion amounted to only a few percent or a trace.[66] Blank or control feces and urine samples were collected from each subject before they entered the study. When incubated with digoxin in TG or BHIC medium, as described in a previous section, 10 of the 11 fecal suspensions were able to reduce digoxin to 20*R*-dihydrodigoxin, most giving quantitative conversion. The factors which cause some persons to reduce digoxin in vivo, while others do not, may be many and complex. One factor which is being explored is the effect of bacterial overgrowth in the upper GI tract on digoxin metabolism. It is possible that in conditions such as achlorhydria, blind loops, and diverticulitis, some of the increased bacterial population of the upper GI tract might include bacteria which can reduce digoxin. It is also logical that persons suffering from GI diseases characterized by malabsorption might absorb digoxin incompletely in the upper GI tract, allowing more of the drug to reach the lower regions where populations of *E. lentum* may be much higher. In some individuals, other resident microorganisms or dietary factors might inhibit the bacteria which reduce digoxin.

2. *Digitoxin and Other Cardiac Glycosides*

While digoxin is the most frequently used cardiac glycoside in the U.S., digitoxin is still used by some physicians, and additional cardiac glycosides are used around the world. A major question concerning the reduction of the cardiac glycosides by *E. lentum* is the

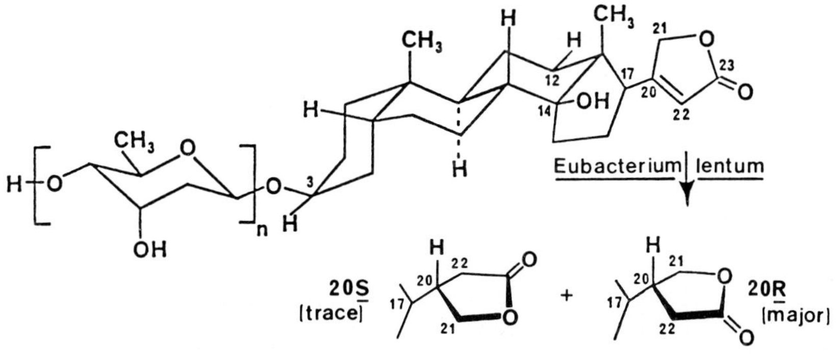

DIGITOXIN SERIES

UNSATURATED LACTONE		REDUCED LACTONE	
DT3	n=3 (DIGITOXIN) →	DHDT3	n=3
DT2	n=2 →	DHDT2	n=2
DT1	n=1 →	DHDT1	n=1
DT0	n=0 →	DHDT0	n=0

FIGURE 4. Metabolism of digitoxin by sugar hydrolysis and *Eubacterium lentum*-catalyzed lactone ring reduction.

specificity or generality of the reaction with respect to substrate. Therefore, the reduction of digitoxin was also examined.

a. Known Routes of Digitoxin Metabolism in Man

Like digoxin, digitoxin was once thought to escape extensive metabolic degradation and to be mostly excreted unchanged in the urine. Reduction of the C-20 — C-22 double bond in the cardenolide ring results in the formation of the dihydro metabolite, similar to that of digoxin. The specific knowledge of digitoxin metabolism is limited due to the complexities of the processes involved. Hepatic microsomal enzymes can hydroxylate digitoxin at the 12β-position of the steroid nucleus to form digoxin.[67,68] The sugar side-chains of the resulting pair of glycosides (digitoxin and digoxin) are hydrolyzed to give the respective bis (DT2 and DG2) and mono (DT1 and DG1) digitoxosides and genins (DT0 and DG0).[69,70] Although evidence for conjugates and other polar metabolites has been reported, the products have not been fully characterized.[67] Dihydrodigitoxin has also been found in the urine of patients taking digitoxin maintenance dosage, and it is likely that the sugar-hydrolyzed metabolites may also be found in the reduced form. One question concerning the reduction of the unsaturated lactone ring in digitoxin is whether the same 20*R*-stereochemistry results as in dihydrodigoxin.

b. Production of Digitoxin Reduction Products

When incubated at a concentration of 10 μg/mℓ at 37°C for 24 hr with growing cultures of *E. lentum* ATCC 25559, digitoxin (DT3, Figure 4) was converted to a single major product (with yields approaching 100%). The HPLC retention time (8.6 min) of the product was quite different from the starting material (DT3 at 7.9 min), but coincided with the major peak observed for reference dihydrodigitoxin. As in the studies with digoxin, *E. lentum* did not hydrolyze the sugar side-chain of digitoxin to give bis- and mono-digitoxosides or genin.

In order to produce large milligram quantities of this digitoxin reduction product, a study of culture conditions and a scale-up were undertaken. Similar to previous studies with digoxin, it was found that the minimum time needed for conversion (total) of digitoxin to dihydrodigitoxin is 24 hr and that the digitoxin has to be added at the beginning of the culture growth period. As in the scale-up for dihydrodigoxin production, the facultative anaerobe *E. coli* was added to the larger stationary flask cultures of *E. lentum. E. coli* alone, when incubated with digitoxin, does not perform the reduction reaction; however, the *E. coli* provides a suitable anaerobic environment for the growth and metabolism of the obligate anaerobe, *E. lentum.* After incubating a total of 500 mg of digitoxin with the mixed *E. lentum-E. coli* flask cultures (BHIC medium, 250 mℓ/500 mℓ Erlenmeyer flask) for 48 hr at 37°C, the whole culture broths were extracted with methylene chloride. The dried extracts were subjected to open column silica gel chromatography (methylene chloride + methanol, 10:1 solvent system), and a total of 383 mg of crystalline dihydrodigitoxin was isolated.

The mass spectrum (FAB) of the reduction product indicated a pseudomolecular ion at 789$(DHDT3 + Na)^+$, which corresponds to a molecular weight of 766 for the dihydrodigitoxin. This confirms that the *E. lentum* cultures added two hydrogens, corresponding to the reduction of the C-20 to 22 double bond, the digitoxin molecule. While it would be expected that the stereochemistry of the new chiral center at C-20 would be *R* as in the 20*R*-dihydrodigoxin, it was necessary to prove this conclusively for the digitoxin series. It was evident from HPLC analysis that reduction of digitoxin produced a major (99%+) and a minor (less than 1%) product, probably corresponding to the mixture of 20*R*-DHDT3 and 20*S*-DHDT3 shown in Figure 4.

Since the sugar moieties can interfere with various spectral studies, it was necessary to remove the sugars by hydrolysis and obtain the pure aglycone for stereochemical assignment. The sugar side chain was cleaved by acid hydrolysis (HCl, pH 1, 24 hr), and silica gel chromatography was used to obtain dihydrodigitoxigenin (183 mg) from dihydrodigitoxin (400 mg). The circular dichroism spectrum of the isolated dihydrodigitoxigenin showed a positive curve ($[\Theta]_{218} = 125$). It had been previously shown that for dihydrodigoxigenin, a positive CD curve was associated with the *R* configuration at C-20.[57]

In order to confirm the configuration at C20, the stereochemistry of dihydrodigitoxigenin (DHDT0) was related to that of dihydrodigoxigenin (DHDG0). The only chemical difference between these two aglycones is the presence of a 12β-hydroxy group in dihydrodigoxigenin. A chemical conversion, involving hydroxylation at C-12, would be extremely difficult; however, mammalian liver systems and certain microbial cultures have been reported to perform this hydroxylation. Thirteen organisms, shown in Table 4, were examined for the ability to 12β-hydroxylate digitoxigenin specifically. The organisms were grown in a two-stage culture procedure in a Pharmamedia-yeast extract medium at 25°C on a rotary shaker at 250 rpm.[76] Digitoxigenin (10 mg/100 mℓ culture) was added to 2-day old stage cultures, and the incubation was continued for 1 week. Analysis by thin-layer chromatography (TLC) of the chloroform extracts of the cultures showed that two of the organisms, *Fusarium roseum* OSU and *Penicillium citrinum* ATCC 16040, were able to perform the 12-β-hydroxylation. The product obtained was shown by HPLC and TLC to be identical to reference digoxigenin. *F. roseum* was chosen for scale-up because it produced a high yield of the specific hydroxylation product (90%).

Incubation of dihydrodigitoxigenin (150 mg), obtained by acid hydrolysis of *E. lentum*-produced dihydrodigitoxin, with shaking cultures of *F. roseum* in PY medium at 25°C, produced a single major product. After extraction with chloroform and silica gel chromatography purification, the product (109 mg) showed HPLC retention times (6.5 min) and TLC R_f values similar to those of 20*R*-dihydrodigoxigenin. The mass spectrum and NMR spectrum also were identical to those obtained from 20*R*-dihydrodigoxigenin, whose structure had previously been proven by NMR studies.[76] The 12β-hydroxylation reaction is illustrated in Figure 5.

In order to characterize further the specificity of the cardenolide reduction catalyzed by *E. lentum,* the sugar-hydrolyzed metabolites (DT2, DT1, and DT0) were incubated indi-

Table 4
MICROORGANISMS SCREENED AND RESULTS OBTAINED FOR THE CONVERSION OF DIGITOXIGENIN TO DIGOXIGENIN (12β-HYDROXYLATION)

Organism screened	ATCC No.	Result obtained[a]
Actinomucor elegans	6476	−
Aspergillus flavipes	11013	−
Aspergillus niger	9142	−
	11394	−
Aspergillus ochraceus	18500	−
Botrytis allii	9435	−
Cunninghamella blakesleeana	8688a	−
Cunninghamella elegans	9245	−
Cylindrocarpon radicicola	11011	−
Fusarium roseum	OSU	+
Penicillium citrinum	16040	+
Streptomyces aureus	15437	−
Syncephalastrum racemosum	18192	−

[a] A plus (+) sign denotes conversion detected by thin-layer chromatographic analysis of culture broth extracts. A minus (−) sign indicates no conversion.

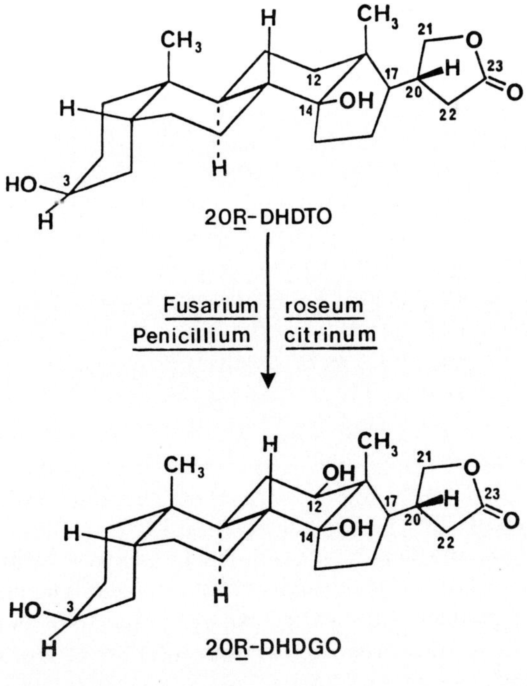

FIGURE 5. Biotransformation of 20*R*-dihydrodigitoxigenin (20*R*-DHDT0) to 20*R*-dihydrodigoxigenin (20*R*-DHDG0) by cultures of *Fusarium roseum* and *Penicillium citrinum*.

vidually with the bacterial cultures, at a concentration of 10 μg/mℓ. In each case, reduction gave almost quantitative yields of the 20*R*-dihydro metabolite. The stereochemistry was confirmed by acid hydrolysis of the glycosides to give the aglycone, whose HPLC, TLC, and NMR characteristics were compared with those of the aglycone (20*R*-DHDT0) of known stereochemistry.

In summary, incubation of digitoxin or its sugar-hydrolyzed metabolites at a concentration of 10 μg/mℓ at 37°C for 24 hr (TG or BHIC medium) results in the formation of the corresponding 20*R*-dihydro metabolite from each of the substrates.

c. Other Cardiac Glycosides

There is a wide variety of cardioactive glycosides in nature. In *Digitalis* species alone, there are the following aglycones: digitoxigenin (3β,14β-dihydroxy); digoxigenin (3β,12β,14β-trihydroxy); gitoxigenin (3β,12β,16β-trihydroxy); dignatigenin (3β,12β,14β,16β-tetrahydroxy); and gitaloxigenin (16β-formyloxy-3β,14β-dihydroxy). In addition, each of these aglycones can be substituted with digitoxose sugars, as well as glucose and possibly acetate functions. In addition, other plant species produce such cardioactive glycosides as ouabain and proscillaridin A (which contains the six-membered diunsaturated bufadienolide ring). It will be interesting to discover whether other naturally occurring or semisynthetic cardiac glycosides are equally as susceptible as digoxin and digitoxin to cardenolide ring reduction, and hence to bacterial inactivation, by *E. lentum* in some patients. If reductive inactivation by *E. lentum* is perceived by the clinical community as a problem of great magnitude, the design of cardiac glycoside derivatives resistant to the reduction may be a worthwhile goal.

C. Additional Transformation Reactions Performed by *Eubacterium* Species

Eubacterium species are Gram-positive nonspore-forming anaerobic bacilli which possess great variability in morphology and biochemical properties. They are also recognized as common members of the human intestinal microflora. In addition to the reduction of the C20-22 double bond in digoxin and other cardenolides, they are capable of performing a number of other biotransformation reactions, and these reactions may be very species- and/or strain-specific. Of 400 organisms isolated from the feces of dihydrodigoxin formers, only two isolates were capable of reducing digoxin, and both were identified as *E. lentum*. *E. lentum* alone among 13 *Eubacterium* species examined was capable of performing the reduction, and only 18 of 28 *E. lentum* strains examined could reduce digoxin.[56]

Some of the other transformation reactions which have been attributed to *E. lentum* and other *Eubacterium* species are listed in Table 5, and most of these have been discovered or investigated only within the past 5 to 10 years. Probably because of the interests of individual investigators, most of the reactions studied have involved endogenous bile acids, sterols, or steroid hormones. The types of reactions include double bond reduction, carbonyl reduction, reductive dehydroxylation, oxidation (dehydrogenation), and epimerization. Additional study of this important genus may indicate whether the specificity for transformations is actually limited mainly to steroidal molecules, or whether a wider range of drugs and exogenous chemicals may serve as substrates. *Eubacterium* species may, therefore, be an important factor in the intestinal transformation of a number of different drugs, much in the same manner as *E. lentum* catalyzes the reductive inactivation of digoxin and related cardiac glycosides.

III. SUMMARY AND CONCLUSIONS

The increasing understanding of microbial transformations continues to contribute to the knowledge of mammalian drug metabolism in two important ways. First, there is growing recognition of the importance of the gastrointestinal microflora in the transformation and

Table 5
SUMMARY OF TRANSFORMATION REACTIONS CATALYZED BY *EUBACTERIUM* SPECIES

Substrates	Products	Reaction type	Ref.
21-Hydroxy-preg-4-ene-3,20-dione (deoxycorticosterone)	Pregn-4-ene-3,20-dione (progesterone)	Dehydroxylation	14, 72
3α,21-Dihydroxy-5β-pregn-20-one	3α-Hydroxy-5β-pregn-20-one	Dehydroxylation	14
16-Dehydro-progesterone	17-Isoprogesterone	Reduction	12
Progesterone	5β-Pregnadione (major) and 5α-pregnadione (minor)	Reduction	12
5β-Pregnadione	5-Hydroxy-pregn-20-one	Reduction	12
3α,21-Dihydroxy-5β-pregnan-20-one	21-Hydroxy-5β-pregnane-3,20-dione	Oxidation	14
3α-Hydroxy-5β-pregnan-20-one	5β-Pregnane-3,20-dione	Oxidation	14
Cholest-5-en-3β-ol (cholesterol)	5β-Cholestan-3β-ol	Reduction	13
Cholest-4-en-3-one	5β-Cholestan-3β-ol	Reduction	13
(24*R*)-Ergost-5-en-3β-ol (campesterol)	(24*R*)-5β-Ergostan-3β-ol	Reduction	13
Stigmast-5-en-3β-ol (β-sitosterol)	5β-Stigmastan-3β-ol	Reduction	13
3β-Hydroxy-androst-5-en-17-one	3β-Hydroxy-5β-androstan-17-one	Reduction	13
Pregn-5-ene-3β,20β-diol	5β-Pregnane-3β,20β-diol	Reduction	13
Ursodeoxycholic acid	Lithocholic acid	Dehydroxylation	73
Linoleic acid	*trans*-Vaccenic acid	Reduction	74
3α, 12α-Dihydroxy-5β-cholan-24-oic acid	12α-Hydroxy-5β-cholan-24-oic	Dehydroxylation	74
3α,7α,12α-Trihydroxy-5β-cholanoic acid	3α,12α-Dihydroxy-5β-cholanoic	Dehydroxylation	75
3α,7α-Dihydroxy-5β-cholanoic acid	3α-Hydroxy-5β-cholanoic acid	Dehydroxylation	75
Chenodeoxycholic acid	Ursodeoxycholic acid	Epimerization	65

disposition of drugs in man and animals. The recent example of the inactivation of digoxin and other cardiac glycosides by the *E. lentum*-catalyzed reduction of the cardenolide ring has been discussed in some detail. Additional investigations will be needed to show why the reductive inactivation occurs in only a minority of the patients who receive digoxin therapy. Second, microbial transformations may serve as convenient models for mammalian metabolic conversions, and as very useful synthetic tools for the preparation of substantial quantities of drug metabolites. For example, the 12β-hydroxylation of dihydrodigitoxigenin to form dihydrodigoxigenin, which occurs in high yields in *F. roseum* cultures, allowed the unambiguous assignment of the 20*R*-stereochemistry to the dihydrodigitoxin series of reduction products. In addition, the scaled up in vitro incubation of digoxin, digitoxin, and their sugar-hydrolyzed metabolites made possible the production of large milligram quantities of the specific 20*R* metabolites for further chemical and biological studies, and for use as chromatographic reference standards in clinical studies.

ACKNOWLEDGMENTS

For the studies on digoxin described in this chapter, the author would like to acknowledge his graduate student, Appavu Chandrasekaran, and valued colleagues, Richard Reuning, James Hui, Douglas Geraets, James Caldwell, and Roy Donnerberg. The digoxin project is supported by a research grant from the National Institute on Aging (NIH 1 RO1 AG04119). The mass spectra were obtained at the Ohio State University Chemical Instrumentation Center (CCIC) by Alan Marshall, Ronald Shomo, and Richard Weisenberger.

REFERENCES

1. **Charney, W. and Herzog, H. L.,** *Microbial Transformations of Steroids, A Handbook,* Academic Press, New York, 1967.
2. **Iizuka, H. and Naito, A.,** *Microbial Transformations of Steroids and Alkaloids,* State College, University Park, Pa., 1967.
3. **Iizuka, H. and Naito, A.,** *Microbial Transformations of Steroids and Alkaloids,* University of Tokyo Press, Springer-Verlag, New York, 1981.
4. **Kieslich, K.,** *Microbial Transformations of Non-Steroid Cyclic Compounds,* Georg Thieme Verlag, Stuttgart, 1976.
5. **Jones, J. B., Sih, C. J., and Perlman, D., Eds.,** *Applications of Biochemical Systems in Organic Chemistry,* John Wiley & Sons, New York, 1976.
6. **Robertson, L. W.,** Microbial transformations of cannabinoids, in *Microbial Transformations of Physiologically Active Compounds,* Rosazza, J. P., Ed., CRC Press, Boca Raton, Fla., 1982, chap. 4.
7. **Rosazza, J. P., Ed.,** *Microbial Transformations of Bioactive Compounds,* Vol. 1 and 2, CRC Press, Boca Raton, Fla., 1982.
8. **Goldman, P.,** Transformations of xenobiotics by the intestinal microflora, in *Microbial Transformations of Bioactive Compounds,* Vol 2, Rosazza, J. P., Ed., CRC Press, Boca Raton, Fla., 1982, chap. 2.
9. **Goldman, P.,** Role of the intestinal microflora, in *Metabolic Basis of Detoxication,* Jakoby, W. B., Bend, J. R., and Caldwell, J., Eds., Academic Press, New York, 1982, chap. 17.
10. **Smith, R. V. and Rosazza, J. P.,** Microbial models of mammalian metabolism, *J. Pharm. Sci.,* 64, 1737, 1975.
11. **Simon, G. L. and Gorbach, S. L.,** Intestinal flora in health and disease, *Gastroenterology,* 86, 174, 1984.
12. **Glass, T. L. and Burley, C. Z.,** Stimulation of 16-dehydroprogesterone and progesterone reductases of *Eubacterium* sp. Strain 144 by hemin and hydrogen or pyruvate, *Appl. Environ. Microbiol.,* 49, 1146, 1985.
13. **Eyssen, H. J., Parmentier, G. G., Compernolle, F. C., DePauw, G., and Piessens-Denef, M.,** Biohydrogenation of sterols by *Eubacterium* ATCC 21408 — Nova Species, *Eur. J. Biochem.,* 36, 411, 1973.
14. **Bokkenheuser, V. D., Winter, J., Dehazya, J., and Welly, W. G.,** Isolation and characterization of human fecal bacteria capable of 21-dehydroxylating corticosteroids, *Appl. Environ. Microbiol.,* 34, 571, 1977.
15. **Dayman, J. and Jepson, J. B.,** The metabolism of caffeic acid in humans: the dehydroxylating action of intestinal bacteria, *Biochem. J.,* 113, 11P, 1969.
16. **Peppercorn, M. A. and Goldman, P.,** Caffeic acid metabolism by bacteria of the human gastrointestinal tract, *J. Bacteriol.,* 108, 996, 1971.
17. **Sandler, M., Goodwin, B. L., Ruthven, C. R. J., Hunter, K. R., and Stern, G. M.,** Variation of levodopa metabolism with gastrointestinal absorption site, *Lancet,* 1, 238, 1974.
18. **Goldin, B. R., Peppercorn, M. A., and Goldman, P.,** Contributions of host and intestinal microflora in the metabolism of L-dopa by the rat, *J. Pharmacol. Exp. Ther.,* 186, 160, 1973.
19. **Gingell, R., Bridges, J. W., and Williams, R. T.,** The role of the gut flora in the metabolism of prontosil and neoprontosil in the rat, *Xenobiotica,* 1, 143, 1971.
20. **Peppercorn, M. A. and Goldman, P.,** The role of intestinal bacteria in the metabolism of salicylazosulfapyridine, *J. Pharmacol. Exp. Ther.,* 181, 555, 1872.
21. **Glazko, A. J., Wolf, L. M., Dill, W. A., and Bratton, A. C.,** Biochemical studies on chloramphenicol (Chloromycetin). II. Tissue distribution and excretion studies, *J. Pharmacol. Exp. Ther.,* 96, 445, 1949.
22. **Glazko, A. J., Dill, W. A., and Wolf, L.,** Observations of the metabolic disposition of chloramphenicol (Chloramphenicol) in the rat, *J. Pharmacol. Exp. Ther.,* 104, 452, 1952.
23. **Koch, R. L. and Goldman, P.,** The anaerobic metabolism of metronidazole forms N-(2-hydroxyethyl)-oxamic acid, *J. Pharmacol. Exp. Ther.,* 208, 406, 1979.
24. **Rosenkranz, H. S. and Speck, W. T.,** Mutagenicity of metronidazole: activation by mammalian liver microsomes, *Biochem. Biophys. Res. Commun.,* 66, 520, 1975.
25. **Wheeler, L. A., Soderberg, F. B., and Goldman, P.,** The relationship between nitro group reduction and the intestinal microflora, *J. Pharmacol. Exp. Ther.,* 194, 135, 1975.
26. **Reddy, B. G., Pohl, L. R., and Krishna, G.,** The requirement of the gut flora in nitrobenzene-induced methemoglobinemia in rats, *Biochem. Pharmacol.,* 25, 1119, 1976.
27. **Wheeler, L. A., Soderberg, F. B., and Goldman, P.,** The reduction of N-hydroxy-4-acetylaminobiphenyl by the intestinal microflora of the rat, *Cancer Res.,* 35, 2962, 1975.
28. **Scheline, R. R.,** Metabolism of foreign compounds by gastrointestinal microorganisms, *Pharmacol. Rev.,* 25, 451, 1973.
29. **Clark, A. G., Fischer, L. J., Millburn, P., Smith, R. L., and Williams, R. T.,** The role of the gut flora in the enterohepatic circulation of stilbesterol in the rat, *Biochem. J.,* 112, 17P, 1969.

30. **Reddy, B. S., Weisburger, J. H., and Wynder, E. L.,** Fecal bacterial β-glucuronidase: control by diet, *Science,* 183, 416, 1974.
31. **Spatz, M., Smith, D. W. E., McDaniel, E. G., and Lacquer, G. L.,** Role of intestinal microorganisms in determining cycasin toxicity, *Fed. Proc., Fed. Am. Soc. Exp. Biol.,* 124, 691, 1967.
32. **Carter, J. H., McLafferty, M. A., and Goldman, P.,** Role of the gastrointestinal microflora in amygdalin (laetrile)-induced cyanide toxicity, *Biochem. Pharmacol.,* 29, 301, 1980.
33. **MacDonald, I. A., Bussard, R. G., Hutchison, D. M., and Holdeman, L. V.,** Rutin-induced beta-glucuronidase activity in *Streptococcus faecium* VGH-1 and *Streptococcus* sp. strain FRP-17 isolated from human feces: formation of the mutagen quercetin, from rutin, *Appl. Environ. Microbiol.,* 47, 350, 1984.
34. **Renwick, A. G. and Williams, R. T.,** The fate of cyclamate in man and other species, *Biochem. J.,* 129, 869, 1972.
35. **Bickel, M. H., Burkard, B. Meier-Strasser, and Van Den Broek-Boot, M.,** Entero-bacterial formation of cyclohyxylamine in rats ingesting cyclamate, *Xenobiotica,* 4, 425, 1974.
36. **Forth, W. Nell, G., Rummel, W. and Andrews, H.,** The hydrogogue and laxative effect of the sulfuric acid ester and the free diphenol of 4-4′-dihydroxy-diphenyl-(pyridyl-2)-methane, *Nauyn-Schmiedeberg's Arch. Pharmacol.,* 274, 46, 1972.
37. **MacDonald, I. A., Rochon, Y. P., Hutchison, D. M., and Holdeman, L. V.,** Formation of ursodeoxycholic acid from chenodeoxycholic acid by a 7β-hydroxysteroid dehydrogenase-elaborating *Eubacterium aerofaciens* strain cocultured with 7α-hydroxysteroid dehydrogenase-elaborating organism, *Appl. Environ. Microbiol.,* 44, 1187, 1982.
38. **Harris, B., Manning, B., Federle, T., and Diasio, R.,** Conversion of 5-fluorocytosine to 5-fluorouracil by human intestinal microflora, *Fed. Proc., Fed. Am. Soc. Exp. Biol.,* 70th Annu. Meeting, Abstr. 1681, St. Louis, April 13 through 16, 1986.
39. **Klubes, P. and Jondork, W. R.,** Dimethylnitrosamine formation from sodium nitrite and demethylamine by bacterial flora of rat intestine, *Res. Commun. Chem. Pathol. Pharmacol.,* 2, 24, 1971.
40. **Hill, M. J.,** The role of colon anaerobes in the metabolism of bile acids and steroids, and its relation to colon cancer, *Cancer,* 36, 2387, 1975.
41. **Norseth, T.,** Biotransformation of methyl mercuric salts in germfree rats, *Acta Pharmacol. Toxicol.,* 30, 172, 1971.
42. **Abdulla, M., Arnesto, B., and Ihse, I.,** Methylation of inorganic mercury in experimental jejunal blind-loop, *Scand. J. Gastroenterol.,* 8, 565, 1973.
43. **Reamer, D. C. and Zoller, W. H.,** Selenium biomethylation products from soil and sewage sludge, *Science,* 208, 500, 1980.
44. **Tyler, V. E., Brady, L. R., and Robbers, J. E.,** *Pharmacognosy,* 8th ed., Lea & Febiger, Eds., Philadelphia, 1981, 170.
45. **Hoffman, B. F. and Bigger, J. T., Jr.,** Digitalis and allied cardiac glycosides, in *The Pharmacological Basis of Therapeutics,* 6th ed., Gilman, A. G., Goodman, L. S., and Gilman, A., Eds., MacMillan, New York, 1980, 747.
46. **Smith, T. W. and Harber, E.,** Digitalis, *N. Engl. J. Med.,* 289, 945, 1973.
47. **Lindenbaum, J., Rund, D. G., Butler, V. P., Jr., Tse-Eng, D., and Saha, J. R.,** Inactivation of digoxin by the gut flora: reversal by antibiotic therapy, *N. Engl. J. Med.,* 305, 789, 1981.
48. **Reuning, R. H. and Geraets, D. R.,** Digoxin, in *Applied Pharmacokinetics: Principles of Therapeutic Drug Monitoring,* Evans, W. E., Schentag, J. J., and Jusko, W. J., Eds., Applied Therapeutics, San Francisco, in press.
49. **Gault, M. H., Longerich, L. L., Loo, J. C. K., Ko, P. T. H., Fine, A., Vasdev, S. C., and Dawe, M. A.,** Digoxin biotransformation, *Clin. Pharmacol. Ther.,* 35, 74, 1984.
50. **Lage, G. L. and Spratt, J. L.,** Structure-activity correlation of the lethality and central effects of selected cardiac glycosides, *J. Pharmacol. Exp. Ther.,* 152, 501, 1966.
51. **Gault, H., Kalra, J., Ahmed, M., Kepkay, D., and Barrowman, J.,** Influence of gastric pH on digoxin biotransformation. I. Intragastric hydrolysis, *Clin. Pharmacol. Ther.,* 27, 16, 1980.
52. **Gault, H., Kalra, J., Ahmed, M., Kepkay, D., Longerich, L., and Barrowman, J.,** Influence of gastric pH on digoxin biotransformation. II. Extractable urinary metabolites, *Clin. Pharmacol. Ther.,* 29, 181, 1981.
53. **Watson, E., Clark, D. R., and Kalman, S. M.,** Identification by gas chromatography-mass spectrometry of dihydrodigoxin-a metabolite of digoxin in man, *J. Pharmacol. Exp. Ther.,* 184, 424, 1973.
54. **Heinz, N. and Flasch, H.,** Comparison of the pharmacokinetics of digoxin and dihydrodigoxin in cats in single-dose studies, *Nauyn-Schmiedeberg's Arch. Pharmacol.,* 303, 181, 1978.
55. **Dobkin, J. F., Saha, J. R., Butler, V. P., Jr., Neu, H. C., and Lindenbaum, J.,** Inactivation of digoxin by *Eubacterium lentum,* an anaerobe of the human gut flora, *Trans. Assoc. Am. Physicians,* 95, 22, 1982.
56. **Dobkin, J. F., Saha, J. R., Butler, V. P., Jr., Neu, H. C., and Lindenbarm, J.,** Digoxin-inactivating bacteria identification in human gut flora, *Science,* 220, 325, 1983.

57. **Bockbrader, H. N. and Reuning, R. H.,** Spectral analysis of the configuration and solution conformation of dihydrodigoxigenin epimers, *J. Pharm. Sci.*, 72, 271, 1983.
58. **Bockbrader, H. N. and Reuning, R. H.,** Digoxin and metabolites in urine: a derivatization-high-performance liquid chromatographic method capable of quantitating individual epimers of dihydrodigoxin, *J. Chromatogr.*, 310, 85, 1984.
59. **Reuning, R. H., Shepard, T. A., Morrison, B. E., and Bockbrader, H. N.,** Formation of (20*R*)-dihydrodigoxin from digoxin in humans, *Drug Metab. Dispos.*, 13, 51, 1985.
60. **Shepard, T. A., Hui, J., Chandrasekaran, A., Sams, R., Reuning, R. H., Robertson, L. W., Caldwell, J. H., and Donnerberg, R. L.,** Digoxin and metabolites in urine and feces: a fluorescence derivatization-HPLC technique, *J. Chromatogr.*, in press.
61. **Gault, H., Kalra, J., Longerich, L., and Dawe, M.,** Digoxigenin biotransformation, *Clin. Pharmacol. Ther.*, 31, 695, 1982.
62. **Robertson, L. W., Chandrasekaran, A., Reuning, R. H., Hui, J., and Rawal, B. D.,** Reduction of digoxin to 20*R*-dihydrodigoxin by cultures of *Eubacterium lentum, Appl. Environ. Microbiol.*, in press.
63. **Bokkenheuser, V. D., Winter, J., O'Rourke, S., and Ritchie, A. E.,** Isolation and characterization of fecal bacteria capable of 16α-dehydroxylating corticoids, *Appl. Environ. Microbiol.*, 40, 803, 1980.
64. **Chandrasekaran, A., Robertson, L. W., Geraets, D. R., Hui, J., Caldwell, J. H., Donnerberg, R. L., and Reuning, R. H.,** Identification of 20*R*-dihydrodigoxin as the major reduced digoxin metabolite in humans, *Drug Metab. Dispos.*, in press.
65. **Shomo, R. E., Chandrasekaran, A., Marshall, A., Reuning, R. H., Robertson, L. W., and Weisenberger, C. R.,** Laser desorption/fourier transform/ion cyclotron resonance vs. fast atom bombardment/magnetic sector mass spectrometry: digoxin, digitoxin, and their reduced and sugar-hydrolyzed metabolites, *Biomed. Mass Spectrom.*, in press.
66. **Geraets, D., Hui, J., Chandrasekaran, A., Caldwell, J., Robertson, L. W., Donnerberg, R., and Reuning, R.,** Influence of hypochlorhydria on excretion of digoxin and metabolites in aged humans, *Clin. Pharmacol. Ther.*, in press.
67. **Storstein, L.,** Pharmacokinetics of digitoxin, in *Handbook of Experimental Pharmacology,* Vol. 56, (Part 2), Greeff, K., Ed., Springer-Verlag, New York, 1981, 31.
68. **Vohringer, H. F. and Rietbrock, N.,** Pharmacokinetics and metabolism of digitoxin in the human, in *Cardiac Glycosides,* Dengler, H. J., Ed., Springer-Verlag, New York, 1978, 64.
69. **Perrier, D., Mayersohn, M., and Marcus, F. I.,** Clinical pharmacokinetics of digitoxin, *Clin. Pharm.*, 2, 292, 1977.
70. **Storstein, L.,** Studies on digitalis. VII. Digitoxin metabolism on a maintenance regimen and after a single dose, *Clin. Pharmacol. Ther.*, 21, 125, 1977.
71. **Bodem, G. and Unruh, E. V.,** Dihydrodigitoxin, a metabolite of digitoxin in humans, in *Cardiac Glycosides,* Bodem, G. and Dengler, H. J., Eds., Springer-Verlag, New York, 1978, 74.
72. **Bokkenheuser, V. D., Winter, J., Finegold, S. M., Sutter, V. L., Ritchie, A. E., Moore, W. E. C., and Holdeman, L. V.,** New markers for *Eubacterium lentum, Appl. Environ. Microbiol.*, 37, 1001, 1979.
73. **White, B. A., Fricke, R. J., and Hylemon, P. B.,** 7β-Dehydroxylation of ursodeoxycholic acid by whole cells and cell extracts of the intestinal anaerobic bacterium, *Eubacterium* species V.P.I. 12708, *J. Lipid Res.*, 23, 145, 1982.
74. **Eyssen, H. and Verhulst, A.,** Biotransformation of Linoleic Acid and Bile Acids by *Eubacterium lentum, Appl. Environ. Microbiol.*, 47, 39, 1984.
75. **Masuda, N., Oda, H., Hirano, S., Masuda, M., and Tanaka, H.,** 7α-Dehydroxylation of bile acids by resting cells of a *Eubacterium lentum-* like intestinal anaerobe, strain C-25, *Appl. Environ. Microbiol.*,47, 735, 1984.
76. **Chandrasekaran, A. and Robertson, L. W.,** Reductive inactivation of digitoxin by cultures of *Eubacterium lentum, Appl. Environ. Microbiol.*, in press.

Antibiotic Formulations

Chapter 5

FORMULATIONS AND MANUFACTURING OF ANTIBIOTICS

Tarun R. Patel and Walter G. Chambliss

TABLE OF CONTENTS

I. INTRODUCTION

The development of an antibiotic formulation involves the same routine steps as the development of any other drug product. However, there are a few characteristics of antibiotics which make the formulation task a little more difficult. First, therapeutic doses are usually quite high, and in a particular formulation, the bulk of the dosage form consists of the active ingredient. Therefore, the physical characteristics of the antibiotic, such as the particle shape and size, dominate the overall formulation. The second characteristic is poor stability. These compounds generally exhibit a poor stability profile; therefore, particular attention has to be paid to the other ingredients in the formula in order to develop a stable product. Third, they are usually very expensive, especially during the early stages of development when they are newly discovered. This becomes an inhibiting factor when large-scale production experiments are planned.

Antibiotics are available in a wide variety of dosage forms such as tablets, capsules, oral suspensions, topical powders, semisolids, and injectables. Most are marketed in the form of tablets, capsules, oral suspensions, and injections (intramuscular and intravenous). Each dosage form requires a unique formulation and manufacturing process. However, every antibiotic dosage form must meet the same basic requirements, as shown in Table 1.

The antibiotic must be effective, identifiable, possess the designated potency, be sufficiently pure, and have the necessary quality attributes as defined by the finished product specifications. It also has to be safe, bioavailable, and maintain a reasonable stability profile throughout its shelflife. In order to assure that the final marketed product meets all of these requirements, a massive development program needs to be undertaken. Such a program involves drug discovery research, pharmacological experiments, animal toxicity testing, clinical studies, and formulation development.

II. DEVELOPMENT STUDIES

Once a therapeutically active antibiotic is identified, the product development group begins the development work. The studies can be divided into four stages: preformulation, formulation, process development, and process validation.

A. Preformulation

During the preformulation stage, physicochemical tests are conducted in order to characterize the bulk antibiotic. Several different batches of the material are tested in order to determine the batch-to-batch reproducibility of the chemical synthesis or extraction process. The stability of the antibiotic alone, and in the presence of ingredients which are commonly used in formulation, is determined. Supplies for animal studies are also prepared during this stage. The animal studies are designed to determine safety as well as possible routes of administration.

Another major function of the preformulation group is to enhance the stability and bioavailability of the antibiotic. Many antibiotics are not absorbed after oral administration due to instability in acidic pH or poor solubility. Techniques such as enteric coating, prodrugs, and solid dispersions have been used to develop oral antibiotics.

B. Formulation

The first step of formulation development is to define the product. For example, before any formulation work begins, the route of administration must be decided. Is it an oral, vaginal, sublingual, or injectible dosage form? Next, which dosage form is best? If oral administration is possible, do we market a tablet, capsule or liquid? This decision is not only based on the pharmaceutical characteristics of the antibiotic, but also on marketing

Table 1
BASIC DOSAGE FORM REQUIREMENTS

Efficacy	Quality
Identity	Safety
Potency	Bioavailability
Purity	Stability

strategy. Marketing also decides on the color, and this should be done as early as possible because color often affects stability, elegance, assay methodology, etc. The flavor should also be selected as early as possible, and the choice should be in tune with the choice of color. The shape of a tablet can affect various physical parameters and biological properties and also can cause some production problems if not selected early enough.

In the initial stages of formulation development, several different excipient combinations are screened for feasibility. Factors such as ease of manufacture, stability, cost, availability, and bioavailability are critical. Long-term stability studies are then conducted on primary and secondary formulations to determine the optimum level of each ingredient and the expected shelflife of the product.

C. Process Development

Once a formulation is finalized, it undergoes process development studies. These studies are conducted in order to develop an economical and reproducible means of manufacturing the product in production. The batch size of product manufactured during formulation development is very small, and the equipment used may not resemble that used in production. In process development, the product is first made in the pilot plant using equipment similar to that which will be used in production. The batches made in the pilot plant are smaller than production size, but can be used to establish the manufacturing procedure. Process characterization studies are conducted to define the critical processing steps and to set acceptable ranges for each variable. Stability studies should be conducted to determine the effect of scale-up on the physical and chemical stability.

D. Process Validation

The final stage is process validation and subsequent transfer of the technology to routine production. Once a process has been developed and optimized at the pilot scale, full-scale production size batches are manufactured and validated by adequate in-process sampling and testing. The process controls are established and the finished product specifications are finalized. Detailed cleaning procedures must also be validated for the equipment and facility to prevent cross-contamination.

III. ORAL DOSAGE FORMS

All types of dosage forms undergo these stages of development. The remainder of this chapter will describe some development problems that the authors have experienced with oral dosage forms.

A. Capsules

Capsules are popular dosage forms for antibiotics because the capsule shell masks the unpleasant taste of the antibiotic. A typical formulation consists of the antibiotic powder, a filler-diluent-like lactose, and a lubricant such as magnesium stearate. Although the formulation appears very simple, production problems can occur if the process and formulations are not optimized.

Table 2
PHYSICAL PROPERTIES OF MAGNESIUM STERATE

Type	Particle shape by SEM	Particle size (VMD) (μm)	Specific surface area (m^2g^{-1})	Bulk density (g $m\ell^{-1}$)	Moisture (%)	Melting range (°C)
A	Modular	15.27	3.53	0.45	4.2	115—120
B	Flaky	16.19	6.02	0.25	3.4	115—120
C	Flaky	10.68	18.03	0.24	4.0	120—127

The physical characteristics of the raw material such as particle size, particle shape, moisture content, and bulk density can affect flow properties of the blend. This can cause the material to fill poorly in the capsule and cause weight variation problems. The capsule fill weight is directly related to the potency of the antibiotic. If the potency drops below a critical value, then the amount of powder required to give the desired activity will exceed the capacity of the capsule. The material then has to be densified before it can be used.

The lubricant is generally less than 5% of the formula, but if it is not properly distributed, then poor flow, slow dissolution, and low bioavailability may occur. The lubricant coats the surface of the antibiotic powder, forming a water barrier. The degree of coating depends upon the amount of lubricant, the surface area of the lubricant and the antibiotic, and also the blending time.

A recent article by Frattini and Simioni[1] demonstrates this phenomenon. The article stressed that the surface area of magnesium stearate, the most commonly used lubricant, varies from lot to lot. Therefore, simply controlling the amount of lubricant in a formulation is not sufficient. The authors evaluated three types of magnesium stearate which differed in specific surface area, as shown in Table 2.

Tablets lubricated with different types of magnesium stearate differed significantly in dissolution rate even though the same amounts of the lubricant were used (shown in Figure 1). The time it takes for 50% of the drug to dissolve, T_{50}, increased proportionally to the surface area of magnesium stearate. However, when an equivalent amount of each type of magnesium stearate (i.e., equivalent lubricating areas) was used, the T_{50} did not change (shown in Figure 2).

The degree of lubricant/antibiotic powder mixed may be very difficult to control because mixing occurs not only when the lubricant is physically mixed with the antibiotic in a blender, but also in the encapsulation machine itself. This phenomenon has been observed during the development of new antibiotic capsules. It is not unusual to find that the dissolution rate is slower when the product is filled on a high-speed encapsulation machine than when it is hand filled.

In order to simulate this machine effect, laboratory scale experiments may be conducted in which the antibiotic/lubricant blend is repeatedly forced through a 200-mesh screen. In general, the dissolution rate decreases as the shear increases. Since the degree of shear occurring in the machine cannot be controlled, it may be necessary to reduce the amount of lubricant in the formulation. Sufficient amounts of lubricant must be present, however, in order for the material to flow in the machine.

B. Tablets

There are three major methods of preparing granulation for tablet compression. These can be described as wet granulation, dry granulation, and direct compression. The most common method of preparing tablets is by wet granulation. The basic steps of this method are listed in Table 3.

In this method, a granulating solution is used to impart adhesiveness to the ingredients

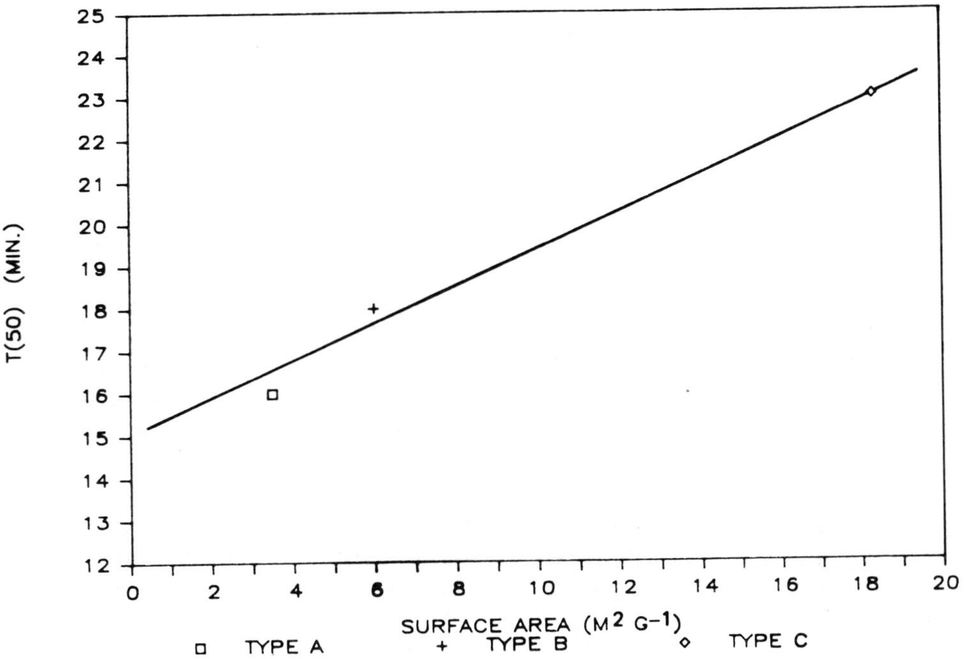

FIGURE 1. Influence of 0.5% (W/W) magnesium stearate on the dissolution rate. Reprinted from Reference 1, p. 1126, by courtesy of Marcel Dekker, Inc.)

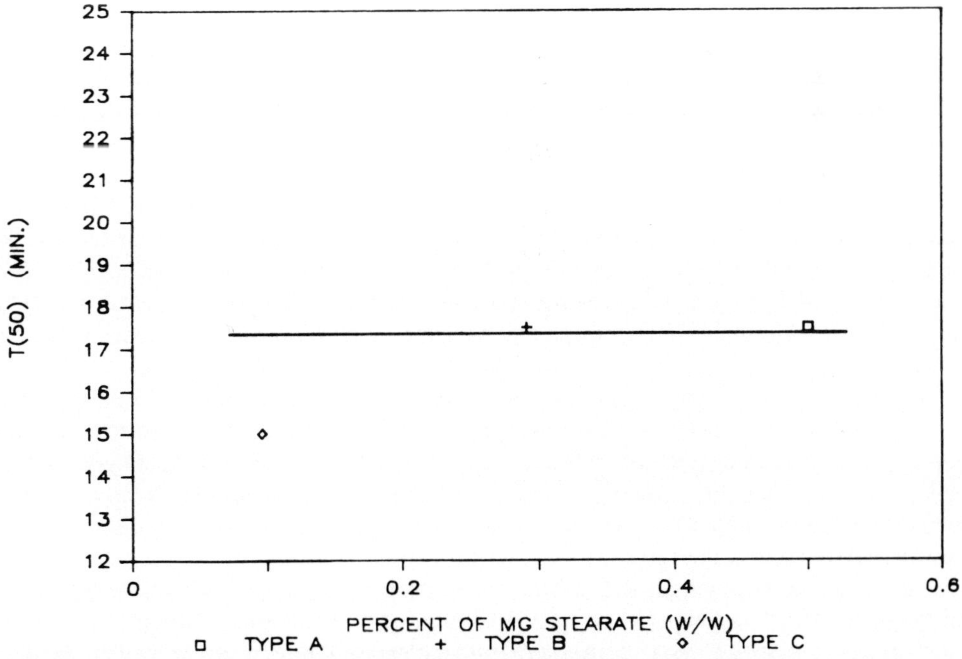

FIGURE 2. Influence of equivalent areas of magnesium stearate on the dissolution rate. (Reprinted from Reference 1, p. 1126, by courtesy of Marcel Dekker, Inc.)

Table 3
WET GRANULATION PROCESS

Weighing
Preblending
Granulation
Drying
Milling
Blending
Compression

Table 4
DRY GRANULATION PROCESS

Weighing
Preblending
Chilsonating
Milling
Blending
Compression

so that they will form granules and a firm tablet when compressed. The wet granulation method, however, cannot be used for antibiotics which are sensitive to heat or moisture.

The basic steps of dry granulation are given in Table 4. In a dry granulation process, the antibiotic is blended with excipients such as diluents and disintegrants and then compacted into a dry mass. The mass, which may be in the form of a large tablet (slug) or a ribbon of material, is milled to the desired granule size, lubricated, and then compressed into tablets. This technique converts generally fluffy antibiotic powder to a granular form, which is compressible without subjecting the drug to conditions which could affect the chemical stability.

Because this method can be used for heat- and moisture-sensitive drugs, it is a popular method of preparing antibiotics for tablet compression. However, the conditions used to process the materials must be optimized in order to prevent production problems. A general description of the chilsonation process will be used as an example. The material passes through rollers and is compressed into a ribbon. The ribbon then falls into a mill where it is reduced to the desired particle size. The pressure exerted by the rollers and the rate of feed are critical processing variables which affect the quality of the granulation. Some antibiotic powders are self-compressing and very cohesive. This may lead to selective chilsonation and a nonuniform granulation. For example, in one lot of antibiotic, the potency of the granulation ranged from 280 μg/mg in the fine powder to 720 μg/mg in the coarse particles. Tablets compressed from the granulation showed an interesting effect in that there was a correlation between potency and tablet weight as shown in Figure 3. The higher the tablet weight, the higher the potency in μg/mg. Optimization of the feed rate and roller pressure in the chilsonator corrected the problem. The results obtained from a subsequent lot are presented in Figure 4. There was no correlation between potency and fill weight. The individual data points were scattered around the average potency and tablet weight as expected.

Antibiotic tablets are often coated to mask the unpleasant taste of the drug. There are four types of coated tablets: film coated, sugar coated, enteric coated, and press coated. Examples of film-coated tablets include Bristol Laboratory's, Pen VK. and Cefadroxil tablets. Antibiotic tablets may also be enteric coated, because the drug is acid labile and therefore easily destroyed by stomach acid. The coating polymer is pH sensitive and dissolves only after it is emptied from the stomach.

If tablet coating is required, the tablet must be formulated in such a manner that it can withstand the abrasion which occurs in the coating pan. In addition, certain logos on the tablet are very difficult to coat. Extra stability studies are required to ensure that the film coat does not alter the stability profile.

C. Powders for Oral Suspension

Oral suspensions generally consist of bulk antibiotic powder, wetting agents, antioxidants, suspending agents, sweeteners, flavors, and dyes. The powder is reconstituted by the phar-

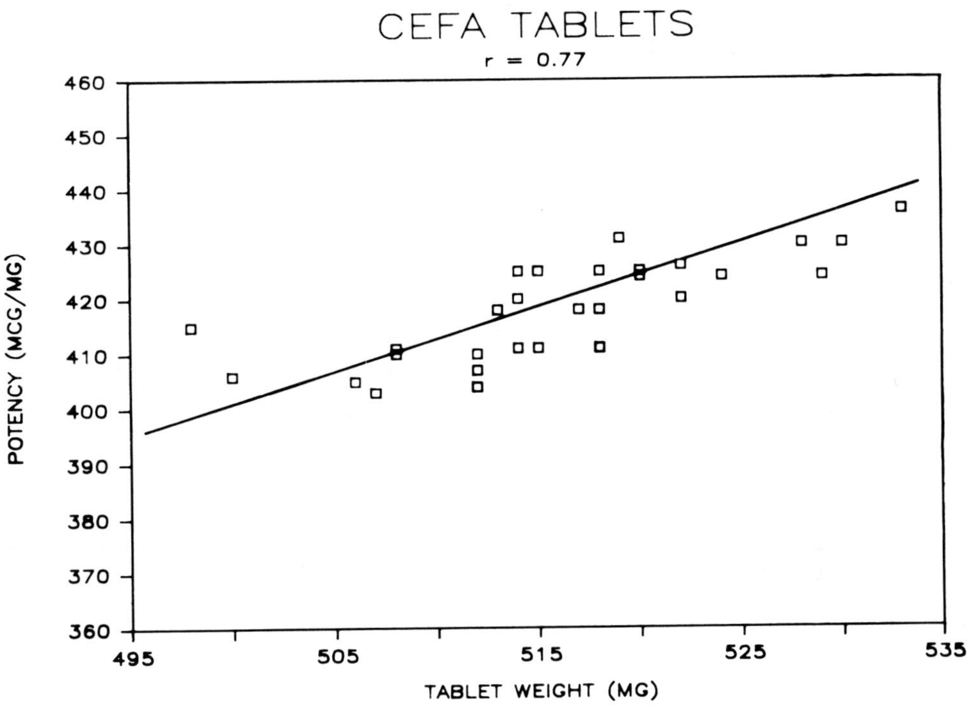

FIGURE 3. Effect of tablet weight on potency: nonoptimized chilsonation process.

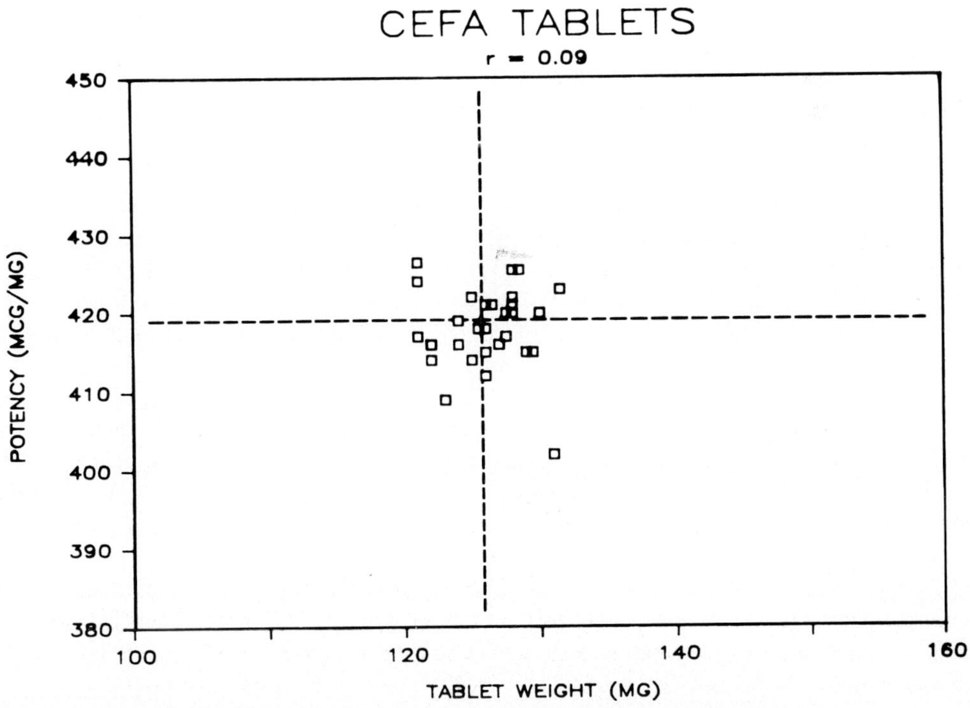

FIGURE 4. Effect of tablet weight on potency: optimized chilsonation process.

macist, using distilled water prior to dispensing. Formulation development studies concentrate on the physical characteristics of the suspension, bioavailability, and palatability.

Production problems usually center around the uniformity of the powder blend, especially for potent antibiotics, and the fill weight of the bottles due to variability in the powder flow. The particle size of the bulk antibiotic is often the most critical variable because segregation can occur during filling if the particle size of the antibiotic differs significantly from the excipients.

IV. CONCLUSION

Antibiotics are available in a wide variety of dosage forms, each requiring extensive formulation and process development studies prior to full-scale production. Typical problems which occur during development and production have been discussed. Some of the critical variables which must be controlled during production have been emphasized. Production problems will always occur, but a carefully planned and systematic approach to product development can minimize the incidence of many problems.

REFERENCE

1. **Frattini, C. and Simioni, L.,** Should magnesium stearate be assessed in the formulation of solid dosage forms by weight or by surface area?, *Drug Dev. Ind. Pharm.*, 10 (7), 1117, 1984.

Chapter 6

PARTICULATE MATTER IN PARENTERAL PRODUCTS

Joseph R. Tricome, Ronald J. Sax, and C. Martin Parkinson

TABLE OF CONTENTS

I. INTRODUCTION

The presence of particulate matter in large and small volume parenteral products is a major concern for both the medical profession and the pharmaceutical industry.[1-3] To maintain proper quality control of particulate matter levels in parenteral solutions, a comprehensive particulate matter program must be undertaken. The program consists of sizing and counting of particulates, identification of particulates, and ultimately determining the sources of particulate matter. This paper examines one approach to a particulate matter program, including the techniques and instrumentation employed for particulate matter analysis. In addition, government regulations which were implemented January 1, 1986 for small volume parenterals will be discussed.

II. PARTICULATE MATTER AND SIZE

Particulate matter is defined as "...extraneous, mobile, undissolved substances, other than gas bubbles, unintentionally present in parenteral solutions".[4]

The definition of particulate matter not only applies to parenteral products but also to other things. The air we breathe, for example, contains particulates from the environment. The dust from roads and construction produces airborne particles which are easily inhaled. Particulate matter is common in the water we drink and as many as 10 ppm is allowed in purified water USP as dissolved solids used for manufacturing of pharmaceutical products.

The size of particulates the human eye can see without magnification is about 40 to 50 μm (micrometers). (A micrometer is defined as 1/1000 of a millimeter.) If we were to equate the size of particulates to the volume of space the particle would occupy, then a 50-μm particle would occupy a volume of 1.25×10^5 μm^3 assuming a unit density. If we had 1 mg of particulates, the space it would occupy is about 10^{12} μm^3, and 1 ppb would contain about 8000 particles per liter.[5] Considering our earlier example of 10 ppm of solids in purified water USP, the number of particulates in this level of solids, using a unit density, is extremely large.

III. PARTICULATE MATTER SIZING AND COUNTING TECHNIQUES

Particulate matter in a parenteral product needs to be closely monitored to maintain effective quality control of a manufacturing process. Because of the size of the particulates, in most cases less than 1000 μm, accurate measurement and counting of particulate matter can be very difficult. Various methods have been developed to accomplish this task. Methods commonly employed to size and count particulates include use of (1) optical microscopy, (2) HIAC® analyzer, (3) Coulter Counter®, and (4) image analyzer.

A. Optical Microscopy

The optical microscope has been the preferred instrument for sizing and counting particulates, because of its cost in comparison to more expensive instrumental techniques and the ability to further identify the particulates if one desires.

The technique of sizing and counting particulates first involves the isolation of the particulates onto a membrane filter. The particulate matter is usually suspended in an aqueous solution of the parenteral product. In the case of a dry-powder fill, 0.22-μm-filtered distilled water is added, and the solution is shaken to dissolve the solid and suspend the particulates. The solution containing the particulates is filtered under Class 100 laminar flow conditions and isolated on a 0.45-μm membrane filter. Once the particulates are isolated, they are sized and counted under the optical microscope.

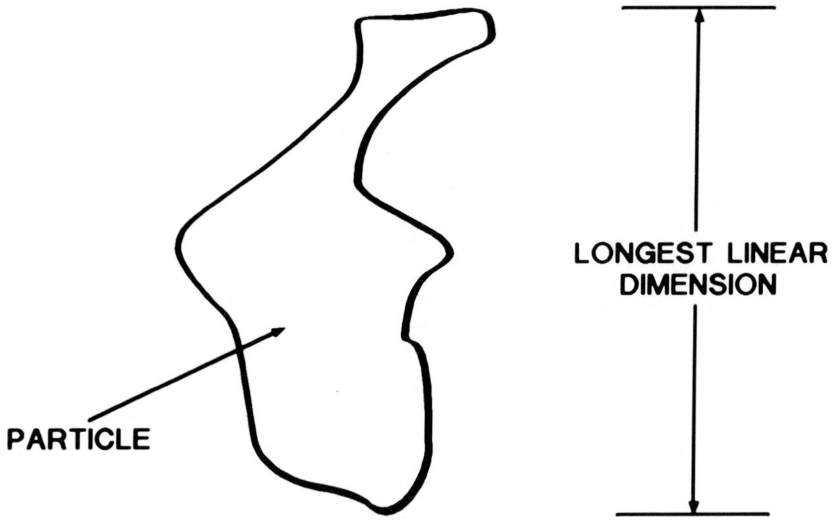

FIGURE 1. A Drawing illustrating the longest linear dimension of a nonspherical particle.

The optical microscope is equipped with an X-Y mechanical stage, illumination source, objective lens and eyepiece capable of 100× magnification and an eyepiece micrometer scale. The eyepiece micrometer scale is calibrated against a stage micrometer. The stage micrometer scale has divisions of 100 μm with a subdivision of 10 μm for each one of the major 100-μm divisions. The eyepiece micrometer scale for each objective used for sizing and counting is calibrated against the stage micrometer.

Each particulate isolated on the filter is sized and counted, utilizing oblique illumination at about a 20° angle. As each field underneath the microscope objective is examined, the counting is accomplished with the aid of a tally counter used frequently for counting blood cells. The particle size is determined by measuring the longest linear dimension of the particle as illustrated in Figure 1.

Once the particles are sized and counted, the total number of particles ⩾10 μm and ⩾25 μm are the normal ranges used for determining the level of particulates present in the product.

B. HIAC® Analyzer

The HIAC® analyzer is electronic instrumentation, based on the principle of light blockage, capable of sizing and counting particles suspended in a liquid medium. The instrumentation, manufactured by HIAC/Royco® (Menlo Park, California), is composed of an automatic bottle sampler (ABS) and electronic control unit with adjustable channels used to select the threshold settings for the various particle size ranges. The size ranges capable of being measured depend on the size of the sensor employed for analysis (Figure 2). The sensors used in the authors' laboratory were 2 to 120 μm and 5 to 150 μm.

A typical sensor is designed as a rectangular flow cell which liquid passes through, based on a predetermined volume or time interval. A light beam passing through the sensor is detected by a photodiode (Figure 3). When a particle passes through the sensor, it interrupts the light. This interruption of the light produces a change in the voltage which is proportional to the projected area or "shadow" of the particle.

The ABS is the part of the HIAC® which controls the vacuum or pressure of the system for the flow of the liquid going through the volume metering tube. If the viscosity of the liquid being tested is sufficiently high, pressure may have to be applied to pump the liquid through the HIAC® sensor. However, for most pharmaceutical parenteral solutions, the vacuum is used to control the liquid flow.

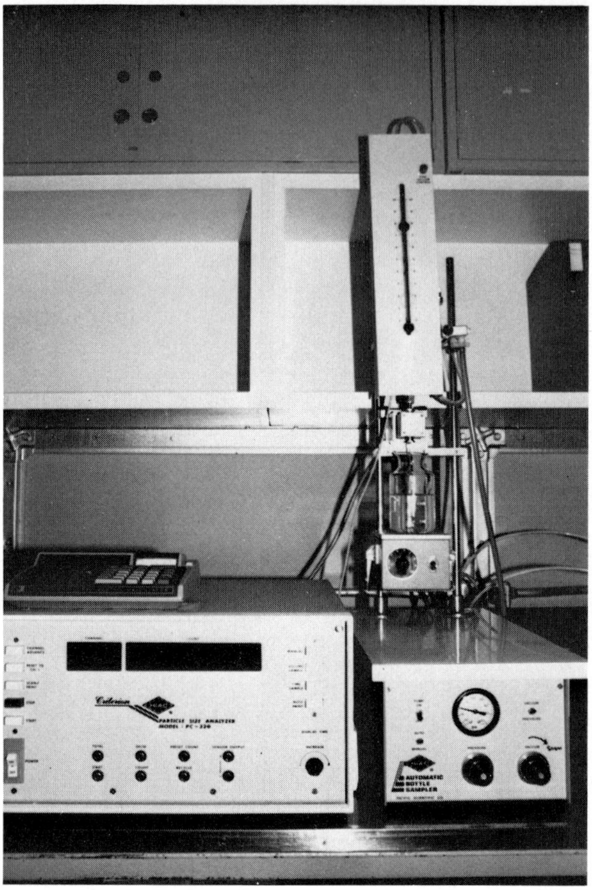

FIGURE 2. Photograph of the HIAC® instrumentation in the authors' laboratory, showing the automatic bottle sampler on the right and the control unit on the left.

C. Coulter Counter®

The Coulter Counter®, manufactured by Coulter Electronics (Luton, England), is used extensively in medical laboratories for counting blood cells. The principle of operation is based on the volume displaced by a particle flowing through an aperture, as determined by the change in voltage between two electrodes immersed in an electrolytic solution. The major limiting factor to the technique is the use of electrolytic solution for particle measurement.[6]

D. Image Analyzer

One way of overcoming the very labor-intensive method of manual microscopic particle size analysis is by using an automated image analysis system.

The Bausch and Lomb Omnicon® 3500, manufactured by Bausch and Lomb (Rochester, New York), is (1) a computer-based system composed of a vidicon camera mounted on top of an optical microscope, (2) a motorized microscopic stage to advance the fields of view containing the particulates on a membrane filter, (3) a video display to show the particulates as they are counted, (4) an input terminal, and (5) a sophisticated software to measure and size the particulates.

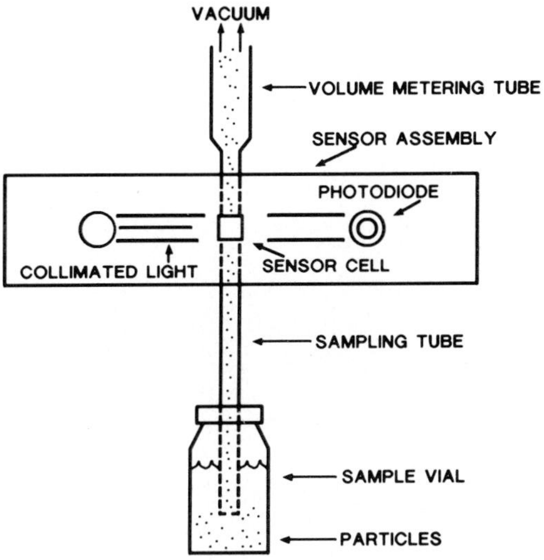

FIGURE 3. A diagram of the particle-counting device in operation. The particles suspended in a liquid are drawn by a vacuum and passed through the sensor where they interrupt a beam of collimated light which is monitored by a photodiode. The drop in potential is measured by the particle counter and is proportional to the diameter of a sphere based on the area of the particle.

The method of operation is based on the vidicon camera receiving an image from the particles isolated on the filter membrane. The image received by the camera is converted to a series of electrical pulses whose width and height are proportional to the size and gray-level of the particles and filter membrane. Adequate gray-level separation between the particulates and membrane filter is essential for proper sizing and counting of particles. This gray-level separation is accomplished by proper adjustment of the instrument threshold above and below the background gray-level.[7]

The system provides a very quick and efficient means of particle size analysis. The information retrieved from the system can be expressed as particle size ranges, histograms, and particle shapes.

The major point to consider when sizing and counting particulates is that results are dependent on the method used; the results from an instrumental method may be different from a manual microscopic method. To illustrate the comparison, let's examine a sphere and a fiber from a theoretical point of view. A manual microscopic method using the longest linear dimension of a sphere having a diameter of 100 μm would be sized at 100 μm. If this same particle was analyzed by HIAC® using a light-blockage principle, the sphere would be sized as 100 μm. A fiber having a length of 100 μm and a width of 5 μm would be sized as 100 μm, using the longest linear dimension. The fiber, when sized by the HIAC®, however, would be sized as about 25 μm, based on its equivalent spherical diameter.

Each method of particle size analysis has its advantages and disadvantages. Each method, depending on the analyst's needs, provides valuable information for monitoring the quality of the parenteral product.

IV. PARTICULATE MATTER IDENTIFICATION

A comprehensive particulate matter program must not only concern itself with the counting of particles, but also with the identification of particles. Identification of particles is important because information obtained concerning the nature of the particles may ultimately lead to the discovery of the source of the problem and its eventual eradication. Some of the most common identification techniques are (1) optical microscopy (polarized), (2) scanning electron microscopy, (3) energy dispersive X-ray analysis, (4) image processing, and (5) chemical microscopy.

A. Optical Microscopy (Polarized)

1. Sample Preparation

In order to identify particles accurately, it is first necessary to remove the particles from the medium (product) in which they are present. Samples which are not already in a liquid state should be reconstituted to a label-specified volume with particle free water.

In the microscopy lab, the optimum method for particle isolation is by filtration through an appropriate filter, i.e., methyl cellulose (porosity 0.45 to 0.80 μm). Sample preparation and filtration should be conducted under laminar flow, Class 100 conditions in order to reduce extraneous particulate contribution from the environment.

Filtration is accomplished by passing the liquid through a filter placed between a glass filtration funnel and supporting glass frit, while being held securely together with a clamp. The filtration unit is then placed on top of a side-arm flask attached to a vacuum source. Strict procedures must be followed at all times in order not to contaminate the sample with extraneous particles.[8]

After filtration, the filter should be placed in a clean Petri slide and dried under laminar flow before proceeding further. Once isolated on a filter, a particle may be removed, mounted on a slide with immersion oil, and protected with a glass cover slip. In the case of some filters, i.e., white methyl cellulose, the whole filter may be mounted in immersion oil of the same refractive index, 1.515, thus causing the filter to "disappear", leaving only the particles visible for microscopic examination.

2. Sample Examination

The next phase of particle identification is accomplished by optical evaluation utilizing a polarizing light compound microscope. The polarizing light microscope differs from a standard biological microscope in that light passing through a polarizing microscope is reduced to a single vibrational plane in order to enhance the optical properties of the specimen being studied.

As light passes through a polarizing microscope it first must pass through a polarizer, which reduces the light from many vibrational planes to a single vibrational plane, perpendicular to the path the light is traveling. This polarized light passes through the specimen mounted on a 360° rotating graduated stage and onto the analyzer, which is in essence another polarizer. When the analyzer is crossed, or at 90° to the polarizer, light transmitted from the polarizer is blocked, thus producing a dark field. When the analyzer is parallel to the polarizer, light continues to be transmitted, thus producing a bright field.

When the analyzer is disengaged or parallel to the polarizer, a specimen illuminated by polarized light appears similar to a specimen being viewed under a standard biological microscope. It is at this time that particles being analyzed are characterized as to size, shape, color, luster, and surface texture. At this time the refractive index of the particle may also be determined.

a. Refractive Index

The refractive index is a characteristic which aids in the identification of a particle. Refractive index is a unitless ratio of the speed of light in a vacuum to its speed in some medium (particles). Most common particles have a refractive index between 1.50 to 1.72.[9]

The refractive index of a particle may be determined by mounting the particle on a glass slide, covering it with a glass cover slip, and immersing the particle in a medium of known refractive index. Careful observation of the particle in plane-polarized light will reveal a halo close to the particle's border. This halo is known as the Becke line. Adjustment of the fine focus up will cause the Becke line to travel either towards the center of the particle or towards the mounting medium. The Becke line will always travel towards the medium of greater refractive index.[10] The object is to match the unknown refractive index of the particle to a known refractive index by successive mountings of the particle in various mounting mediums of known refractive index, using the Becke line to determine which direction, i.e., higher or lower refractive index, to follow. Eventually a match will be found where the particle will become indistinguishable from the surrounding medium; thus the refractive index of the particle will be known. Knowledge of the refractive index of a particle is extremely useful in identifying particulate matter.

b. Birefringence

Another optical property which will aid in the identification of nonopaque particles is that of birefringence. Nonopaque particles can generally be placed into two groups according to their appearance in crossed polarized light. One group will remain dark when the microscope stage is rotated 360°. This group is called isotropic. The other group will become alternately light and dark as the stage is rotated 360°. This group is called anisotropic. Particles behave isotropically or anisotropically according to their atomic arrangement. Isotropic particles belong to the isometric or cubic system of atomic arrangement where the spacing of atoms is alike in all three axial directions. They may also belong to the amorphous group of materials. Anisotropic particles belong to the tetragonal, hexagonal, rhombohedral, orthorhombic, monoclinic, and triclinic crystallographic systems where atoms of the crystals are arranged in either a single optical axis (uniaxial) or two optical axes (biaxial).[11,12] The result of this atomic ordering is that a particle will exhibit two refractive indices when uniaxial and three or more refractive indices when biaxial.

When dealing with uniaxial, anisotropic particles, one will note that as the particle is rotated between crossed polarizers it will disappear, or become extinct, four times in 360° rotation. These points at which the particles disappear are called extinction positions, and are 90° apart. Locating the extinction positions of a particle will reveal the vibrational directions (refractive indices) of each crystal, which will parallel the vibrational directions of the two polars when the crystal is extinct.[10]

When polarized light enters an anisotropic crystal, the light is split into two vibrational components (double refraction). The two components follow the two major vibrational directions which have different refractive indices. The greater the refractive index, the slower the polarized light will travel through a particle. As both components emerge from the particle, one will be retarded by a definite amount. The greater the difference between the two refractive indices ($N_2 - N_1$), birefringence, the greater the retardation. Particle thickness also plays a role in determining retardation.[10] Interference due to retardation between the two components will produce a color (interference color) when both are recombined by the analyzer. By employing a color chart called the Michel-Levy chart and by determining the particle thickness in nanometers (nm), the birefringence of a particle may be determined, as in McCrone et al,[10]

$$N_2 - N_1 = \frac{r}{t}$$

where r = the retardation interference color in nanometers, t = the particle thickness in nanometers, and $N_2 - N_1$ = birefringence. Another method for determining birefringence is by calculating retardation by employing a Berek compensator. The Berek compensator, inserted after the analyzer, may give a more precise numerical retardation value than determined by color matching on the Michel-Levy chart.

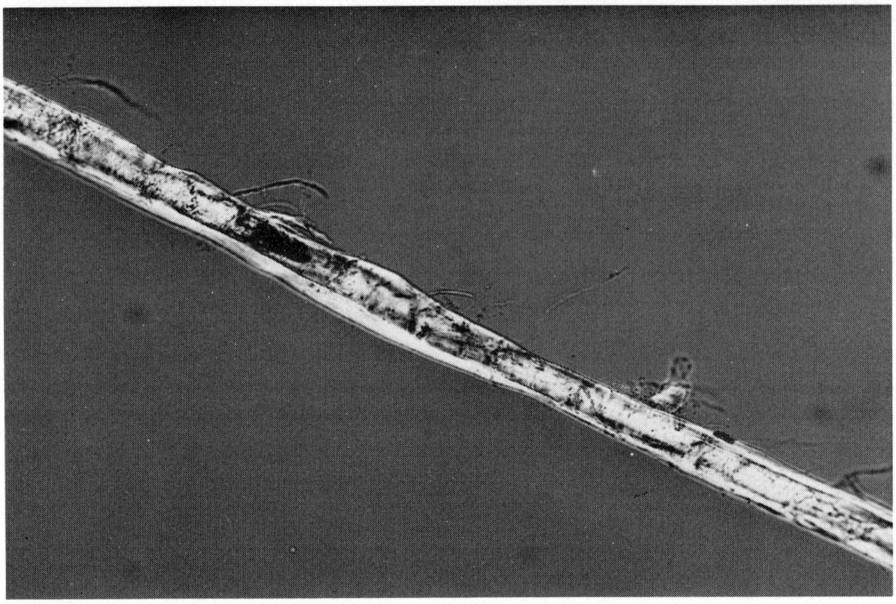

FIGURE 4. Optical photomicrograph of a paper fiber in plane-polarized light at 200× magnification.

3. Fiber Identification

Fibers from many sources may be found in pharmaceutical products.[13] Optical microscopy employing polarized light has proven very useful in determining what fiber type is present and ultimately the source of contamination. Three common fibers found are (1) paper fibers, (2) cotton fibers, and (3) synthetic fibers.

a. Paper Fibers

Paper fibers are primarily composed of cellulose, whose atoms are arranged in a crystalline lattice. The crystalline structure of the fiber exhibits birefringence under polarized light. Paper fibers are long, flat, and tapered, resembling translucent ribbons (Figure 4). They are generally colorless unless the paper has been dyed. Refractive index in the long direction is 1.53 to 1.58.[14] Paper fibers exhibit total extinction when rotated 360° under crossed polars.

b. Cotton Fibers

Cotton fibers are similar to paper fibers in that they, too, are mainly composed of cellulose and are ribbon-like in appearance. Like paper fibers, cotton fibers exhibit birefringence due to their crystalline structure. Cotton fibers may be colorless or dyed. Refractive index in the long direction is 1.544 to 1.578.[15] The key in distinguishing paper fibers from cotton fibers is that the cellular structure of cotton causes the fiber to be twisted (Figure 5) instead of relatively flat, as is the case with paper fibers. Because the fiber structure is twisted, the fiber does not exhibit total extinction. Instead, extinction may be observed as a black band traveling up and down the fiber as the microscope stage is rotated 360° under crossed polars.[13]

c. Synthetic Fibers

Synthetic fibers such as polyamides (nylon) appear as smooth, transparent, continuous cylinders under the microscope (Figure 6). They may appear colored if they have been dyed. The refractive index of nylon is about 1.580 in the long direction.[16] Nylon fibers exhibit a high degree of birefringence (0.060).[13] Nylon fibers exhibit complete extinction when rotated 360° under crossed polars.

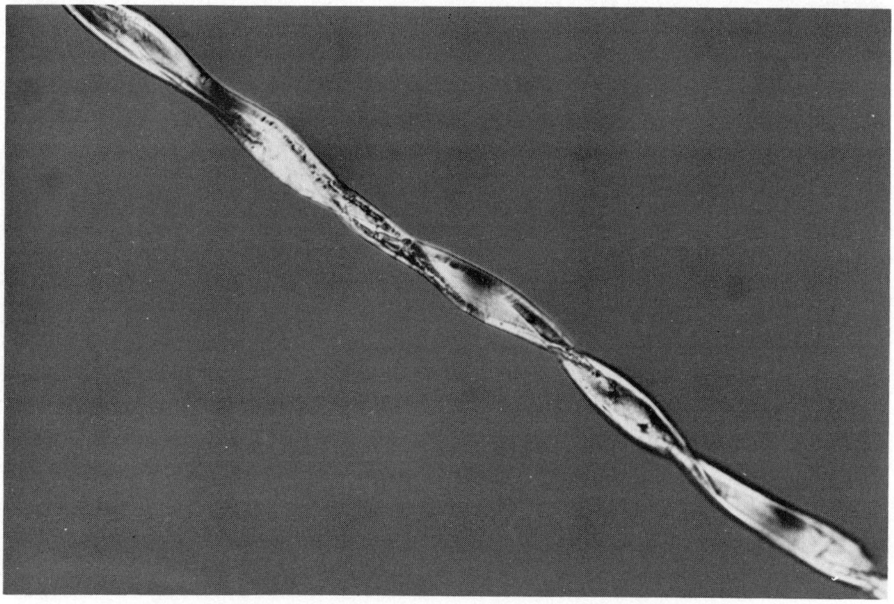

FIGURE 5. Optical photomicrograph of a cotton fiber in plane-polarized light at 200 × magnification.

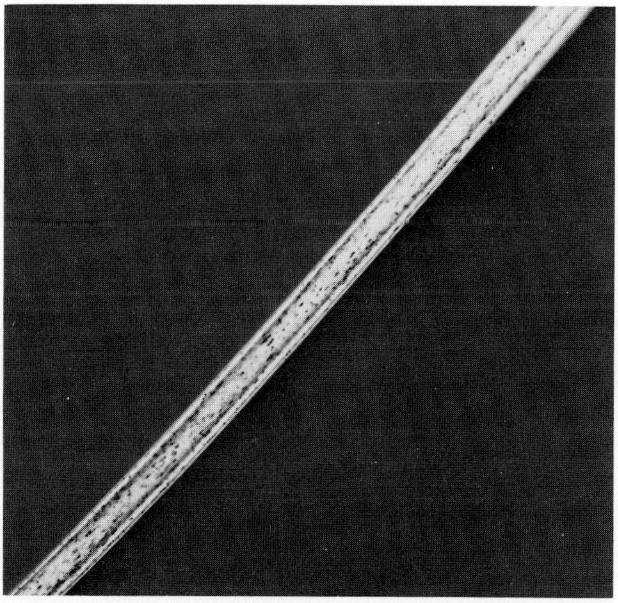

FIGURE 6. Optical photomicrograph of a nylon fiber in plane-polarized light at 200 × magnification.

B. Scanning Electron Microscopy

Once optical microscopy has been performed, further identification of the particle may be necessary to support the optical data or provide additional information. One of the primary tools used in identification is the scanning electron microscope (SEM, Figure 7). This microscope is capable of enhanced magnification as compared to an optical microscope and

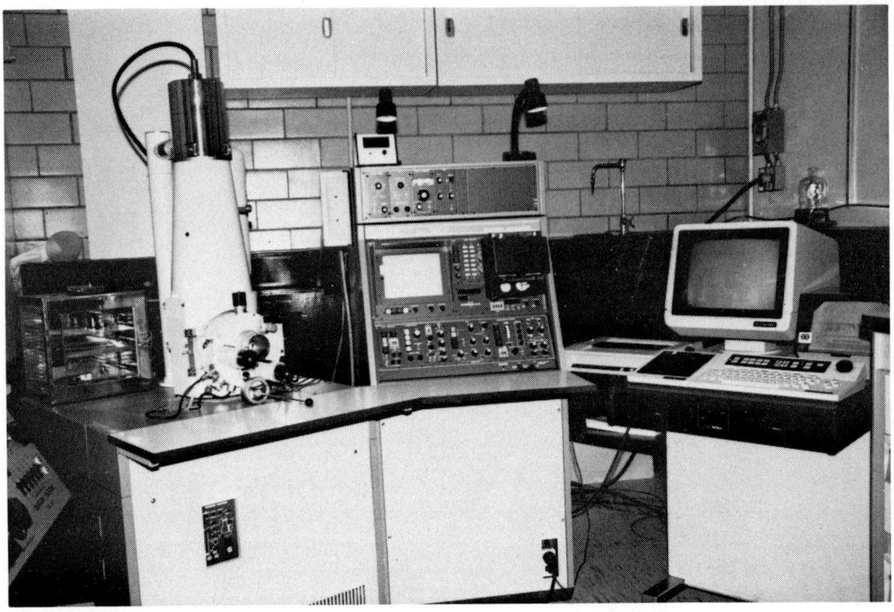

FIGURE 7. Photograph of the scanning electron microscope and energy dispersive X-ray spectrometer in the authors' laboratory. The SEM is on the left and the EDS is on the right.

in addition greatly improves the depth of field. "Depth of field is the extent of the zone on a specimen which appears acceptably in focus."[17] A properly mounted specimen can be examined at magnifications well over 100,000 times, allowing morphological characteristics to be observed. Typical resolution of a modern SEM is about 7 nm. Resolution is defined as "the smallest separation at which two points can be seen as distinct entities."[17]

The principle of operation of the scanning electron microscope is that a pinpoint beam of high-energy electrons produced by a tungsten filament is focused by an electromagnetic field across a specimen in a systematic scan pattern. The beam-specimen interaction produces signals which are translated into a video image and used to study the characteristics of the particle to be identified.

The electron beam and specimen interactions produce various types of radiation. The various types of radiation are Auger electrons, secondary electrons, back-scattered electrons, X-rays, bremssträlung, cathodoluminescence, elastically scattered electrons, and transmitted electrons (Figure 8).[19,20] The signals which are used by the particle analyst most frequently are secondary electrons, back-scattered electrons, and X-rays.

1. Secondary Electrons

Secondary electrons (SE) originate from the primary electron beam striking the specimen and producing low energy electrons from near the sample surface. The secondary electrons are drawn to a positively biased detector, converted to an electronic signal, and displayed on a cathode ray tube (CRT)(TV screen). Since the secondary electron image is generated from near the surface of the sample, topographical information may be gathered, and sizing and counting of particles may be performed (Figure 9).

2. Backscattered Electrons

Another signal which is utilized by the microscopist is the backscattered electrons. A backscattered electron is one which has been scattered one or more times within the specimen and then escapes back through the surface. Thus, the backscattered electron, being more

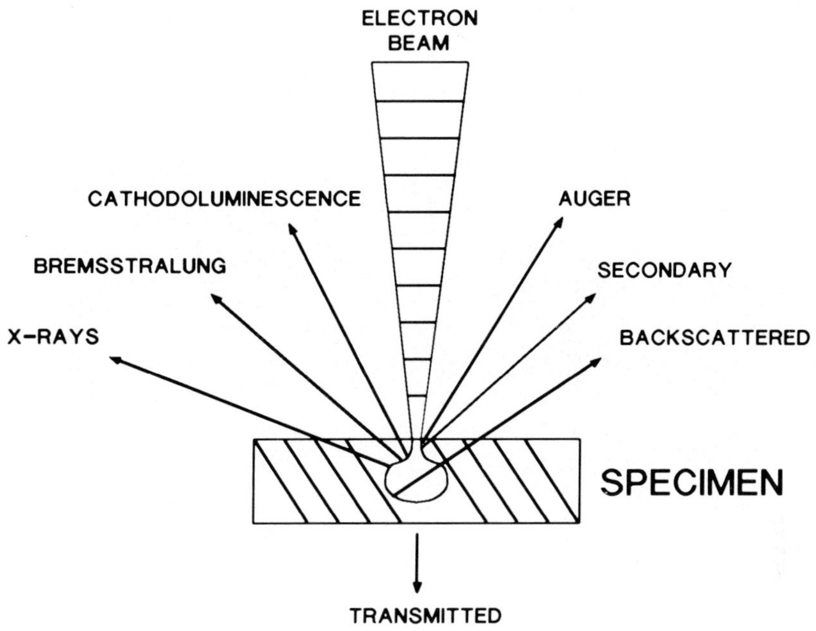

FIGURE 8. An illustration of primary electron beam specimen interaction and the major types of radiation produced.

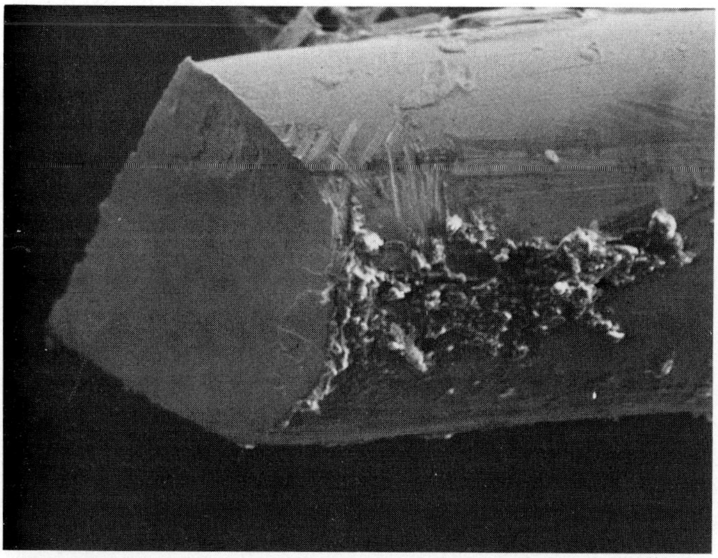

FIGURE 9. Scanning electron micrograph of a secondary electron image of a piece of copper wire.

energetic than a secondary electron, has less information about topography and more information about composition. The backscattered electron travels in a straight line and is not affected by the low voltage of the secondary electron collector. The higher the atomic number of the element present, the more backscattering will be detected. This atomic number phenomenon is useful in contrasting different elements on single or on multiple particles. The elements of higher atomic number will exhibit stronger emission which causes brighter

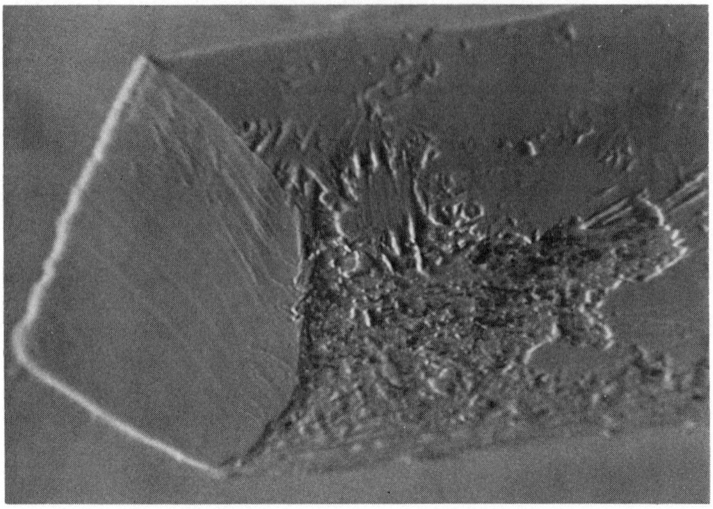

FIGURE 10. Scanning electron micrograph of a backscattered electron image of a piece of copper wire.

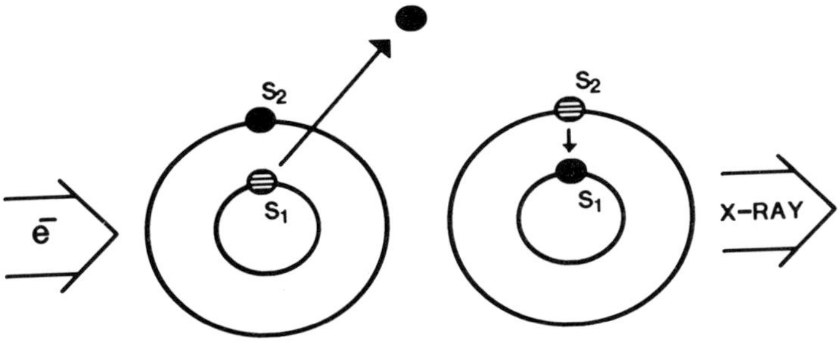

FIGURE 11. A drawing illustrating X-ray generation. The primary electron from the SEM ejects an electron from a low energy shell (S_1). An electron from a higher energy shell (S_2) fills the gap in the lower energy shell. The energy lost in the process is emitted as radiation.

areas on the CRT, thereby creating an image which is similar to a map (Figure 10). The map may be useful in identification of particles which are similar or dissimilar to each other on a membrane filter, thereby increasing the microscopist's knowledge of the compositional distribution of particulates. Another use for the backscattered electron is the shadowing effect due to the straight line path which they follow. Minor imperfections, in a seemingly smooth surface, are enhanced by the shadowing effect.

3. X-Rays

When the particle is bombarded by electrons, one type of energy that is released is the X-ray (Figure 11). The primary source of information relating to the elemental composition of the particle are the X-rays. Each specific element emits characteristic X-rays. The X-rays strike a semiconductor crystal at the detector and the energy is absorbed by electrons in the crystal. The energy is converted to an electrical signal which is then amplified and changed to a digital signal to be collected by a multichannel analyzer (MCA). The MCA sorts the signals and counts the X-rays at the characteristic energy level; the information gathered is displayed as a spectrum on a CRT (Figure 12).

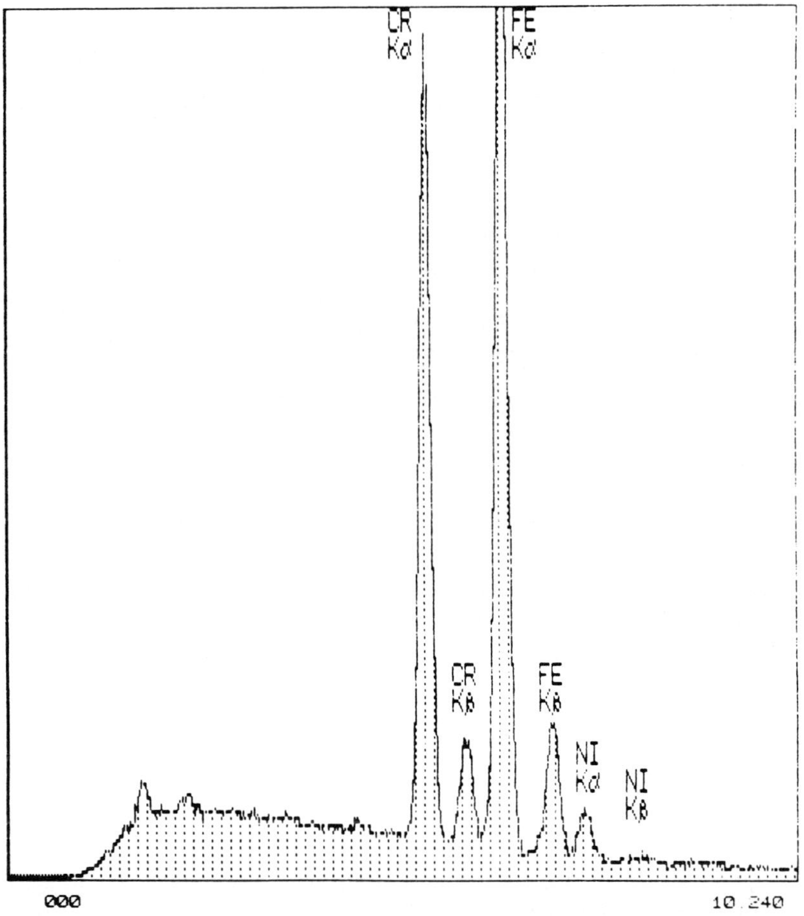

FIGURE 12. An X ray spectrum from 0 to 10 keV of stainless steel showing the Kα and Kβ lines of chromium, iron, and nickel.

C. Energy Dispersive X-Ray Analysis

The X-ray data are analyzed by a computer equipped with sophisticated software to give both qualitative and quantitative information relating to the particle's elemental composition. Due to limitations discussed below, only elements with atomic number 11 or higher can be detected by most energy dispersive X-ray spectrometer systems (EDS).[20] X-ray analysis is limited due to detector window thickness and by the accelerating voltage used to bombard the sample.

The upper limit of detected X-rays is set by the accelerating voltage which as a rule of thumb is of about one and one-half times over voltage to excite the X-ray of a specific element. Therefore, certain elements may be excluded by low accelerating voltages either by choice or by instrument limitations.

The lower limit of detection is set by the absorption effect of the detector window. Lower atomic-number elements such as boron or below generate low energy X-rays which are easily absorbed by the beryllium window, the specimen coating, or by the specimen electrons.

D. Image Processing

Another useful tool associated with the X-ray system is image processing. X-rays and secondary electrons can be processed into various types of useful information depending on the computer software available. Another way of looking at a sample is the processed X-ray map as compared to an SE image (SEI). An image processor will assign colors to selected elements and display individual maps on the CRT. This is a very powerful tool for the microscopist in identifying an unknown. When both topography and elemental composition of a particle are analyzed simultaneously, more useful information may be obtained than when the two are analyzed separately.

E. Chemical Microscopy

Chemical spot tests have been proven useful as quick methods which give important information. One such test for the presence of iron involves placing a few drops of acid, i.e., hydrochloric acid, on a suspected particle and following it with a few drops of potassium ferrocyanide. If iron is present, bubbling plus a yellow color will appear after the addition of hydrochloric acid. The particle will then turn blue when the potassium ferrocyanide is added.[21] Another useful chemical spot test is accomplished by adding Herzberg's reagent to a fiber to determine if it is of cellulosic origin. If the fiber is cellulosic, the Herzberg's reagent will cause the fiber to turn brown.[22]

V. SOURCES OF PARTICULATE MATTER

After identification of the particulates has been accomplished, the microscopist has the difficult task of identifying the sources of the particulates, if possible. The major concept behind determining the origin of the particulates is either the elimination of the particulate matter source or reduction in the level of particulate matter in the product. If the source cannot be eliminated, then measures need to be taken to minimize the contribution of the source to the product.

Major particulate matter sources are the (1) environment, (2) manufacturing process of bulks, (3) packaging components, (4) personnel, and (5) filling equipment. These sources are by no means the origin of all particulate matter present in products. However, they are the sources which can create a large majority of the particulates. Let's examine some of the sources and the types of particulate matter produced.

A. Environment

The environment consists of the area in which the work is being performed as well as material present in the area. The types of particulates most prevalent are dust-related particles, natural and synthetic fibers, and in some cases, fiberglass.

Dust particles of varying elemental composition can be introduced into the product through the environment and the conditions in which the product is being manufactured. The cleanliness of the area and handling involved with the manufacturing and filling are major considerations when these dust particulates are found.

Natural fibers, sometimes referred to as cellulosic fibers, may enter into a product through the environment. Paper in the areas where product is being manufactured and filled as well as improper cleaning of the packaging components will result in the introduction of fibers into an area. Common sources of paper fibers are the boxes and cardboard dividers in which containers are received. Synthetic fibers originate from filters or building material. These fibers are easily distinguishable from natural fibers, as discussed earlier.

B. Manufacturing Process

The manufacturing process of bulk pharmaceuticals, due to its complexity, can result in particulate matter. The steps of a bulk process will vary depending on whether it is a liquid or a dry powder. In most cases, a liquid can be terminally filtered to reduce or eliminate most particulates inherent in the manufacturing process. In the case of a dry powder, however, particulates may be produced by the process equipment, personnel handling, and raw materials.

C. Packaging Components

Particulate matter from packaging components can originate from the glass container, closure, and product/container interaction. Glass chips from manufacturing of the container or mishandling can be easily produced. The size of the chips can be several millimeters and will be readily apparent in the solution. Rubber closures will introduce particulates depending on the type of rubber employed. Natural rubber in comparison to synthetic elastomers has low-coring characteristics when penetrated by a needle. However, when a synthetic elastomer such as butyl rubber is penetrated by a hypodermic needle, coring of the stopper could result, thus introducing particles of the stopper material.[23] Draftz and Graf[9] reported elastomer fragments ranging in size from 5 to 50 μm in studies of saline and dextrose intravenous solutions.

D. Personnel

Personnel in the areas where manufacturing and filling are taking place may introduce particulates. One common source of particulate matter is latex gloves coated with starch. Starch granules show a black cross when viewed between crossed polars under a polarizing microscope. The purpose of the starch on the gloves is to aid personnel inserting their hands into the gloves. Cotton fibers shed from the clothing of manufacturing personnel may also contribute particulates to the product.

E. Filling Equipment

The filling equipment may introduce particulates into the product. Any time moving parts are involved, abrasion takes place which will result in metal shavings as well as various plastics being introduced into the product.

VI. GOVERNMENT REGULATIONS

The *United States Pharmacopeia* established particulate matter limits for large volume parenterals (LVPs) in 1980.[24] The limits were a maximum of 50 particles per milliliter $\geqslant$ 10 μm and maximum 5 particles per milliliter $\geqslant$ 25 μm based on a microscopic method of analysis. Particulate matter limits for small volume parenterals went into effect on January 1, 1986.

The limits for small volume parenterals (SVPs) were first presented in the November-December 1983 Pharmacopeial Forum. The attempt of the USP subcommittee on parenteral products was to establish a "patient oriented" standard utilizing a one fifth rule to set particulate matter limits arbitrarily in small volume parenterals.[25]

Starting with the existing limits for LVPs extrapolated to a 1ℓ volume, the LVP limits would be a maximum 50,000 particles per container $\geqslant$ 10 μm and maximum 5,000 particles per container $\geqslant$ 25 μ. Based on the use of SVPs and assuming a maximum of 5 SVPs could be added to an LVP, the USP subcommittee did not want to more than double the particle loading given to a patient. Therefore, taking one fifth of the LVP limit, a maximum of 10,000 particles per container $\geqslant$ 10 μm and a maximum of 1,000 particles per container $\geqslant$ 25 μm were established as limits for SVPs. The method of analysis, however, will be based on instrumentation employing light-blockage and not microscopic principles.[28] The instrument most commonly used is the HIAC®.

VII. CONCLUSION

The limits for the particulate matter, regardless of how they were established, will provide a means of assessing the quality of small volume parenterals. Once a comprehensive program for monitoring particulate levels has been established, only through identification of the particulates and their sources can reduced levels be achievable in parenteral products.

ACKNOWLEDGMENTS

We thank Dr. Naseem Muhammad for his valuable comments and suggestions in reviewing this manuscript.

REFERENCES

1. **Turco, S. and Davis, N. M.,** Clinical significance of particulate matter: a review of the literature, *Hosp. Pharm.*, 8, 8, 1973.
2. **Rapp, R. P. and Bivins, B. A.,** Support for in-line terminal I.V. filtration, *Particulate Microb. Control,* 2, 12, 1983.
3. **Knapp, J. Z. and Kushner, H. K.,** Particulate inspection of parenteral products: from biophysics to automation, *J. Parenter. Sci. Technol.*, 36, 121, 1982.
4. United States Pharmacopeial Convention, Inc., *United States Pharmacopeia,* 21st rev. ed., Mack Publishing, Eastman, Pa., 1985, 1257.
5. **DeLuca, P. P.,** Particulate Matter Monitoring Workshop, presented at the Center for Professional Advancement, East Brunswick, N.J., November 28 to 30, 1979.
6. **Graf, J.,** The "electrozone" counter: applications to non-aqueous particle-fluid systems, in *Particle Size Analysis,* Stockman, J. D. and Fochtman, E. G., Eds., Ann Arbor Science Publishers, Ann Arbor, 1977, chap. 7.
7. **Graph, J.,** Sizing with modern image analyzers, in *Particle Size Analysis,* Stockman, J. D. and Fochtman, E. G., Eds., Ann Arbor Science Publishers, Ann Arbor, 1977, chap. 4.
8. **Pfizer,** *Determination of Particulate Matter,* Pfizer Quality Control Division, New York, 1976.
9. **Draftz, R. G. and Graf, J.,** Identifying particulate contaminants, *Bull. Parenter. Drug Assoc.*, 28(1), 35, 1984.
10. **McCrone, W. C., McCrone, L. B., and Delly, J. G.,** Crystal optics, in *Polarized Light Microscopy,* Ann Arbor Science Publishers, Ann Arbor, 1978, chap. 9.
11. **West, P. W.,** Chemical identification by means of polarized light microscopy, in *Chemist-Analyst,* J. T. Baker Chemical Co., Phillipsburg, N.J., June, September and December 1954, reprint 5-6.
12. **Mason, C. W.,** Relation of optical properties to aggregates and crystals, in *Handbook of Chemical Microscopy,* Vol. 1, 4th ed., John Wiley & Sons, New York, 1983, chap. 12.
13. **Draftz, R. G.,** Artifacts and their sources, *Particle Analyst,* 1(5), 33, 1968.
14. **McCrone, W. C. and Delly, J. G.,** *The Particle Atlas,* Vol. 2, 2nd ed., Ann Arbor Science Publishers, Ann Arbor, 1973, 356.
15. **McCrone, W. C. and Delly, J. G.,** *The Particle Atlas,* Vol. 2, 2nd ed., Ann Arbor Science Publishers, Ann Arbor, 1973, 352.
16. **McCrone, W. G. and Delly, J. G.,** *The Particle Atlas,* Vol. 2, 2nd ed., Ann Arbor Science Publishers, Inc., Ann Arbor, 1973, 365.
17. **Postek, M. T., Howard, K. S., Johnson, A. H., and McMichael, K. L.,** Advances in microscopy, in *Scanning Electron Microscopy, A Student's Handbook,* Ladd Research Industries, Burlington, Vt., 1980, chap. 1.
18. Kevex Corporation, The physics of the process, in *Energy Dispersive X-Ray Microanalysis, An Introduction,* Kevex Corporation, Foster City, Calif., 1983, chap. 2.
19. **Goldstein, J. I., Newburg, D. E., Echlin, P., Flori, C., Joy, D. C., and Lifshin, E.,** Electron-beam-specimen interactions, in *Scanning Electron Microscopy and X-Ray Microanalysis,* Plenum Press, New York, 1981, chap. 3.
20. **Postek, M. T., Howard, K. S., Johnson, A. H., and McMichael, K. L.,** X-ray analysis, in *Scanning Electron Microscopy, A Student's Handbook,* Ladd Research Industries, Burlington, Vt., 1980, chap. 4.

21. **Feigl, F.,** *Spot Tests in Inorganic Analysis,* 6th ed., Elsevier, New York, 1972, 269.
22. **Clark, G. L.,** *The Encyclopedia of Microscopy,* Reinhold Publishing Corp., New York, 1961, 346.
23. **Wang, Y. J. and Chien, Y. W.,** Primary packaging systems, in *Sterile Pharmaceutical Packaging: Compatibility and Stability, Tech. Rep. No. 5,* Parenteral Drug Association, Philadelphia, 1984, chap. 2.
24. United Pharmacopeial Convention, Inc., *United States Pharmacopeia,* 20th rev. ed., Mack Publishing, Eastman, Pa., 1980, 863.
25. United States Pharmacopeial Convention, Inc., Particulate matter in small volume parenteral injections, in *Pharmacopeial Forum,* United States Pharmacopeial Convention, Inc., Rockville, Md., November to December, 1983, 3729.

Bioavailability and Pharmacokinetics

Chapter 7

INTRODUCTION TO BIOAVAILABILITY AND PHARMACOKINETICS

P. L. Madan

TABLE OF CONTENTS

I. BIOAVAILABILITY

Bioavailability is loosely defined as the rate and exent of absorption of an administered dose that reaches the site of action in the form in which it is effective. Since it is difficult, if not impossible, to measure the concentration of the drug at the site of action, one usually measures the drug concentration in the blood or in the plasma. This is based upon the premise that most drugs reach the site of action, or the biophase, through systemic circulation. The venous and arterial blood is considered systemic circulation during the absorptive phase after oral administration and blood in the portal vein is excluded.

Both the extent and the rate of absorption are characterized by the following five parameters:

1. Area under the curve (AUC),
2. Actual peak height (C_{max}),
3. Time to reach peak (t_{max}),
4. Rate of excretion, and
5. Extent of excretion.

The area under the curve and the extent of excretion relate to the extent of absorption, and C_{max}, t_{max}, and rate of excretion are indicative of rate of absorption.

Figure 1 shows the hypothetical blood level curve following oral administration of a drug. The shaded area in this figure represents the area under the curve and the peak height and time of peak are identified as C_{max} and t_{max}, respectively.

In a bioavailability study it is customary to determine AUC, C_{max}, and t_{max}. The importance of determining these three parameters is obvious from Figure 2 which shows hypothetical blood level curves upon oral administration of three different products containing identical amounts of the same drug. As can be seen in this figure, all three products give the same areas under the curves. However, drug products A and B are clearly superior since their peaks are above the minimum effective concentration (MEC), whereas the drug product C does not even reach the minimum effective concentration. Products A and B differ in C_{max} and t_{max} and, from a clinical standpoint, this difference would translate in terms of their onset of action and duration of activity. Product A would exhibit its action earlier than product B and its duration of action would also be shorter than product B. C_{max} is particularly important for potent drugs, since these drugs in general have a low therapeutic index.

A. Methods of Estimating AUC

Several methods exist for the measurement of area under the curve. Some of these are discussed here.

1. Planimeter Method

Planimeter is an instrument which mechanically measures the area of plane figures. Although the use of a planimeter for determining AUC is the simplest approach, this method is not in common use since most research facilities do not have this instrument.

2. Counting Squares

In this method, a plot of plasma level-time curve is prepared on a rectilinear graph paper (Figure 3) and the number of squares within the curve is counted. Knowing the area of one square (height × width), the area under the curve from time 0 to time t is calculated from the relationship:

$$AUC = (\# \text{ of squares})(\text{area of one square}) \tag{1}$$

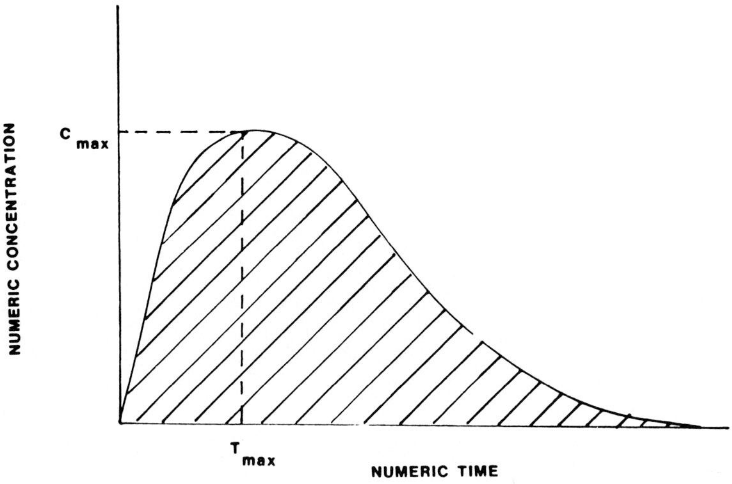

FIGURE 1. Typical rectilinear plot of drug concentration in the plasma following an oral dose, indicating area under the concentration-time curve, peak concentration (C_{max}), and time of peak concentration (t_{max}).

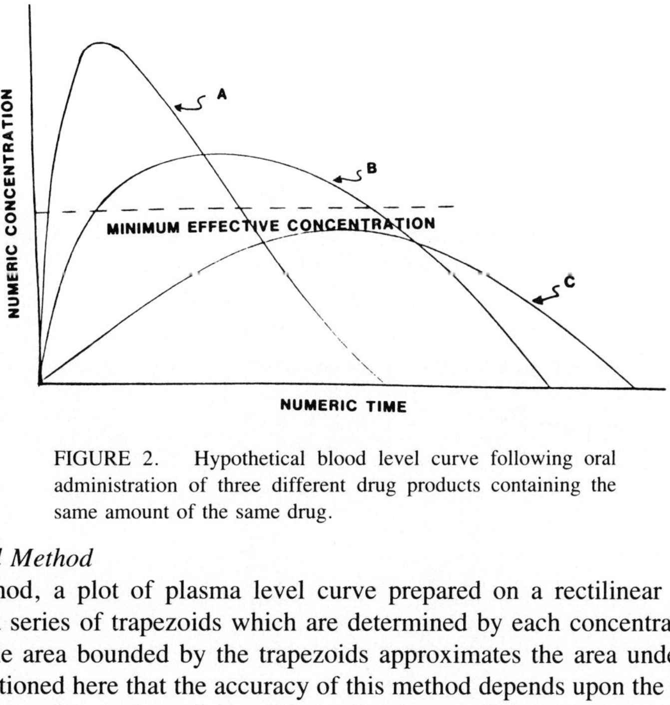

FIGURE 2. Hypothetical blood level curve following oral administration of three different drug products containing the same amount of the same drug.

3. Trapezoidal Method

In this method, a plot of plasma level curve prepared on a rectilinear graph paper is described by a series of trapezoids which are determined by each concentration-time point (Figure 4). The area bounded by the trapezoids approximates the area under the curve. It should be mentioned here that the accuracy of this method depends upon the number of data points: the greater the number of data points, the closer is the approximation of area under the curve. The area of a trapezoid is equal to one half the product of the sum of the height times the width. The area under the curve from time 0 to time t is the sum of the areas of all trapezoids (Table 1).

4. Cutting and Weighing

In this method the area under the curve from time 0 to time t on a rectilinear graph paper is cut out and weighed on an analytical balance. The weight thus obtained is converted to the proper units by dividing it by the weight of a unit area of the same paper.

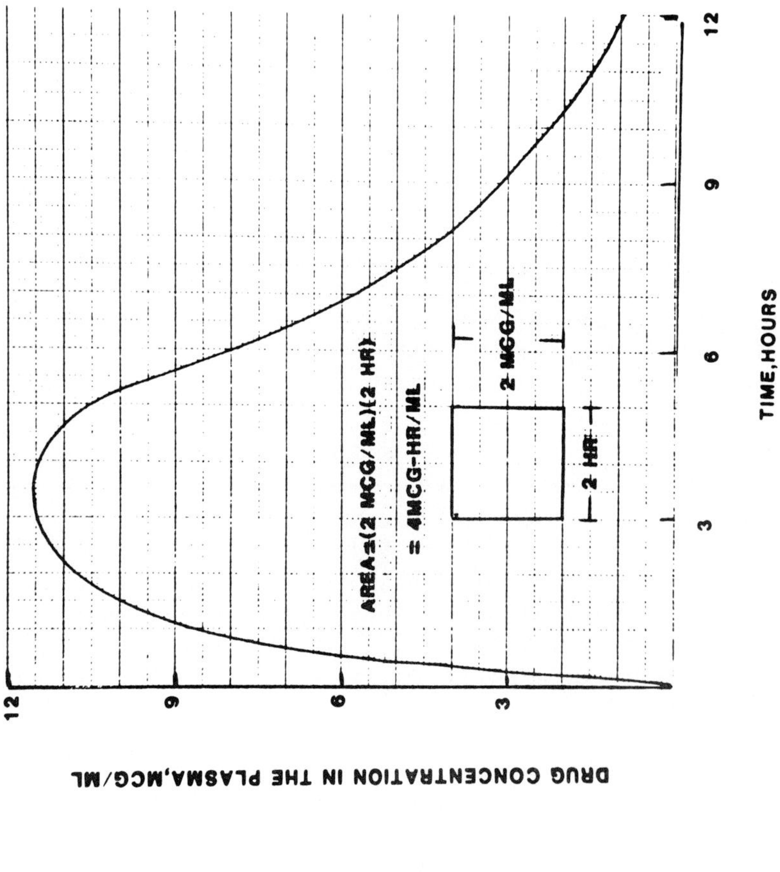

FIGURE 3. Typical rectilinear plot of drug concentration in the plasma following an oral dose.

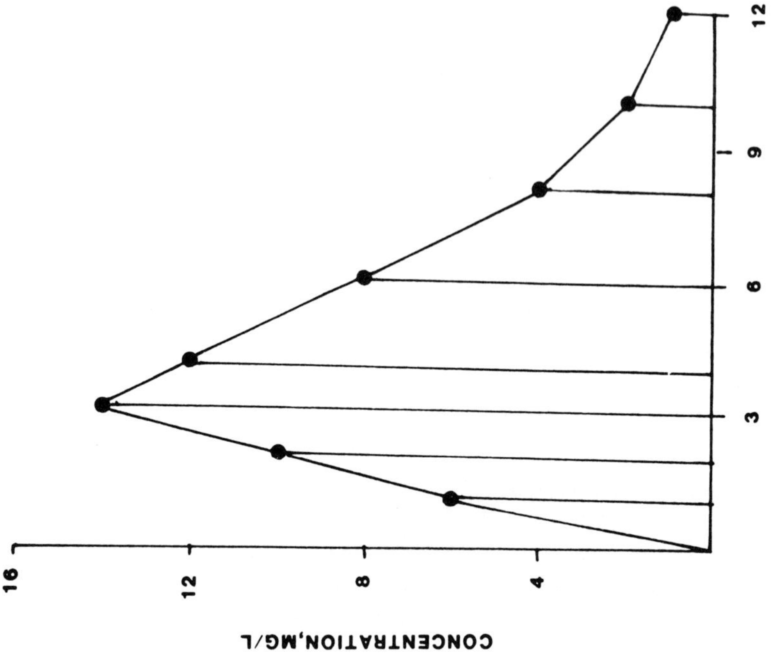

FIGURE 4. Trapezoidal method of calculating the area under the curve. The data points are connected by straight line segments rather than a smooth curve. The area of each trapezoid is delineated.

Table 1
CALCULATION OF AUC BY TRAPEZOIDAL METHOD

Time (hr)	Concentration (mg/ℓ)	Area (mg hr/ℓ)
0	0	0
1	6	3
2	10	8
3	14	12
4	12	13
6	8	20
8	4	12
10	2	6
12	1	3
	Total AUC from 0 to 12 hr	77

B. Total AUC

Pharmacokinetic studies are seldom carried out long enough to allow the drug concentrations to fall to undetectable levels. Thus the entire curve from time zero to time infinity is usually not available for estimating the total area under the curve. However, if the study is carried out sufficiently beyond the absorptive phase, then the area under the curve for the portion not studied, i.e., from time t to time infinity, is readily estimated from the equation:

$$\text{AUC from t to } \alpha = C_t/K \qquad (2)$$

where C_t is the concentration at time t (the last point in the AUC graph) and K is the apparent first order elimination rate constant. This area added to the area calculated from time 0 to time t gives the total area under the curve.

II. PHARMACOKINETICS

Pharmacokinetics is a major branch of pharmaceutical sciences concerned with the study and characterization of the time course of drug absorption, distribution, metabolism, and excretion, as well as the relationship of these processes to the intensity and time course of therapeutic and adverse effects of drugs. In other words, pharmacokinetics involves the application of mathematics and biochemistry in a physiologic and pharmacologic context. Thus, the science of pharmacokinetics deals with the study of the time course of drug and metabolite levels in different fluids, tissues, and excreta of the body, and the mathematical relationships required to develop models to interpret such data. The ultimate goal of pharmacokinetics has different meaning to different scientists. To some, the study of kinetics of absorption, distribution, metabolism, and excretion of drugs or chemicals may enhance their understanding of the basic mechanisms underlying these processes. To others, it is the elucidation of relationships between the pharmacologic response and levels of drug or their metabolites. To still others, this science offers an approach to improving and/or optimizing the therapeutic management of individual patients.

At one time it was customary to assume that the response to a drug or a drug product was simply a function of the pharmacologic activity of the drug. It was also usual to define the potency of the drug in terms of a milligram per kilogram dose. It was equally usual to compare the so-called potencies of drugs with similar pharmacologic effects, without a frame of reference. It is now evident that for most drugs a more appropriate assessment of potency or the therapeutic effect would be realized by considering drug concentration as a function of response rather than the dose as a function of response relationship.

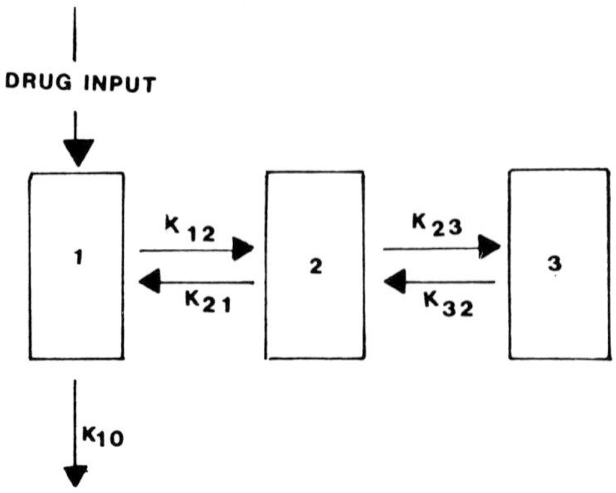

FIGURE 5. Example of a three-compartment caternary model consisting of a central compartment and two peripheral compartments.

The most commonly employed approach to the pharmacokinetic characterization of a drug is to represent the body as a system of compartments, even though these compartments may have no apparent physiologic or anatomic reality. A compartment is considered as a tissue or group of tissues which have similar blood flow and drug affinity. Within each compartment the drug is considered to be uniformly distributed and homogeneously mixed so that the drug concentration represents an average concentration. Compartment models are based on linear assumptions using linear differential equations.

A. Compartment Models

Three types of compartment models are recognized. These are:

1. Mammillary Model

This model consists of one or more peripheral compartments connected around a central compartment like satelites. The central compartment represents plasma and highly perfused tissues which rapidly equilibrate with the drug.

2. Caternary Model

The caternary model consists of compartments joined to one another in a series like the compartments of a train (Figure 5). Since the caternary model does not apply to the way most functional organs in the body are directly connected to the plasma, it is not used in pharmacokinetics as often as the mammillary model.

3. Perfusion Model

This model, also known as the physiologic model or blood flow model, is based upon known anatomic or physiologic data. An example of a perfusion model is shown in Figure 6. The number of tissue compartments in a perfusion model varies with the drug. Each box represents a tissue compartment. Typically, organs of major importance in drug distribution are considered separately and the tissues or organs that have no affinity for the drug are excluded from consideration. To date, numerous drugs, including digoxin, thiopental, methotrexate, and lidocaine, have been described with perfusion models. Tissue levels of some of these drugs cannot be accurately predicted from compartment models, although they generally describe blood levels well.

The mammillary model is the most common compartment model used in pharmacokinetics. It may be considered as a strongly connected system since one can estimate the amount of

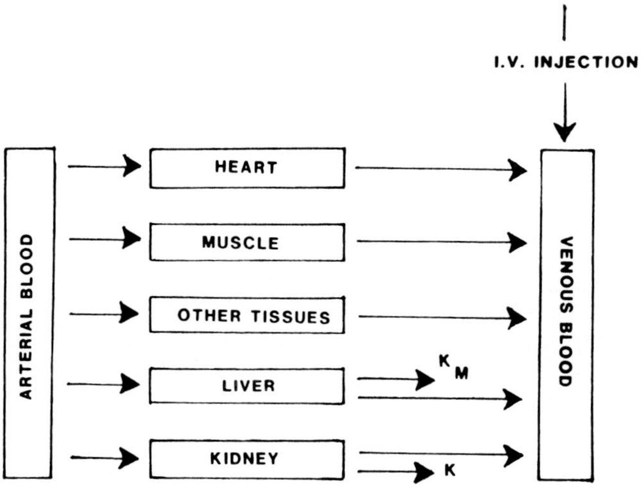

FIGURE 6. A typical pharmacokinetics model of drug perfusion.

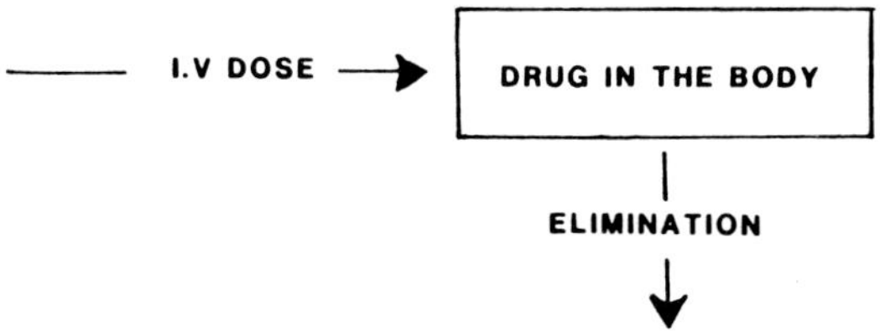

FIGURE 7. One-compartment open model.

drug in any compartment of the system after the drug is introduced into a given compartment. Elimination of the drug occurs from the central compartment since the organs involved in drug elimination, primarily kidney and liver, are well-perfused tissues.

B. One-Compartment Model

The one-compartment model is the most simple model and depicts the body as a single homogeneous unit. In this model the drug concentrations in all body tissues at any given time are not necessarily the same but any changes that occur in the plasma quantitatively reflect changes occurring in tissue drug levels.

1. Intravenous Injection

Following rapid intravenous injection (bolus dose) of a drug that distributes in the body according to a one-compartment model (Figure 7) and is eliminated by apparent first-order kinetics, the rate of loss of drug from the body, dc/dt, is given by

$$dc/dt = -KC \tag{3}$$

where C is the concentration of drug in the body at time t after injection, K is the apparent first order elimination rate constant for the drug and includes the rate constants of such individual processes as renal excretion (K_e), biliary excretion (K_b), metabolism (K_m), etc. The negative sign in Equation 3 indicates that the drug is being lost from the body.

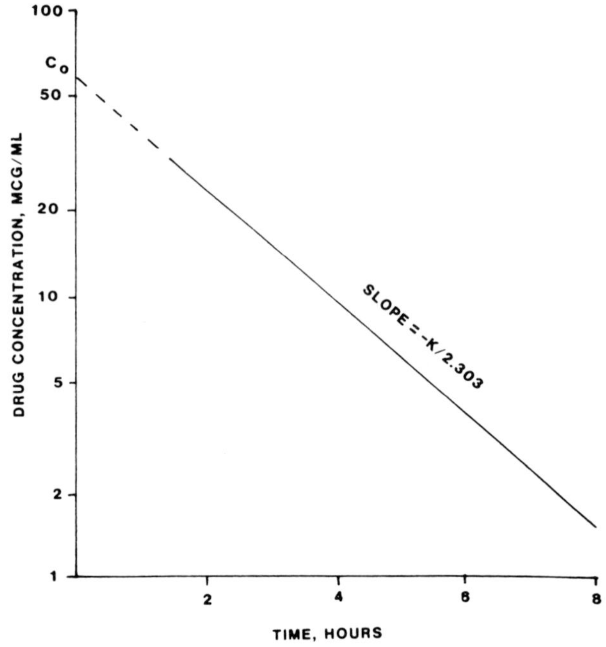

FIGURE 8. Semilogarithmic plot of drug concentration in the plasma as a function of time.

Integration of Equation 3 from time 0 to time t describes the time course of the concentration of drug in the body after injection

$$\ln C_t = \ln C_0 - Kt \tag{4}$$

where C_t is the plasma drug concentration at time t and C_o is the plasma drug concentration immediately after injection. Based on the relationship

$$\ln x = 2.303 \log x \tag{5}$$

Equation 4 can be converted to common logarithms

$$\log C_t = \log C_0 - Kt/2.303 \tag{6}$$

Equation 6 shows that a plot of log C_t vs. t will be linear with a slope equal to $-K/2.303$ (Figure 8). The apparent first order rate constant of elimination may be calculated as

$$k = (-2.303)(\text{slope}) \tag{7}$$

Extrapolation of the log C_t vs. t plot to time zero gives the intercept from which C_o may be estimated. The antilogarithm of the intercept, C_o, may be used in the calculation of apparent volume of distribution, V_d by the relationship

$$V_d = \text{i.v. dose}/C_0 \tag{8}$$

Apparent volume of distribution is not a "real" volume but an artifact. It is a hypothetical volume of body fluid that would be required to dissolve the total amount of drug at the same

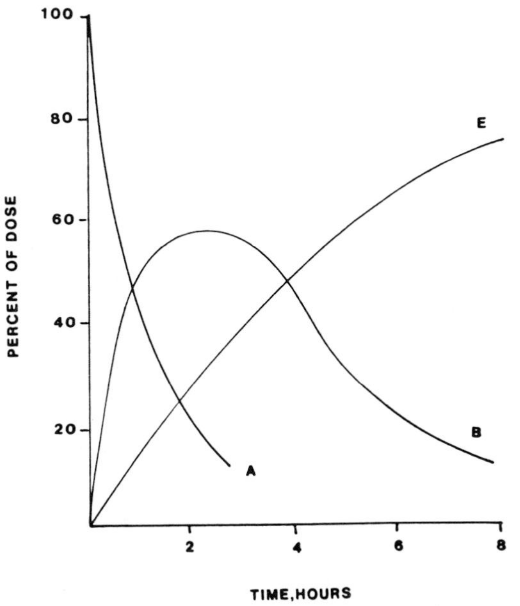

FIGURE 9. Time course of drug disappearance from the absorption site (A) and appearance of the eliminated drug (E). The net result is curve B, which describes the time course of drug in the body.

concentration as that found in the blood. Despite its name, this parameter usually has no direct physiologic meaning because it is usually a characteristic property of the drug rather than of the biologic system. If V_d is found to be a very large number, i.e., $\geqslant$ 100% of body weight, then it may be assumed that the drug is concentrated in certain tissue compartments. In general, drugs with a large apparent volume of distribution are more concentrated in extravascular tissues and less concentrated intravascularly. Similarly, drugs with a smaller volume of distribution tend to be more concentrated intravascularly and less concentrated in extravascular tissues.

2. Extravascular Administration

a. General

A more complex drug concentration-time profile is observed following extravascular administration because absorption from these sites does not occur at a constant rate. This is because the amount of drug at the absorption site after administration continues to decrease with time as the drug is absorbed into the systemic circulation and distributed throughout the body. Also, as soon as the drug reaches systemic circulation it undergoes various elimination processes. The concentration of drug in the plasma therefore depends on the kinetics of absorption, distribution, and elimination processes (Figure 9). The curve B in Figure 9 may thus be considered to describe the relationship between the absorption rate and the elimination rate as a function of time. Four distinct zones can be recognized (Figure 10). Soon after administration (Zone I), the rate of absorption (dA/dt) is much greater than the rate of elimination (dE/dt). As a function of time, the absorption rate starts decreasing somewhat and the elimination rate starts becoming greater and greater. Eventually the two rates become more or less equal (Zone II, $dA/dt \simeq dE/dt$) and after this, the absorption rate starts declining (Zone III, $dA/dt < dE/dt$), eventually reaching a point (Zone IV) when the absorption is more or less complete and $dA/dt \simeq 0$.

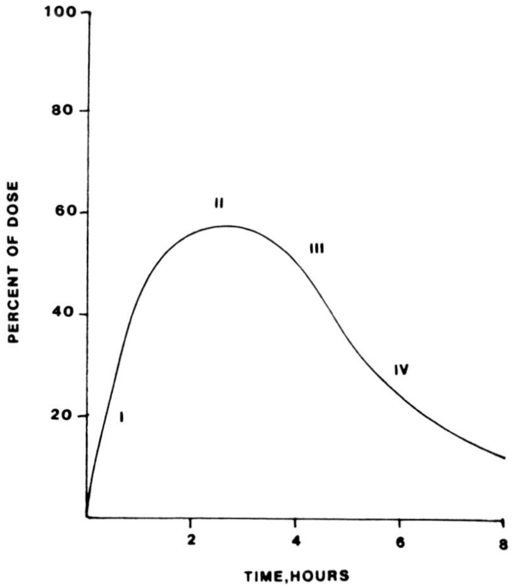

FIGURE 10. Relationship between the rate of absorption and the rate of elimination during the time course of the drug in the body.

Many drugs appear to be absorbed in a first-order fashion according to the following equation

$$dc/dt = K_a C_a - KC \tag{9}$$

where K_a is the apparent first-order absorption rate constant, K is the apparent first-order elimination rate constant, C is the concentration of drug in the body, and C_a is the concentration of drug at the absorption site. Integration of equation (9) from time zero to time t gives

$$\ln C_t = \ln C_0 K_a/(K_a - K) - Kt \tag{10}$$

Equation 10 can be converted to common logarithms,

$$\log C_t = \log C_0 K_a/(K_a - K) - Kt/2.303 \tag{11}$$

Equation 11 shows that a plot of log C_t vs. t will be linear with a slope $= -K/2.303$ (Figure 11), and extrapolation of the terminal linear portion of the plot gives the intercept = log $C_o K_a/(K_a - K)$. If $K_a >>> K$, then the intercept reduces to log C_o, from which C_o can be computed. It should be pointed out here that in those cases where the absorption of a drug is slower than its elimination e.g, rapid metabolism, or administration in a slowly releasing dosage form, the semilogarithmic plot of C_t vs. t cannot be used to estimate K or half-life. In such instances, the drug must be administered either in a more rapidly absorbed form or given intravenously.

b. Determination of K_a

The most commonly used method of determining the apparent first-order rate constant of absorption, K_a is the feathering technique, also known as the method of residuals. This method is based upon the assumption that $K_a >>> K$. The procedure is as follows:

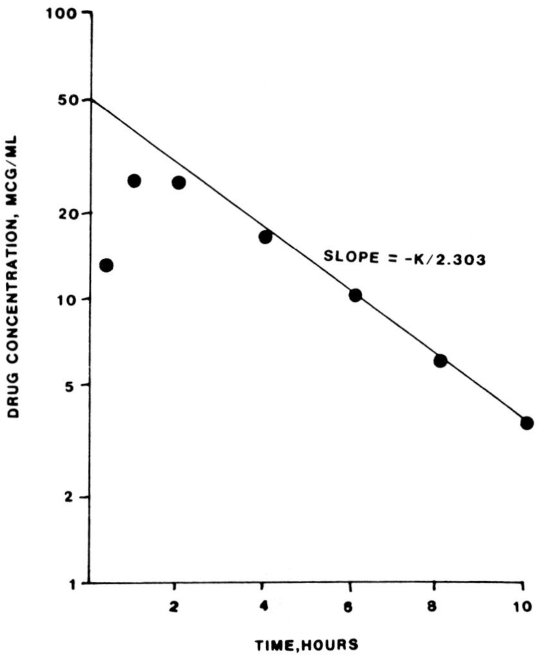

FIGURE 11. Typical semilogarithmic plot of drug concentration in plasma following oral administration.

The drug plasma concentration vs. time graph is plotted on a semilogarithmic paper with the concentration on the logarithmic y-axis and time on the x-axis. The slope of the terminal (elimination) phase is obtained by extrapolation to time zero. Any points on the upper part of the elimination line (e.g., y'_1, y'_2, y'_3---) are taken and dropped vertically to obtain the corresponding points on the absorptive phase of the curve (e.g., y_1, y_2, y_3---). The values of y_1 and y'_1, y_2, y'_2, y_3 and y'_3 etc. are read from the graph and the values of the differences Δ_1, Δ_2, Δ_3, etc. are determined, i.e., $\Delta_1 = y'_1 - y_1$, $\Delta_2 = y'_2 - y_2$, $\Delta_3 = y'_3 - y_3$, etc. The values of Δ_1, Δ_2, Δ_3, etc. are plotted at the corresponding time points to obtain a straight line with a slope of $-K_a/2.303$ (Figure 12).

In using this method, one should use the terminal points to describe the straight line representing elimination and not use the data points shortly after t_{max} since drug absorption is still occurring at that time. Similarly, y_1, y_2, y_3, etc. should be taken from the pre-t_{max} portion of the elimination line since absorption is more predominant during the earlier part of the curve and then starts declining as a function of time.

If drug absorption begins immediately after extravascular administration, the residual line obtained by feathering the plasma level-time curve (Figure 12) will intersect on the y-axis. The value of this intercept is $K_aC_o/(K_a - K)$. The value of C_o in this intercept, as well as in Equation 11, is a function of dose, V_d, and F, the fraction of dose absorbed, given by

$$C_0 = (F)(\text{dose})/V_d \tag{12}$$

knowing the V_d of the drug (e.g., from the intravenous administration) one can calculate the fraction of dose absorbed by rearranging Equation 12

$$F = C_0V_d/\text{dose} \tag{13}$$

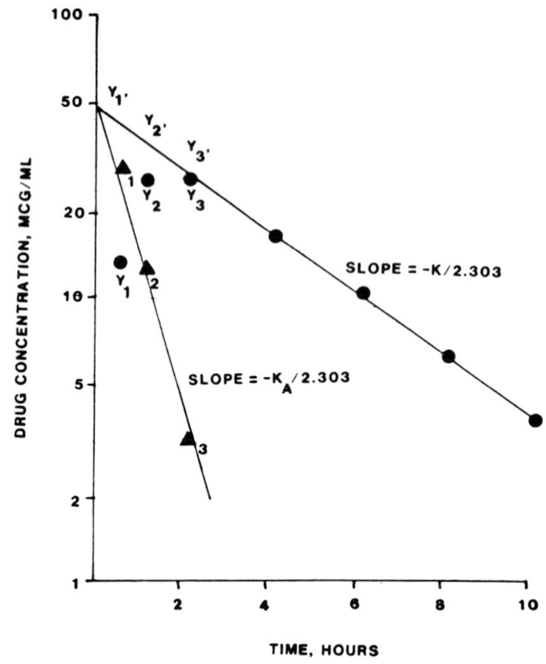

FIGURE 12. Method of residuals to calculate the absorption rate constant in one-compartment open model following oral administration.

C. Multicompartment Models

Multicompartment models are needed to explain the observation that after a rapid intravenous injection (bolus dose) the plasma level vs. time curve does not decline linearly as a single first-order rate process. Most drugs entering the systemic circulation require a finite time to distribute fully throughout the available body space. During this distributive phase, drug concentration in the plasma will decrease more rapidly than in the postdistributive phase. If drug distribution is related to blood flow then the drug distributes at various rates into different tissue groups. Those tissues which have the highest blood flow may equilibrate with the plasma compartments making up what is generally referred to as the central compartment. Similarly all less highly perfused tissues are often ''lumped'' together into a single peripheral compartment, and all poorly perfused tissues are considered as another single peripheral compartment. As with the one-compartment model, drug elimination in multicompartment systems is assumed to occur in a first-order fashion and transfer of drug between various body compartments is also assumed to occur by first-order rate processes.

It appears that many drugs confer upon the body the characteristics of a two compartment model, i.e., a central compartment consisting of the highly perfused tissues connected to a peripheral compartment. There are three possible types of two-compartment systems (Figure 13) which differ in that elimination takes place either from the central compartment, or from the peripheral compartment, or from both of these compartments.

As analytical methods improve and pharmacokineticists treat more rigorously the plasma concentration vs. time curve of a drug, a three compartment model may be required to adequately describe the time course of the drug data. Figure 14 shows the simplest of a three-compartment model in which elimination occurs from the central compartment which is reversibly connected to a ''shallow'' peripheral compartment (compartment 2) and a ''deep'' peripheral compartment (compartment 3).

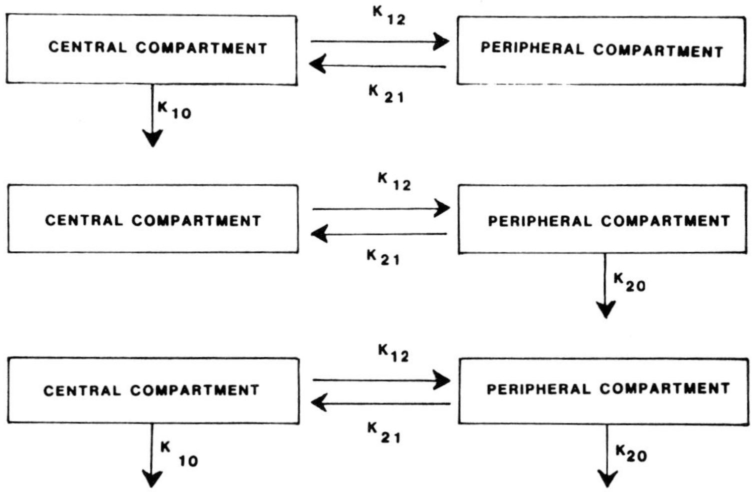

FIGURE 13. Examples of three types of two-compartment open models consisting of a central compartment and a peripheral compartment.

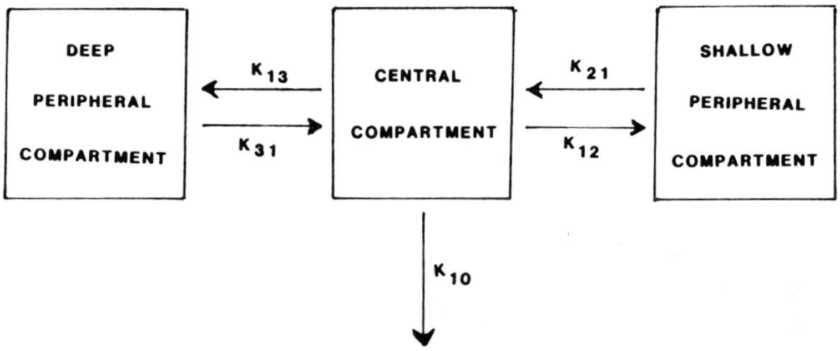

FIGURE 14. A typical three-compartment open model consisting of a central compartment (compartment 1), a "shallow" peripheral compartment (compartment 2), and a "deep" peripheral compartment (compartment 3).

1. Two-Compartment Model

The three types of two-compartment models shown in Figure 13 are mathematically indistinguishable on the basis of the usually available experimental data. Since the major sites of metabolism, i.e., the liver, and elimination, i.e., the kidneys, are well perfused with blood and are readily accessible to the drug and its metabolites it is usually assumed that drug elimination from a two-compartment model takes place exclusively from the central compartment.

Following rapid intravenous injection (bolus dose) of a drug that distributes in the body according to a two-compartment model with elimination occurring from the central compartment, the rate of change in the concentration of drug in the central compartment C can be described by the following differential equation

$$\frac{dC}{dt} = K_{21}\,C_p - K_{12}\,C - K_{10}\,C \tag{14}$$

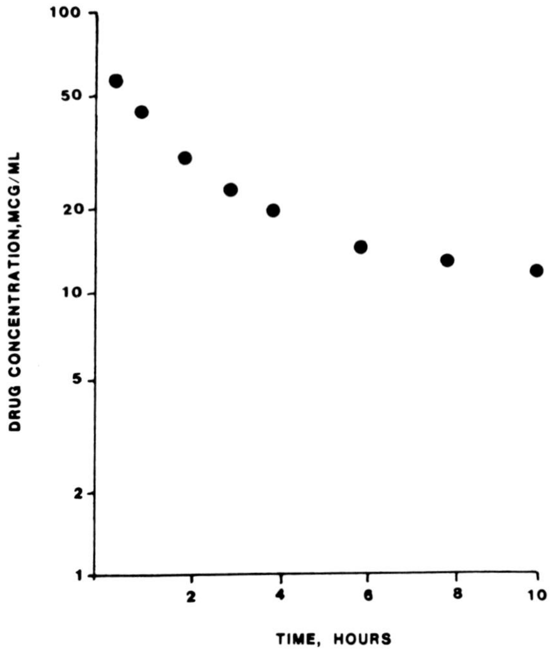

FIGURE 15. Biexponential curve describing a two-compartment open model following intravenous bolus injection.

where C_p is the concentration of drug in the peripheral compartment, K_{21} and K_{12} are the apparent first-order intercompartmental distribution rate constants, and K_{10} is the apparent first-order elimination rate constant from the central compartment.

Upon integration, Equation 14 yields

$$C = Ae^{-at} + Be^{-bt} \tag{15}$$

where C is the concentration of drug in the central compartment at any time t, and A, B, a, and b are complex constants given by

$$a + b = K_{12} + K_{21} + K_{10} \tag{16}$$

$$ab = K_{21} K_{10} \tag{17}$$

$$A = C_0(a - K_{21})/(a - b) \tag{18}$$

$$B = C_0(K_{21} - b)/(a - b) \tag{19}$$

The term C_o in Equations 18 and 19 is the concentration of drug in the central compartment immediately after intravenous administration. From Equations 18 and 19,

$$C_0 = A + B \tag{20}$$

Thus, a semilogarithmic plot of drug-plasma concentration versus time according to Equation 15 will yield a biexponential curve as shown in Figure 15. In deriving the equations for a two-compartment model, the constant a is by definition larger than the constant b and,

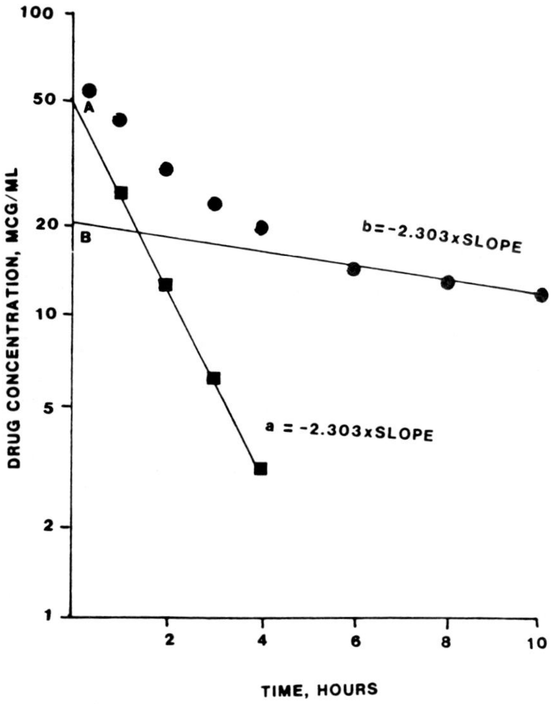

FIGURE 16. Application of method of residuals for the determination of constants A, B, a, and b in a two-compartment open model following intravenous bolus dose.

therefore, at some time the term Ae^{-at} in Equation 15 will approach zero, while Be^{-bt} will still have a finite value. Equation 15 will then be reduced to a monoexponential form given by

$$C = Be^{-bt} \tag{21}$$

which in common logarithms is

$$\log C = \log B - bt/2.303$$

An estimate of b can thus be obtained from the slope of the terminal exponential phase, and the biological half-life can be determined by the relationship

$$t_{1/2} = 0.693/b \tag{22}$$

Extrapolation of the terminal linear phase to time zero gives the intercept B, and application of the method of residuals yields a second linear segment with a slope of $-a/2.303$ and an intercept at time zero equal to A (Figure 16). Once the experimental constants A, B, a, and b are known, the pharmacokinetic parameters K_{12}, K_{21}, K_{10} V_c (volume of central compartment), and V_{ss} (volume of distribution at steady state) can be calculated, using the following relationships:

$$K_{12} = a + b - K_{21} - K_{10} \tag{23}$$

$$K_{21} = (Ab + Ba)/(A + B) \tag{24}$$

$$K_{10} = ab/K_{21} \quad (25)$$

$$V_c = dose/(A + B) \quad (26)$$

$$V_{ss} = V_c(1 + K_{10}/b) \quad (27)$$

2. Three-Compartment Model

The three-compartment model is an extension of a two-compartment model, with an additional ''deep'' peripheral compartment. A drug which confers upon the body the characteristics of a three-compartment model is distributed most rapidly to a highly perfused central compartment, less rapidly to the second ''shallow'' peripheral compartment, and very slowly to the third ''deep'' peripheral compartment. There are indications that a three-compartment model may be necessary to characterize the pharmacokinetic profile of digoxin, tubocurarine, and diazepam. The simplest three-compartment model is one where elimination occurs from the central compartment which is reversibly connected to the ''shallow'' and ''deep'' peripheral compartments (Figure 14).

Following rapid intravenous injection (bolus dose) of a drug that distributes in the body according to this model, one can develop the differential equation which describes the rates of flow of drug into and out of the central compartment along lines similar to those used for the two-compartment model. A solution of the differential equation for the three-compartment model gives the following equation

$$C = He^{-ht} + Ae^{-at} + Be^{-bt} \quad (28)$$

where C is the concentration of drug in the central compartment at time t, H, A, and B are the y intercepts of the extrapolated lines for the central, shallow peripheral, and deep peripheral tissue compartments, respectively, and h, a, and b are the first-order rate constants for the central, shallow peripheral, and deep peripheral tissue compartments, respectively.

According to Equation 28, a semilogarithmic plot of concentration vs. time should give a triexponential curve which, following feathering technique, should give three straight lines having intercepts H, A, and B with the slopes $-h/2.303$, $-a/2.303$, and $-b/2.303$, respectively (Figure 17). The values of rate constants K_{10}, K_{12}, K_{21}, K_{13}, and K_{31}, and the volume of central compartment are determined using the following equations:

$$K_{21} = 0.5(m + \sqrt{m^2 - 4n}) \quad (29)$$

where

$$m = (Ha + Hb + Ab + Ah + Ba + Bh)/(H + A + B) \quad (30)$$

$$n = (Hab + Abh + Bah)/(H + A + B) \quad (31)$$

$$K_{31} = 0.5(m - \sqrt{m^2 - 4n}) \quad (32)$$

where m and n are defined in Equations 30 and 31

$$K_{10} = hab/K_{21}\,K_{31} \quad (33)$$

$$= (A + B + H)\,abH/(Abh + Bah + Hab) \quad (34)$$

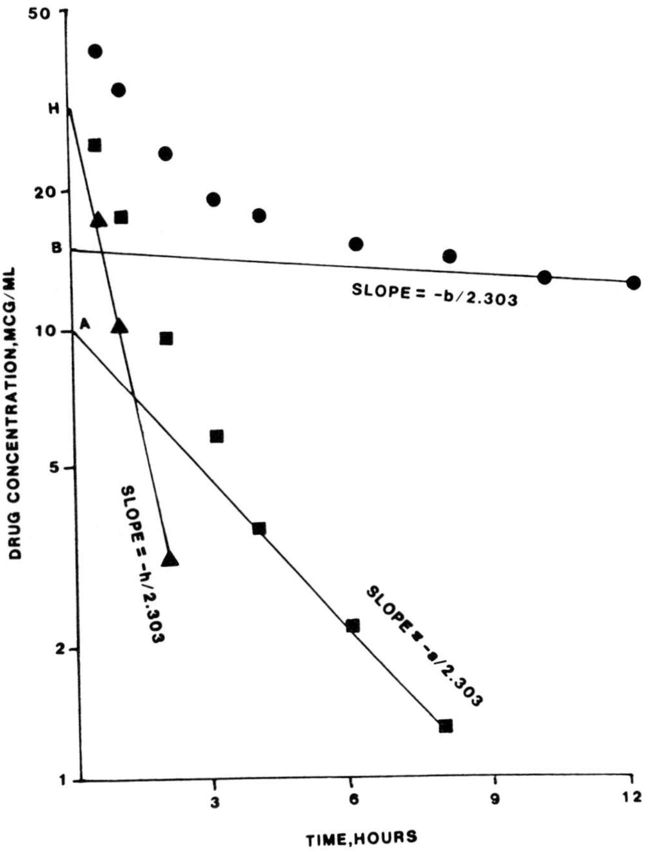

FIGURE 17. Triexponential curve describing a three-compartment open model following intravenous bolus injection.

$$K_{12} = \frac{(ab - ah + bh) - K_{21}(a + b + h) - K_{10}\,K_{31} - K_{21}^2}{K_{31} - K_{21}} \tag{35}$$

$$K_{13} = a + b + h - (K_{10} + K_{12} + K_{21} + K_{31}) \tag{36}$$

$$C_0 = H + A + B \tag{37}$$

$$V_c = \text{dose}/C_0 \tag{38}$$

$$AUC = (A/a) + (B/b) + (H/h) \tag{39}$$

D. Determination of Compartment Models

Models based on compartmental analysis always use the fewest number of compartments necessary to describe the experimental data adequately. However, great care must be exercised in determining the sampling intervals, especially in the case of those drugs which distribute relatively rapidly. If samples are taken too late after administration of the dose, it is possible to miss the distributive phase and hence one or more compartments. For example, if in the data plotted in Figure 18 no samples were taken soon after the drug was administered, one could easily mistake this graph for a one-compartment model, since the

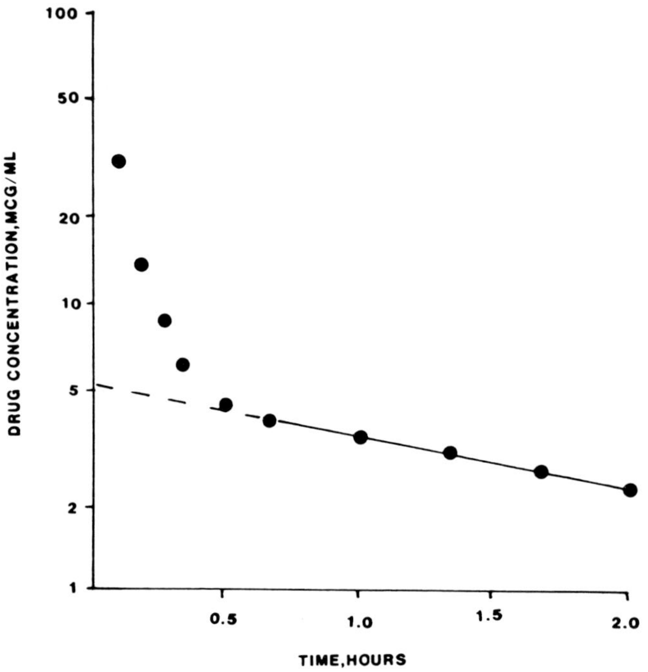

FIGURE 18. Potential error of mistaking a two-compartment open model for a one-compartment open model if no samples were taken soon after the intravenous administration of a bolus injection.

distributive phase could be missed altogether. Also, extrapolation of the data to time zero will give a lower value of C_o than was actually the case.

Compartment models having more than three compartments are rarely of pharmacologic significance. In certain instances, the pharmacokineticist attempts to "lump" a few compartments together to get a smaller number of compartments which together describe the data fairly adequately.

III. NONCOMPARTMENTAL PHARMACOKINETICS

It is frequently seen that drugs requiring multicompartmental analysis after intravenous administration can be described by a one-compartment model after oral administration. In order to avoid the unreconcilable difficulties that may result from pharmacokinetic analysis based on compartmental models, investigators in this field are attempting to utilize the noncompartmental approaches that appear to apply to almost all drugs. These methods permit a comprehensive pharmacokinetic analysis without resort to such tedious methods as curve fitting, mathematical equations, and, in this day and age, sophisticated computers. It should be mentioned here that these methods cannot be applied to all pharmacokinetic problems. However, these methods are useful for the clinical application of pharmacokinetics.

A. Noncompartmental Methods

The noncompartmental methods for calculating the kinetics of absorption, distribution, and elimination parameters of a drug are based on the theory of statistical moments. Two moments are generally used: the zero moment and the first moment of plasma concentration-time profile. The zero moment of the drug concentration in plasma vs. time curve is the total area under the curve from time zero to time infinity. This is the AUC discussed

Table 2
DRUG CONCENTRATION AND DRUG CONCENTRATION × TIME FOLLOWING ORAL ADMINISTRATION OF A SINGLE DOSE

Time (hr)	C (mcg/mℓ)	CT (mcg hr/mℓ)
0	0	0
1	6	6
2	10	20
3	14	42
4	12	48
6	8	48
8	4	32
10	2	20
12	1	12

Note: AUC = 77; AUMC = 348.

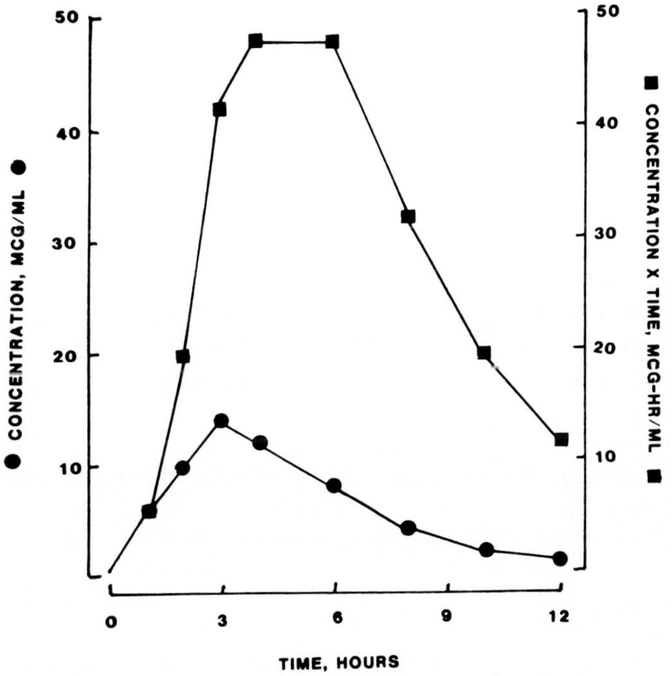

FIGURE 19. Typical plot of drug concentration and drug concentration × time vs. time following oral administration of a single dose.

previously. The first moment of a plasma concentration-time profile is the total area under the curve resulting from a plot of the product of drug concentration and time vs. time. This is the AUMC, the total area under the (C) (t) versus t plot from time zero to time infinity, known as the area under the first moment curve. Table 2 and Figure 19 show typical data for calculation of AUC and AUMC. The ratio AUMC to AUC for any drug is a measure of its mean residence time, MRT. Irrespective of distribution characteristics, MRT represents the time required for 63.2% of an intravenous bolus dose to be eliminated.

Mean residence time is a function of how the dose is administered. The MRT values for noninstantaneous administration will always be greater than the MRT following intravenous bolus administration. The MRT of a drug following intravenous bolus administration provides a useful estimate of the persistence time in the body. When applied to drugs that distribute rapidly it can be shown that

$$MRT_{i.v.} = 1/K \tag{40}$$

where K is the first-order elimination rate constant which is related to the biological half life, $t_{1/2}$, by

$$t_{1/2} = 0.693/K \tag{41}$$

The MRT of a drug that distributes slowly and requires multicompartment characterization is a complex function of the rate constants for distribution and elimination pertinent to the model, given by

$$MRT_{i.v.} = 1/K_1 \tag{42}$$

where K_1 is a rate constant related to the clearance and apparent volume of distribution at steady state, V_{ss}, by the following relationship:

$$K_1 = \text{clearance}/V_{ss} \tag{43}$$

The constant K_1 is usually greater than b (see Equation 15) and in many cases the ratio of 0.693 to b serves as the effective half-life, $t_{1/2}$, of the drug.

B. Drug Clearance

Clearance is the most useful pharmacokinetic parameter for clinical applications and for evaluation of elimination mechanisms of a drug. Clearance is a function of both the intrinsic ability of certain organs, such as the kidneys and liver, to eliminate a drug and the blood flow rate to these organs. It can be shown by means of integral calculus that the ratio of the overall elimination rate of a drug to its concentration in the blood is equal to the ratio of the drug ultimately eliminated to the total area under the drug concentration vs. time curve. If the entire dose administered reaches the systemic circulation, e.g., following intravenous administration, then, the amount eliminated is equal to the dose and clearance can be expressed as

$$Cl = \text{dose}/AUC \tag{44}$$

Equation 44 can also be used for drugs given orally or intramuscularly by substituting the amount of drug absorbed for the term "dose". This equation provides the basis for the routine estimation of the total body clearance of a drug after a single dose.

1. Apparent Volume of Distribution

The most useful volume term in pharmacokinetics is the apparent volume of distribution at steady state or V_{ss}. In noncompartmental methods, estimation of V_{ss} does not require data obtained at steady state, but it can be calculated using the following equation:

$$V_{ss} = \text{i.v. dose } (AUMC)/(AUC)^2 \tag{45}$$

2. Extent of Absorption

Noncompartmental methods for estimating the extent of absorption of a drug after oral or intramuscular administration require a comparison of areas under the curve, i.e., the fraction of the dose that actually reaches the systemic circulation, or F, can be estimated from the following equation

$$F = (AUC)_{test}/(AUC)_{i.v.} \tag{46}$$

where $(AUC)_{test}$ and $(AUC)_{i.v.}$ are areas under the drug concentration vs. time curve, following the administration of equivalent amounts of test product (oral or antramuscular) and the intravenous dose, respectively. Equation 46 can also be used for estimating the extent of absorption of a drug in a test dosage form relative to its absorption from a standard dosage form. In such cases, AUC from the standard dosage form replaces $(AUC)_{i.v.}$ in Equation 46.

3. Rate of Absorption

Noncompartmental methods for estimating the rate of absorption of a drug after oral or intramuscular administration are based upon the mean absorption time, or MAT when absorption is a first-order process.

$$MAT = 1/K_a \tag{47}$$

where K_a is the first-order absorption rate constant. Since, from Equation 47, $K_a = 1/MAT$, it follows that the absorption half-life is given by (0.693)(MAT).

The mean absorption time of a drug given by oral or intramuscular route can be estimated by the following equation:

$$MAT = (MRT)_{test} - (MRT)_{i.v.} \tag{48}$$

where $(MRT)_{test}$ is the mean residence time after oral or intramuscular administration and $(MRT)_{i.v.}$ is the mean residence time after intravenous bolus injection. Since MRT is the ratio of AUMC to AUC, Equation 48 becomes

$$MAT = (AUMC/AUC)_{test} - (AUMC/AUC)_{i.v.} \tag{49}$$

REFERENCES

1. **Clark, B. and Smith, D. A.,** *An Introduction to Pharmacokinetics,* Blackwell Scientific, Oxford, 1981.
2. **Creasy, W. A.,** *Drug Disposition in Humans,* Oxford University Press, New York, 1979.
3. **Curry, S. H.,** *Drug Disposition and Pharmacokinetics,* Blackwell Scientific, Oxford, 1980.
4. **Gibaldi, M.,** *Biopharmaceutics and Clinical Pharmacokinetics,* Lea & Febiger, Philadelphia, 1984.
5. **Gibaldi, M. and Perrier, D.,** *Pharmacokinetics,* Marcel Dekker, New York, 1975.
6. **Goodman, L. S. and Gilman, A.,** *The Pharmacological Basis of Therapeutics,* Macmillan, New York, 1975.
7. **Harvey, S. C.,** Drug absorption, action, and disposition, in *Remington's Pharmaceutical Sciences,* Gennaro, A. R., et al., Eds., Mack Publishing, Easton, Pa., 1985, chap. 37.
8. **Harvey, S. C. and Withrow, C. D.,** Basic pharmacokinetics, in *Remington's Pharmaceutical Sciences,* Gennaro, A. R., et al., Eds., Mack Publishing, Easton, Pa., 1985, chap. 38.
9. **Morselli, P. L.,** *Drug Disposition during Development,* Spectrum Publications, New York, 1977.
10. **Notari, R. E.,** *Biopharmaceutics and Pharmacokinetics. An Introduction,* Marcel Dekker, New York, 1980.

11. **Ritschel, W. A.,** *Handbook of Basic Pharmacokinetics,* Drug Intelligence Publications, Hamilton, Ill., 1976.
12. **Rollins, D. E.,** Principles of clinical pharmacokinetics, in *Remington's Pharmaceutical Sciences,* Gennaro, A. R., et al., Eds., Mack Publishing, Easton, Pa., 1985, chap. 37.
13. **Rowland, M. and Tozer, T. N.,** *Clinical Pharmacokinetics. Concepts and Applications,* Lea & Febiger, Philadelphia, 1980.
14. **Smith, S. E. and Rawlins, M. D.,** *Variability in Human Drug Response,* Butterworth, London, 1973.
15. **Teorell, T., Dedrick, R. L., and Condliffe, P. G.,** *Pharmacology and Pharmacokinetics,* Plenum Press, New York, 1974.
16. **Wagner, J. G.,** *Biopharmaceutics and Relevant Pharmacokinetics,* Drug Intelligence Publications, Hamilton, Ill., 1971.
17. **Wagner, J. G.,** *Fundamentals of Clinical Pharmacokinetics,* Drug Intelligence Publications, Hamilton, Ill., 1975.

Chapter 8

PHARMACOKINETICS OF AZTREONAM: A NEW MONOBACTAM ANTIBIOTIC

Sampat M. Singhvi

TABLE OF CONTENTS

I. ABSTRACT

The monobactam antibiotics, exemplified by aztreonam, represent a recent major development in the field of beta-lactam antibiotics. These synthetic antimicrobial agents have a unique monocyclic beta-lactam ring structure, unlike the fused biannular configuration of penicillins and cephalosporins. Aztreonam, the first monobactam to undergo extensive clinical trials, possesses potent antibacterial activity against a wide range of aerobic Gram-negative organisms.

^{14}C-Aztreonam was administered as single intramuscular, intravenous, or subcutaneous doses to rats, dogs, monkeys, and healthy volunteers. In rats, dogs, and humans, radioactivity was excreted primarily in urine (70 to 90% of the dose); in monkeys, the radioactivity was excreted about equally in urine and feces. Unchanged aztreonam accounted for 77 to 87% of the radioactivity excreted in the urine of all four species; SQ 26,992, the biotransformation product resulting from hydrolysis of the monobactam ring, accounted for 10 to 15%, and minor, unidentified biotransformation products accounted for the remainder.

In a tissue-distribution study in rats, aztreonam and its biotransformation products were found to distribute rapidly into most tissues and were rapidly eliminated from blood into urine and bile. After administration of ^{14}C-aztreonam, radioactivity and unchanged aztreonam were detected in fetuses and amniotic fluid of pregnant rats and in suckling pups, serum, and milk of lactating rats.

In man, the average elimination half-life of aztreonam was about 1.0 hr and the total body and renal clearances were 1.5 and 1.0 mℓ/min/kg, respectively. The apparent elimination half-life of SQ 26,992 was more than 10 times that of aztreonam. SQ 26,992 was tested and found to have no antimicrobial activity.

II. HISTORICAL BACKGROUND AND ANTIBACTERIAL ACTIVITY

Beta-lactam antibiotics (penicillins and cephalosporins) have played a major role in the treatment of bacterial infections (Figure 1). The widespread clinical use of β-lactam antibiotics is based on their excellent clinical efficacy and low frequency of side effects. Traditionally, the naturally occurring penicillins and cephalosporins have been products of fungi, but recently cephalosporins have been found in soil bacteria. The subsequent development of new and more potent penicillins and cephalosporins has been mainly due to the existence of β-lactamases, a class of enzymes which hydrolyze β-lactam antibiotics to inactive products and constitute the major form of resistance to β-lactam antibiotics in the clinic today.

During the 1970s, new generations of penicillins and broad-spectrum cephalosporins were developed. A highly sensitive screening procedure was developed by Squibb scientists in an attempt to discover novel β-lactam antibacterial molecules from bacteria isolated from natural sources, such as soil, water, and vegetation. Using this approach, the first positive culture came in 1978 from a soil sample taken in the Pine Barrens of southern New Jersey. The soil sample contained *Chromobacterium violaceum,* a Gram-negative bacterium, which produced a novel β-lactam antibiotic (Figure 2).[1]

The uniqueness of this β-lactam antibiotic centered on its nucleus. Unlike the bicyclic β-lactam antibiotics derived from fungi, this nucleus possessed a simple and unique monocyclic structure. This provided the basis for the family name monobactam, a monocyclic β-lactam antibiotic.[1] Several more bacterially produced monobactams were subsequently isolated and characterized. Unlike the naturally occurring penicillins and cephalosporins, these new monobactam compounds, isolated from natural sources, exhibited weak antibacterial activity. However, the structural simplicity of these monocyclic compounds offered increased opportunities for chemical modifications and improvements in antimicrobial activity.

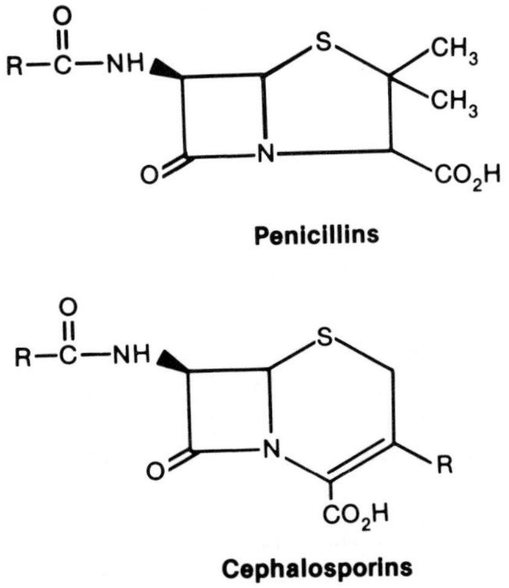

FIGURE 1. Structures of penicillins and cephalosporins.

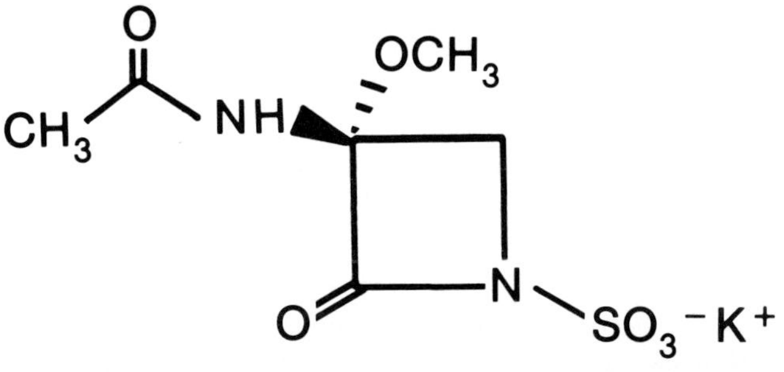

FIGURE 2. Structure of SQ 26,180, the naturally occurring monobactam antibiotic.

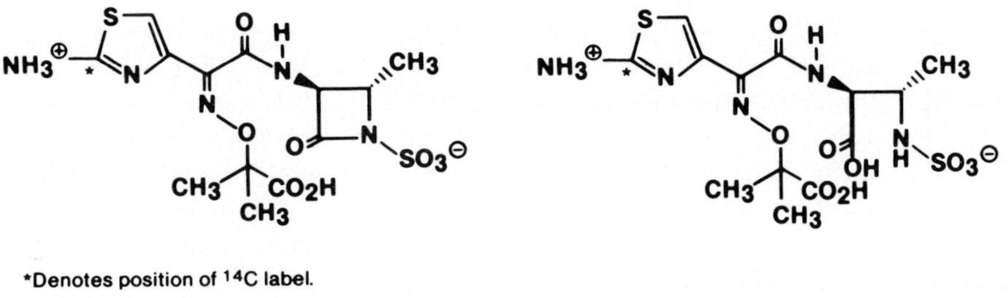

A B

FIGURE 3. Structures of (A) aztreonam and (B) SQ 26,992.

Extensive chemical modifications of these naturally occurring compounds resulted in two major findings. First, substitution of the methyl group at the four position of the β-lactam ring provided increased stability against β-lactamase and chemical stability. Second, addition of an acid group on the acyl side chain increased Gram-negative potency, while lowering activity against Gram-positive organisms.

These efforts led to the development of aztreonam (Figure 3), the first synthetic monobactam antibiotic developed for clinical use. The structure-activity relationships in the

Table 1
ANTIBACTERIAL ACTIVITY OF AZTREONAM

Organism (no. of isolates)	MIC_{90} (μg/mℓ)
Staphylococcus aureus (20)	>100
Streptococcus pyogenes (20)	12.5
Streptococcus faecalis (20)	>100
Bacteroides fragilis (10)	>100
Escherichia coli (80)	0.2
Klebsiella pneumoniae (70)	0.3
Enterobacter cloacae (30)	12.5
Serratia marcescens (120)	1.6
Proteus mirabilis (25)	<0.1
Indole-positive *Proteus* (50)	<0.1
Providencia stuartii (15)	<0.1
Citrobacter freundii (50)	0.7
Pseudomonas aeruginosa (60)	12.0
Haemophilus influenzae (50)	0.2
Neisseria gonorrhoeae (20)	0.2

Data from Reference 3.

subgroups of the aztreonam molecule can be described as follows: (1) The 1-sulfonic acid group attached to the ring nitrogen activates the β-lactam. (2) The α-methyl group at the four position of the β-lactam ring improves the stability of aztreonam to β-lactamase hydrolysis. (3) The aminothiazole oxime moiety on the acyl side chain provides potent activity against aerobic Gram-negative bacteria. (4) Last, the addition of the carboxylic acid group on the oxime side chain is responsible for enhanced activity against *Pseudomonas aeruginosa*.[2]

The antibacterial spectrum of aztreonam is unique among β-lactam antibiotics. It exhibits negligible activity against Gram-positive bacteria such as staphylococci and streptococci and against anaerobic bacteria (Table 1). The potent antibacterial activity of aztreonam is directed specifically against aerobic Gram-negative bacteria, which are commonly encountered in nosocomial infections.[3]

As will be discussed later, aztreonam shows excellent pharmacokinetic properties in experimental animals, i.e., approximately 40 to 50% of the parenterally administered dose is excreted unchanged in the urine, and about 15% is excreted in bile. Also, aztreonam shows good distributional characteristics in animals; it penetrates well into the genital tract, heart, kidney, liver, lung, spleen, and meninges at therapeutic levels. It can protect mice infected systemically with a wide variety of aerobic Gram-negative organisms ranging from *Escherichia coli* and *Haemophilus influenzae* to *Serratia marcescens* and *P. aeruginosa*. In these studies, aztreonam was shown to be more efficacious than cefuroxime, cefaperazone, cefotaxime, or piperacillin.[4]

The efficacy of aztreonam has also been determined in more complex animal models of infectious diseases. For example, aztreonam was highly effective in treating infections in animal models such as an ascending urinary tract infection in mice caused by *E. coli; P. aeruginosa* and *S. marcescens* pneumonia in rats; *H. influenzae* meningitis in neonatal rats; and wound infections caused by *S. marcescens* and *Proteus rettgeri*.[5]

Aztreonam, with its unique single-ring structure, offers an added advantage of being suitable for patients with penicillin-allergic reactions. In a recent clinical study,[6] 41 patients with IgE antibody to one or more penicillin moieties, as determined by positive skin-test reactions, were challenged with aztreonam. No significant reactivity to aztreonam was observed in any patient. In other clinical studies, the drug-specific antibody response among humans receiving repeated therapeutic doses of aztreonam was very low (0 to 3%), indicating that aztreonam is only weakly immunogenic compared to other β-lactam antibiotics.[7]

Thus, preclinical studies of aztreonam have documented excellent antimicrobial and pharmacologic properties, such as antimicrobial activity targeted against Gram-negative bacteria, good stability against a variety of β-lactamase enzymes, lack of any major adverse reactions after parenteral administration, and an excellent safety profile in laboratory animals at considerably higher doses than those used in clinical situations. Clinical studies of aztreonam are being conducted in various centers in the U.S. New drug applications have been approved for marketing of aztreonam as a parenterally administered antibiotic in many countries and are pending approval in the U.S. and other countries.

III. DISPOSITION OF AZTREONAM IN ANIMALS AND HUMANS

All disposition studies were carried out using aztreonam labeled with ^{14}C in the thiazole ring of the side chain (Figure 3). As mentioned earlier, the β-lactamases in microorganisms are primarily responsible for deactivation of β-lactam antibiotics. Although β-lactamase enzymes are not present in mammalian cells, chemical (nonenzymatic) hydrolysis of the β-lactam moiety can occur in mammals. In fact, compounds arising from hydrolysis of the β-lactam ring are often present as minor impurities in the β-lactam antibiotics.[8] The structure of SQ 26,992, the compound formed by hydrolysis of the β-lactam ring, is shown in Figure 3. In this review, the term "biotransformation product" will be used to describe the compounds formed from the β-lactam antibiotics in vivo, whether by enzymatic (metabolic) processes or by nonenzymatic (chemical) processes. Results obtained after parenteral administration of ^{14}C-aztreonam to rats, dogs, monkeys, and humans will be discussed.

^{14}C-Aztreonam was synthesized with a specific activity of 8.0 μCi/mg and a radiochemical purity of more than 96%, as determined by thin-layer chromatography and high-performance liquid chromatography.[9] Solutions containing ^{14}C-aztreonam at desired specific activities were prepared by dissolving appropriate amounts of labeled and unlabeled aztreonam in water along with an equivalent amount of L-arginine.

All samples collected from animals and man after administration of ^{14}C-aztreonam were counted for concentrations of total radioactivity using liquid scintillation spectrometry. In addition, all individual serum, urine, and bile samples collected up to 24 hr after dosing, and pooled urine and fecal samples were also analyzed by a specific thin-layer radio chromatographic (TLRC) procedure for concentrations of unchanged aztreonam.[9] A bioassay procedure was also used for concentrations of unchanged aztreonam in selected serum and urine samples.[10] The TLRC procedure, in combination with HPLC in most cases, was also utilized for characterization of the biotransformation profiles of aztreonam in biological fluids.

A. Disposition in Rats

After single 50-mg/kg intramuscular (i.m.) or intravenous (i.v.) doses, the total recovery of the radioactive doses in 96 hr averaged 70% (i.m.) and 68% (i.v.) in urine, and 26% (i.m.) and 33% (i.v.) in feces, indicating substantial biliary excretion of aztreonam and its metabolites (Table 2). In 0- to 24-hr urine, unchanged aztreonam accounted for about 84% of the radioactivity and SQ 26,992 accounted for about 14% of the radioactivity, regardless of the route of administration. In separate studies in two unanesthetized rats with cannulated bile ducts, an average of 15% of a single 50-mg/kg i.m. dose of ^{14}C-aztreonam was recovered in bile in 24 hr. In the 0- to 24-hr bile, aztreonam accounted for 51% of the radioactivity, SQ 26,992 about 5%, and unknown compounds accounted for the remainder (Table 2).

The distribution of aztreonam and its biotransformation products in several different tissues was determined in groups of three male and three female rats at 0.25, 2, 6, and 24 hr after

Table 2
DISPOSITION OF ^{14}C-AZTREONAM IN RATS

Av % of dose excreted in 0—96 hr (N = 5)

Sample	Intramuscular	Intravenous
Urine	70	68
Feces	26	33
Total	96	101

Av % Relative Distribution of Radioactivity

	Aztreonam	SQ 26,992	Others (unknown)
Urine (0—24 hr)			
Dose, i.m. (N = 5)	83	14	3
Dose, i.v. (N = 5)	85	13	2
Bile (0—24 hr)			
Dose, i.m. (N = 2)	51	5	44

Table 3
AZTREONAM DISTRIBUTION IN RAT TISSUE

	Time (hr)			
Tissue	0.25	2	6	24
Serum	100	100	100	100
Kidney	189	274	987	2904
Liver	132	633	2693	670
Lung	26	27	45	81
Skin	23	32	46	58
Brain	2	2	25	24
Meninges	54	158	130	150

single 50-mg/kg intramuscular doses of ^{14}C-aztreonam.[11] In comparing concentrations of radioactivity in tissues common to both sexes, no major differences were observed at any of the time points studied. Average concentrations of total radioactivity in six tissues of interest, relative to concentrations in serum, are shown in Table 3 (only data from male rats are shown).

In general, concentrations of total radioactivity in all other tissues not shown in this table did not exceed concentrations in serum; exceptions were the small and large intestines (and their contents) and urinary bladder. Measured concentrations of radioactivity were generally highest at the first-time point (0.25 hr), and then declined to low but detectable ($\geq$0.01 μg equivalents of aztreonam per milliliter) levels at 24 hr. Concentrations of unchanged aztreonam were determined in selected tissues. Again, there were no major differences between males and females with regard to concentrations of unchanged aztreonam in these tissues. Between 6 and 24 hr, concentrations of unchanged aztreonam in serum and liver declined at about the same rate, but those in kidneys and lung declined at a somewhat slower rate.[11]

The results of whole-body autoradiography essentially confirmed the results of the distribution of ^{14}C-aztreonam as determined by liquid scintillation counting. A whole-body autoradiogram of a female rat sacrificed at 15 min after dosing is shown in Figure 4. The dark areas in this figure represent the presence of radioactivity. Radioactivity was widely

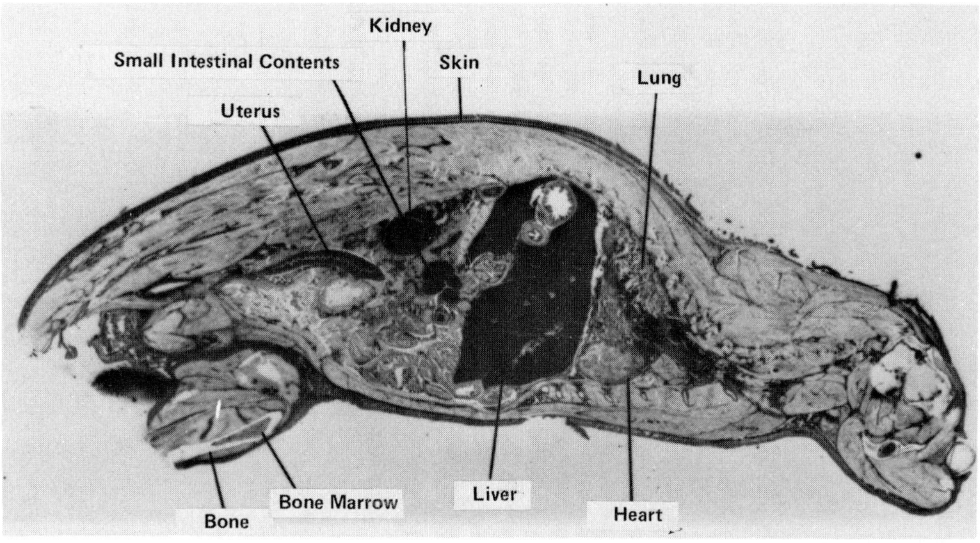

FIGURE 4. Whole-body autoradiograph of a female rat sacrificed 0.25 hr after single intramuscular administration of ^{14}C-aztreonam (50 mg/kg). (From Singhvi, S. M., Ita, C. E., Shaw, J. M., and Migdalof, B. H., *Antimicrob. Agents and Chemother.*, 26, 127, 1984. With permission.)

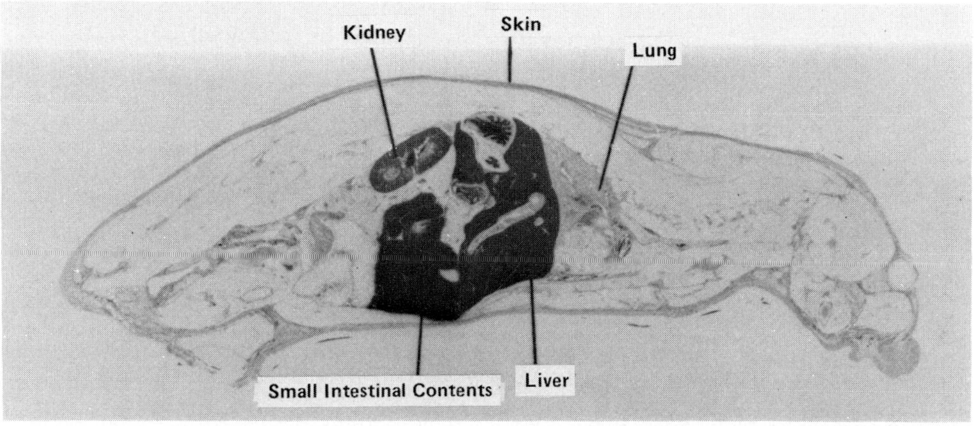

FIGURE 5. Whole-body autoradiograph of a female rat sacrificed 2 hr after single intramuscular administration of ^{14}C-aztreonam (50 mg/kg). (From Singhvi, S. M., Ita, C. E., Shaw, J. M., and Migdalof, B. H., *Antimicrob. Agents Chemother.*, 26, 127, 1984. With permission.)

distributed throughout the body, and was present in high concentrations in organs well perfused with blood, especially liver, heart, kidneys, lungs, and skin. Excretion of radioactivity occurred rapidly, as shown by the presence of radioactivity in the small intestine.

An autoradiogram of a female rat sacrificed 2 hr after dosing is shown in Figure 5. When compared with the 15-min time point, concentrations of radioactivity in the well-perfused organs were lower and concentrations of radioactivity seen in the contents of the small and large intestine were higher, indicating continuation of excretion. At 6 hr, the distribution of radioactivity is similar to that observed at 2 hr in that most of the radioactivity was in liver, kidney, and intestinal contents (Figure 6). At 24 hr, little radioactivity was found in the body, except for the presence of small amounts in liver and kidneys; substantial amounts were seen in the large intestine (Figure 7). At no time were concentrations of radioactivity in brain distinguishable from background in the autoradiographs.

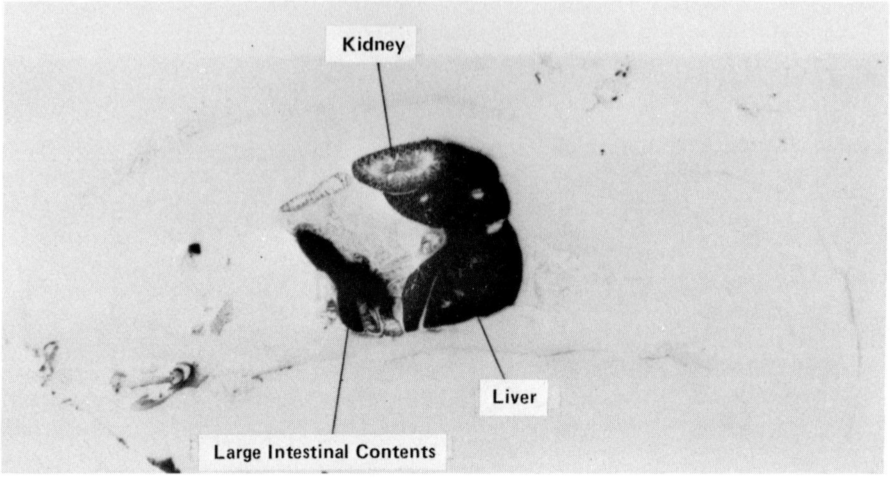

FIGURE 6. Whole-body autoradiograph of a female rat sacrificed 6 hr after single intramuscular administration of ^{14}C-aztreonam (50 mg/kg).

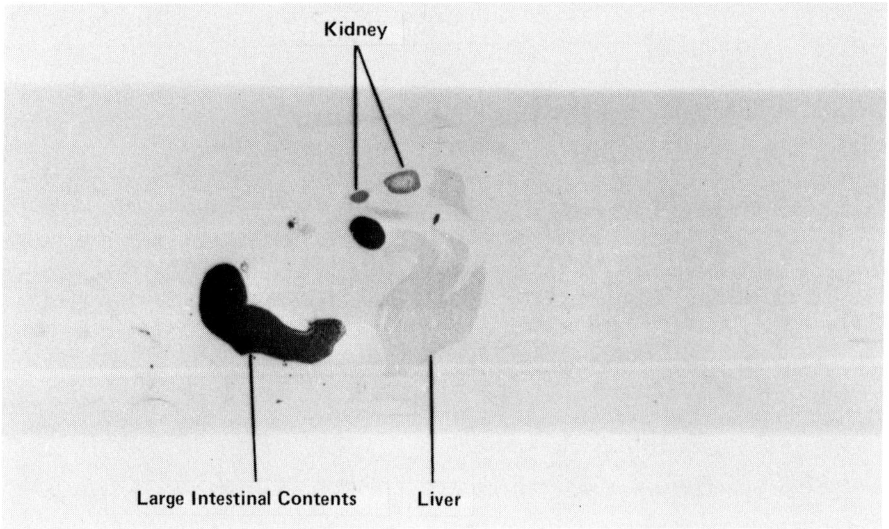

FIGURE 7. Whole-body autoradiograph of a female rat sacrificed 24 hr after single intramuscular administration of ^{14}C-aztreonam (50 mg/kg). (From Singhvi, S. M., Ita, C. E., Shaw, J. M., and Migdalof, B. H., *Antimicrob. Agents and Chemother.*, 26, 127, 1984. With permission.)

Placental transfer was investigated in rats on day 16 of gestation after single doses of ^{14}C-aztreonam were administered subcutaneously at 150 mg/kg and samples of maternal blood, fetuses, placentas, and amniotic fluid obtained at 1, 2, 4, and 8 hr were analyzed.[12] Aztreonam and its biotransformation products passed into fetuses and amniotic fluid; average concentrations for total radioactivity in the tissues studied are shown in Figure 8. For all tissues, maximum concentrations of aztreonam and total radioactivity were attained within 1 hr. In general, concentrations of total radioactivity in maternal serum were higher than in the other three tissues; the only exception was placenta at 8 hr, where the concentrations of radioactivity were higher than in maternal serum. Average concentrations of unchanged aztreonam in these tissues are shown in Figure 9. Again, concentrations of aztreonam were higher in

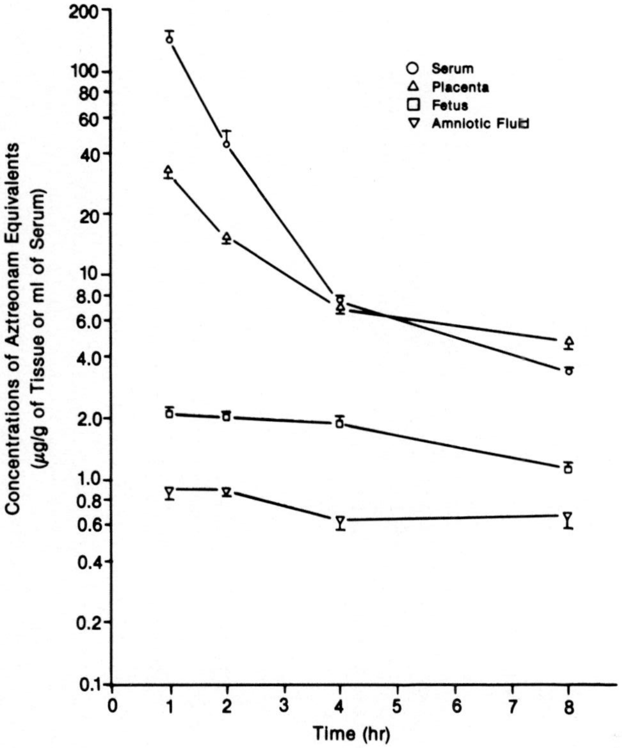

FIGURE 8. Mean (± standard error of the mean) concentrations of total radioactivity after a single subcutaneous dose of ^{14}C-aztreonam (150 mg/kg) to groups of four rats on day 16 of gestation.

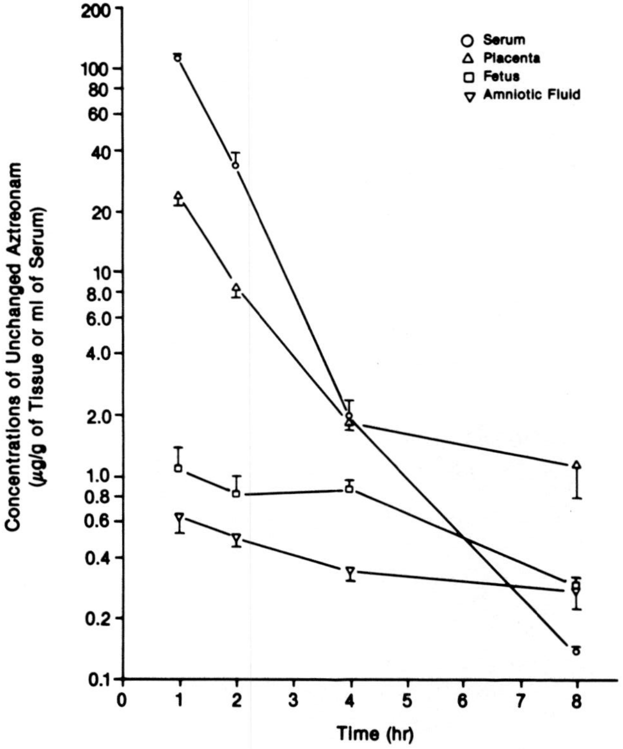

FIGURE 9. Mean (± standard error of the mean) concentrations of unchanged aztreonam after a single subcutaneous dose of ^{14}C-aztreonam (150 mg/kg) to groups of four rats on day 16 of gestation.

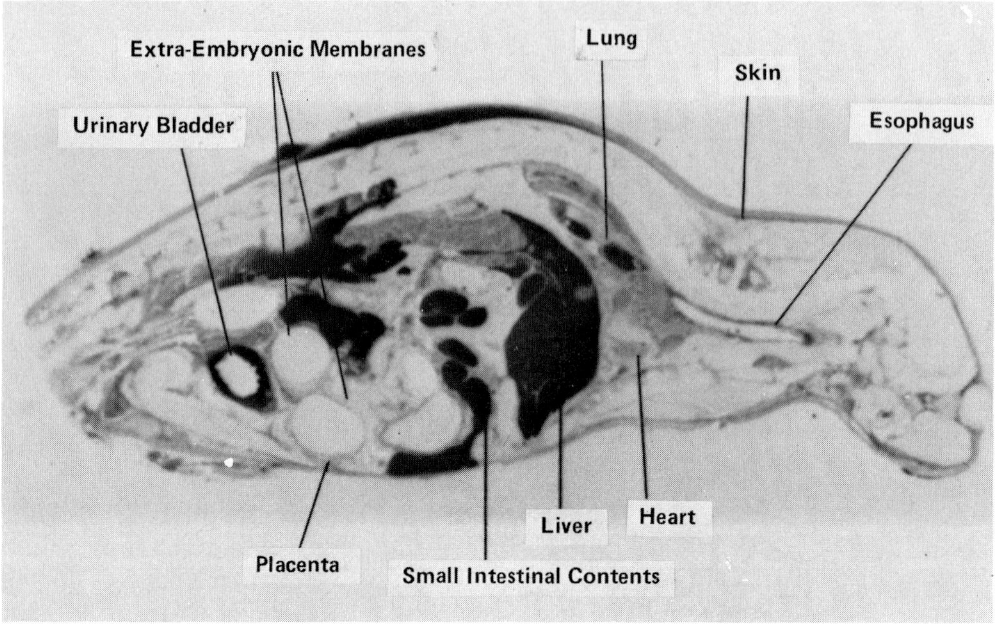

FIGURE 10. Whole-body autoradiograph of a female rat sacrificed 1 hr after single subcutaneous administration of ^{14}C-aztreonam (150 mg/kg) on day 16 of gestation.

maternal serum compared to the other three tissues, except at 8 hr, where the concentrations of aztreonam were higher in all three tissues compared to serum. These results also indicated that the rate of decline of concentrations of both total radioactivity and, to a greater extent, of unchanged aztreonam, was slower in each of three tissues than the rate of decline in maternal serum.

The results of whole-body autoradiography of pregnant rats were generally consistent with the data on tissue distribution obtained by quantitation of radioactivity in excised tissue samples from other animals dosed and sacrificed at the same time points. A typical autoradiogram of a rat killed at 1 hr after dosing is shown in Figure 10. The extensive distribution of radioactivity throughout the body was very similar to the distribution shown earlier in nonpregnant rats. The extraembryonic membranes and the placentas were also found to have detectable amounts of radioactivity. At 2 hr, autoradiograms indicated that concentrations of radioactivity were lower in the well-perfused organs, and radioactivity was also detectable in extraembryonic membranes and placentas. At 4 and 8 hr, detectable amounts of radioactivity were observed in the liver, extraembryonic membranes, and contents of the small intestine, but not in placentas.[12]

The results of this study indicated that aztreonam and its biotransformation products passed from maternal blood through the placenta into the fetus, and that rat fetuses in reproduction studies were exposed to aztreonam and its biotransformation products after subcutaneous administration of aztreonam to dams.

In another study,[12] concentrations of total radioactivity and unchanged aztreonam were determined in milk and serum of lactating female rats, and in their pups, 1, 2, 4, and 8 hr after a single 150-mg/kg subcutaneous dose of ^{14}C-aztreonam (Figure 11). Milk samples were obtained from groups of dams at either 1 and 4 hr or at 2 and 8 hr after dosing. The highest concentrations of total radioactivity and unchanged aztreonam observed were at 1 hr in milk and at 4 hr in suckling pups. Concentrations of both total radioactivity and unchanged aztreonam declined more rapidly in the maternal serum than in the pups and milk. In the pups, most of the radioactivity consisted of unchanged aztreonam, as indicated

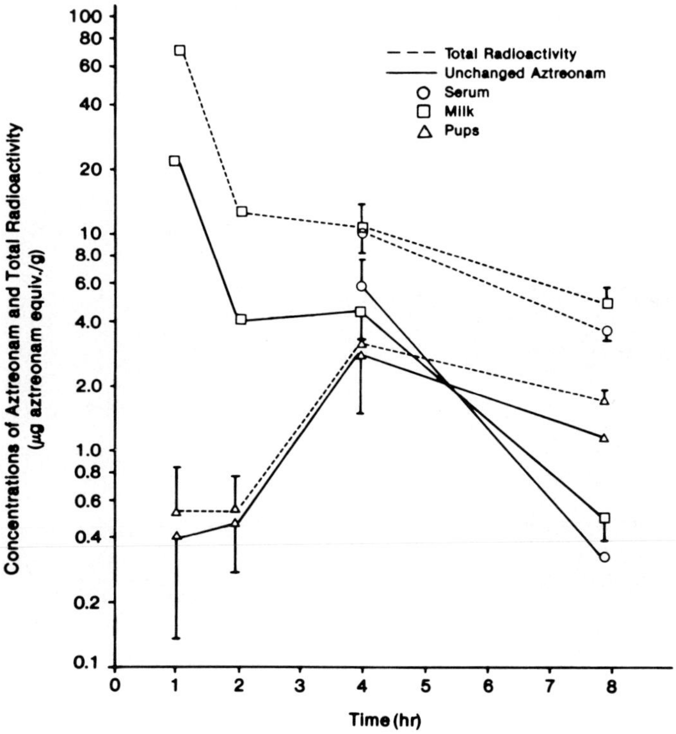

FIGURE 11. Mean (± standard error of the mean) concentration of total radioactivity and unchanged aztreonam after a single subcutaneous dose of ^{14}C-aztreonam (150 mg/kg) to groups of four lactating rats on either day 7 or day 14 of lactation. (From Singhvi, S. M., Ita, C. E., Shaw, J. M., Keim, G. R., and Migdalof, B. H., *Antimicrob. Agents Chemother.*, 26, 132, 1984. With permission.)

by the closeness of the curves for total radioactivity and unchanged aztreonam in Figure 11. These results indicated that the biotransformation processes were not as rapid in the pups.

Whole-body autoradiographs were also prepared for pups that were sacrificed at 1, 2, 4, and 8 hr after dosing of their dams. Radioactivity was detectable at all these time points in the brain, stomach, and intestines of the pups. A representative photograph and an autoradiograph of a pup sacrificed 2 hr after administration of ^{14}C-aztreonam to its dam on day 8 of lactation are shown in Figure 12. Since no radioactivity was detected in the brains of dams and in nonpregnant rats, whereas radioactivity was detectable in the brains of the pups, it appears that the blood-brain barrier was not fully developed in 7-day-old pups. The immaturity of the blood-brain barrier in the fetus and newborn is well documented.[13]

The results of this study demonstrated that rat pups in reproduction studies were exposed to aztreonam and its biotransformation products after subcutaneous administration of aztreonam to the dams.

B. Disposition in Dogs

Disposition studies were conducted in four male dogs after single 25-mg/kg intravenous and subcutaneous doses of ^{14}C-aztreonam.[9] The excretion and biotransformation results of aztreonam in dogs are summarized in Table 4.

In 96 hr, averages of 86 and 7% of the dose were recovered in urine and feces, respectively, after the intravenous dose; corresponding values were 90 and 9% after the subcutaneous dose. Aztreonam was apparently released relatively slowly from the subcutaneous injection

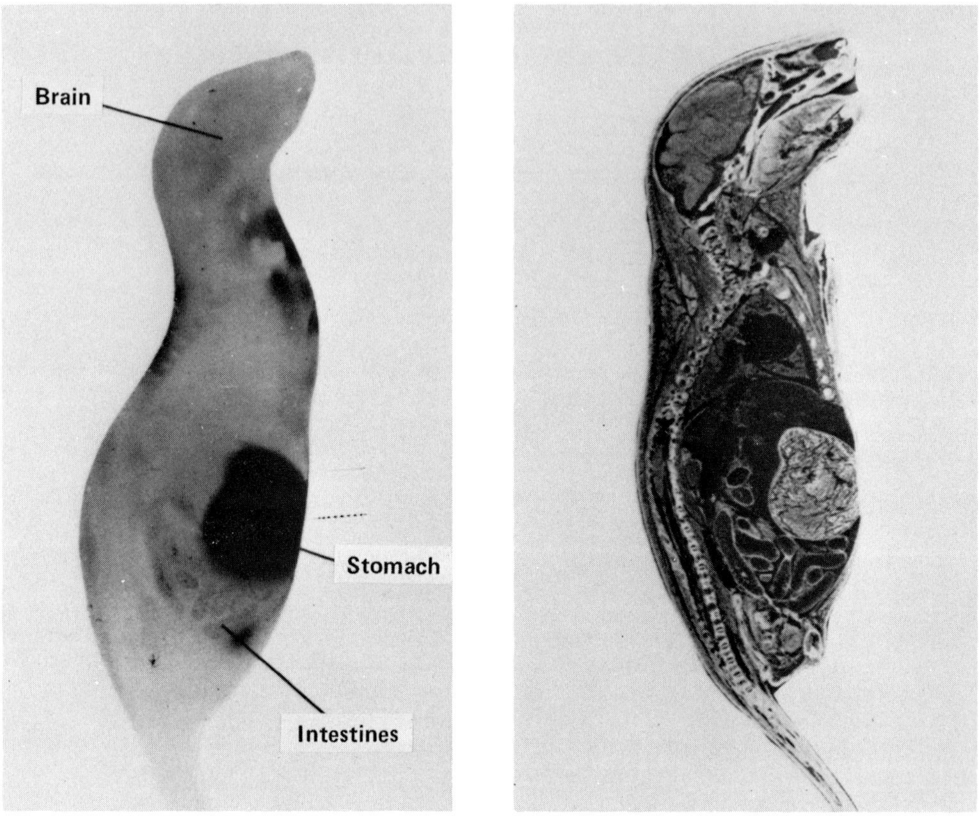

FIGURE 12. Lateral section and whole-body autoradiograph of a pup sacrificed 2 hr after a single subcutaneous dose of ^{14}C-aztreonam (150 mg/kg) to its dam on day 7 of lactation. (From Singhvi, S. M., Ita, C. E., Shaw, J. M., Keim, G. R., and Migdalof, B. H., *Antimicrob. Agents Chemother.*, 26, 132, 1984. With permission.)

Table 4
DISPOSITION OF ^{14}C-AZTREONAM IN DOGS

Av % of Dose Excreted in 0—96 hr (N = 4)

Sample	Intramuscular	Intravenous
Urine	90	86
Feces	9	7
Total	99	93

Av % Relative Distribution of Radioactivity

0—24 hr urine	Aztreonam	SQ 26,992	Others (unknown)
Dose, s.c.	85	11	4
Dose, i.v.	86	10	4

site; excretion in urine after 2 hr averaged 67% after the i.v. dose, but only 24 % after the s.c. dose. After dosing by both routes of administration, 85 to 86% of the radioactivity in 0- to 24-hr urine was attributable to unchanged aztreonam, 10 to 11% to SQ 26,992, and 4% to other unidentified compounds. Thus, unchanged aztreonam accounted for 73 to 75% of the doses in 24-hr urine.[9]

Table 5
EXCRETION OF TOTAL RADIOACTIVITY IN 0 TO 48 HR

	^{14}C-Aztreonam dose			SQ 26,992-^{14}C dose		
Dog no.	**Urine**	**Feces**	**Total**	**Urine**	**Feces**	**Total**
1075	83.4	5.1	88.5	96.2	0.5	96.7
2031	78.3	9.8	88.1	95.1	0.5	95.6
Average	80.8	7.5	88.3	95.6	0.5	96.1

After the s.c. dose, maximum concentrations of total radioactivity and unchanged aztreonam in serum were attained within 1 hr (average of 0.9 hr). Commencing at 1 hr and extending throughout the remainder of the 96-hr study period (for total radioactivity) and to 8-hr (for unchanged aztreonam), the concentrations were greater after the s.c. dose than after the i.v. dose. These results indicate that the absorption from the s.c. site occurred over time (i.e., was not instantaneous), but a comparison of AUC values and urinary excretion data obtained after i.v. and s.c. doses indicated that the absorption of the s.c. dose was complete.[9]

The serum elimination half-life for unchanged aztreonam averaged 0.7 hr after the i.v. dose and the apparent elimination half-life averaged 1.3 hr after the s.c. dose. The apparently longer mean half-life after the s.c. dose was a reflection of the rate of absorption of the s.c. dose from the site of injection, and did not reflect elimination kinetics. Serum protein binding of total radioactivity averaged about 28% at 0.25 and 2 hr after the i.v. dose and 32 and 35% at 0.25 and 2 hr after the s.c. dose. The differences between the two routes were not statistically significant ($p > 0.05$). Thin-layer radiochromatographic profiles were also obtained for serum samples in dogs; radioactivity not accounted for as unchanged aztreonam was present in decreasing order, as components unextractable in methanol, as SQ 26,992, and as other unidentified extractable biotransformation products.[9]

In disposition studies of aztreonam in rats and dogs, the elimination of SQ 26,992, the most prominent biotransformation product of aztreonam, appeared to be much slower than the elimination of aztreonam. In order to study the apparent difference in elimination rates between aztreonam and SQ 26,992, the pharmacokinetics of the two compounds were studied in dogs after equimolar i.v. doses of ^{14}C-aztreonam and SQ 26,992-^{14}C in two dogs. The excretion data from these studies are summarized in Table 5. Most of the radioactivity was excreted in urine after dosing with both compounds; an average of about 8% of the aztreonam was excreted in feces, while less than 1% of the SQ 26,992 dose was excreted in feces.

The biotransformation profiles of the two compounds in urine are summarized in Table 6. As previously indicated, SQ 26,992 is the most prominent biotransformation product of aztreonam after dosing with aztreonam. When SQ 26,992 was administered i.v., it was eliminated unchanged in the urine.

The serum levels vs. time plots of aztreonam and SQ 26,992 after administration of the corresponding compounds in one of the two dogs are shown in Figure 13. The data for the other dog were very similar. The half-life values for aztreonam and SQ 26,992 both averaged about 1.1 hr after administration of the corresponding compounds. Other pharmacokinetic parameters were also similar for the two compounds (Table 7).

After intravenous administration of aztreonam, the apparent average elimination half-life of SQ 26,992 was about 7 to 8 hr (Table 7 and Figure 14), when determined from plots of amounts of aztreonam or SQ 26,992 remaining to be excreted versus time. However, after the intravenous dose of SQ 26,992, the apparent elimination half-life of SQ 26,992 was only about 1.1 hr (Figure 14), a value similar to that obtained for serum data. Based on the results of this study, it appears that either the elimination of SQ 26,992 is formation-rate

Table 6
BIOTRANSFORMATION PROFILE IN URINE

^{14}C-Aztreonam Dose

	Dose (%) excreted in urine as			
Dog no.	**Aztreonam**	**SQ 26,992**	**Others**	**Total**
1075	71.8	8.3	3.3	83.4
2031	66.1	9.7	2.5	78.3
Average	69.0	9.0	2.9	80.8

SQ 26,992-^{14}C Dose

All of the administered SQ 26,992 was eliminated unchanged in urine in both dogs.

limited or the distribution and elimination of SQ 26,992 administered directly is different from that of SQ 26,992 formed in vivo after administration of aztreonam.

Results of this study are similar to those reported for other β-lactam antibiotics; for example, penicilloic acid-biotransformation products of penicillins are eliminated more slowly than the pencillins.

C. Disposition in Monkeys

Disposition studies of ^{14}C-aztreonam were conducted in four female cynomolgus monkeys after single 25-mg/kg intravenous and intramuscular doses.[9] The disposition data in monkeys are summarized in Table 8. Recovery of the radioactive doses in 96 hr after dosing averaged 43% (i.m.) and 44% (i.v.) in urine, and 43% (i.m.) and 46% (i.v.) in feces, with total recovery in the excreta averaging 86% (i.m.) and 90% (i.v.). Recovery was incomplete due, in part, to excretion continuing past the end of the collection period; an average of 2% of the dose was excreted in a 72- to 96-hr period. For both routes of administration, unchanged aztreonam accounted for an average of about 77 to 78% of the radioactivity in 0- to 24-hr urine, and SQ 26,992 accounted for 19 to 20%; unknown compounds accounted for the remainder.

After both routes of administration, the relative percentage of radioactivity in urine attributable to unchanged aztreonam decreased from an average of about 88% in 0- to 4-hr urine to about 25 to 30% in 8- to 24-hr urine. On the other hand, average relative amounts of SQ 26,992 increased from about 10% of the radioactivity in 0- to 4-hr urine to about 65 to 70% in 8- to 24-hr urine.

In the feces, SQ 26,992 was the major radioactive component after an i.m. or i.v. dose of ^{14}C-aztreonam. In the 0- to 72-hr pooled fecal samples, SQ 26,992 accounted for about 46 to 47% of the radioactivity, and unchanged aztreonam for 11 to 17%. Radioactive compounds in feces were not further characterized.

Concentrations of total radioactivity in serum from 0.5 to 24 hr after single i.m. and i.v. doses were comparable. Selected serum samples were analyzed for unchanged aztreonam, and results were similar for both routes of administration. For example, at 1 hr, concentration of total radioactivity in serum averaged 37 and 39 μg equivalents per milliliter for the i.m. and i.v. routes, respectively; average concentrations of unchanged aztreonam were 26 μg/mℓ after i.m. dose and 30 μg/mℓ after the i.v. dose.

Binding of radioactivity to serum proteins at 0.25 and 2 hr averaged 49 and 56% for the i.m. route and 49 and 59 for the i.v. route. Differences between the two routes were not statistically significant.

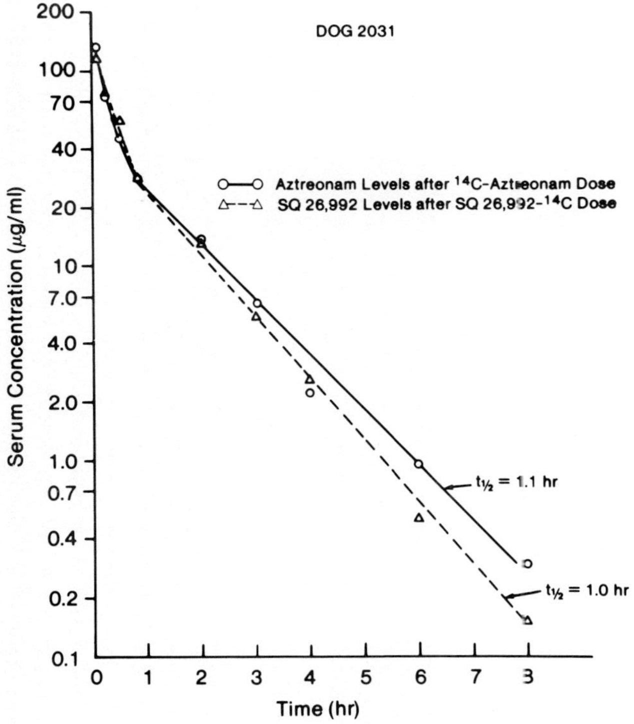

FIGURE 13. Serum concentrations of unchanged aztreonam or unchanged SQ 26,992 in dog 2031 after intravenous doses of ^{14}C-aztreonam or SQ 26,992-^{14}C (equimolar to 25 mg/kg of aztreonam).

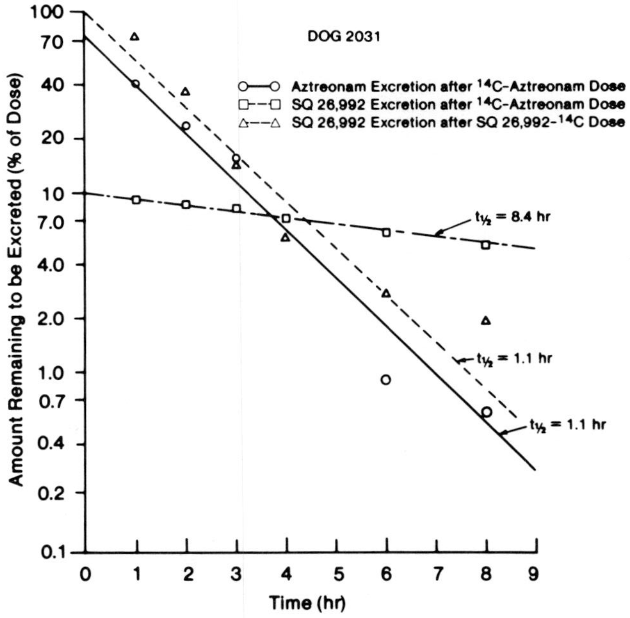

FIGURE 14. Amount of unchanged aztreonam or SQ 26,992 remaining to be excreted in urine of dog 2031 after intravenous doses of ^{14}C-aztreonam or SQ 26,992-^{14}C (equimolar to 25 mg/kg of aztreonam).

Table 7
PHARMACOKINETIC PARAMETERS FOR AZTREONAM AND SQ 26,992

		^{14}C-Aztreonam dose		SQ 26,992-^{14}C dose	
		Dog 1075	Dog 2031	Dog 1075	Dog 2031
Parameter	**Units**				
Body clearance	mℓ/hr/kg	295	267	250	272
Renal clearance	mℓ/hr/kg	212	177	250	272
Vd_{ss}	mℓ/kg	444	319	403	293
Vd_{β}	mℓ/kg	530	397	469	373
$t_{1/2}$	hr	1.0	1.1	1.2	1.0
MRT[a]	hr	1.5	1.2	1.6	1.1
$t_{1/2}$ of SQ 26,992[b]	hr	6.8	8.4	—	—

[a] Mean residence time.
[b] Half-life of SQ 26,992 after elimination of the parent compound (aztreonam), calculated from urinary data.

Table 8
DISPOSITION OF ^{14}C-AZTREONAM IN MONKEYS

Av % of Dose Excreted in 0—96 hr (N = 4)

Sample	Intramuscular	Intravenous
Urine	43	44
Feces	43	46
Total	86	90

Av % Relative Distribution of Radioactivity

	Aztreonam	SQ 26,992	Others (unknown)
Urine (0—24 hr)			
Dose, i.m.	77	20	3
Dose, i.v.	78	19	5
Feces (0—72 hr)			
Dose, i.m.	11	47	42
Dose, i.v.	17	46	37

There was excellent agreement between the results obtained by the TLRC assay and the microbiological assay for concentrations of unchanged aztreonam in selected serum and urine samples (Figure 15). Similar results were observed when serum and urine samples obtained from other species including humans after parenteral administration of ^{14}C-aztreonam were assayed by both methods. These results indicated that SQ 26,992 and other biotransformation products of aztreonam were devoid of any significant antimicrobial activity. This observation was confirmed when SQ 26,992 and three unknown minor biotransformation products of aztreonam isolated from monkey urine were shown to lack any detectable antimicrobial activity.

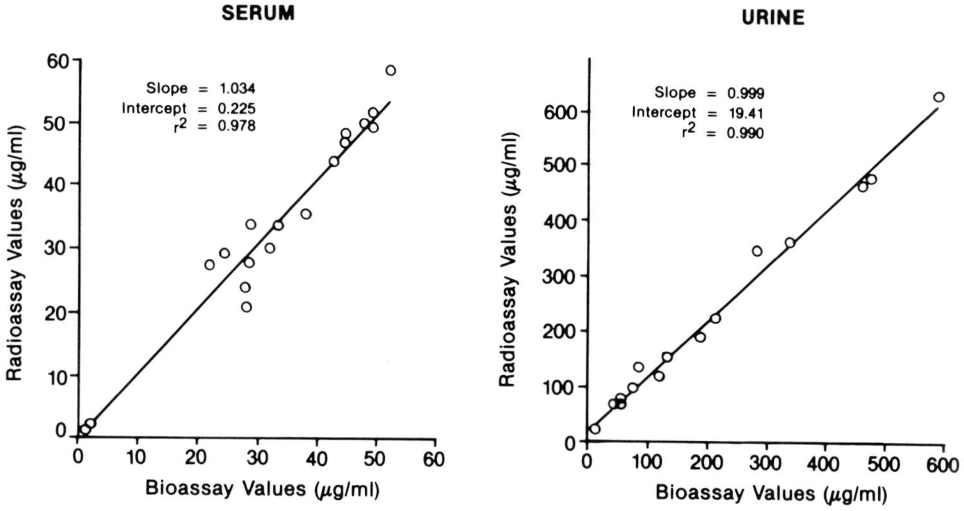

FIGURE 15. Comparison of radioassay and bioassay for concentrations of unchanged aztreonam in selected serum and urine samples of monkeys after single intramuscular and intravenous doses (25 mg/kg) of ^{14}C-aztreonam.

Table 9
DISPOSITION OF ^{14}C-AZTREONAM IN HUMANS

Av % of Dose Excreted in 0—144 hr (N = 4)

Sample	Intramuscular	Intravenous
Urine	78	76
Feces	15	12
Total	93	88

Av % Relative Distribution of Radioactivity

	Aztreonam	SQ 26,992	Others (unknown)
Urine (0—24 hr)			
Dose, i.m.	85	10	5
Dose, i.v.	87	9	4
Feces (0—96 hr)			
Dose, i.m.	7	21	72
Dose, i.v.	11	28	61

D. Disposition in Humans

Disposition studies were also conducted in four healthy male volunteers after single 500-mg i.m. and i.v. doses of ^{14}C-aztreonam.[14] Excretion and biotransformation results of this study are summarized in Table 9.

In 144 hr, the excretion of radioactivity in urine accounted for an average of 78% for the i.m. dose and 76% for the i.v. dose, and fecal excretion accounted for 15% and 12%, respectively. Unchanged aztreonam accounted for 85 to 87% of the radioactivity in 0- to 24-hr urine, SQ 26,992 accounted for 9 to 10%, and the remainder of radioactivity (about 4 to 5%) was unknown compounds. SQ 26,992 was positively identified as the most prominent biotransformation product of aztreonam in urine by chromatographic, spectrometric, and spectroscopic methods.[14]

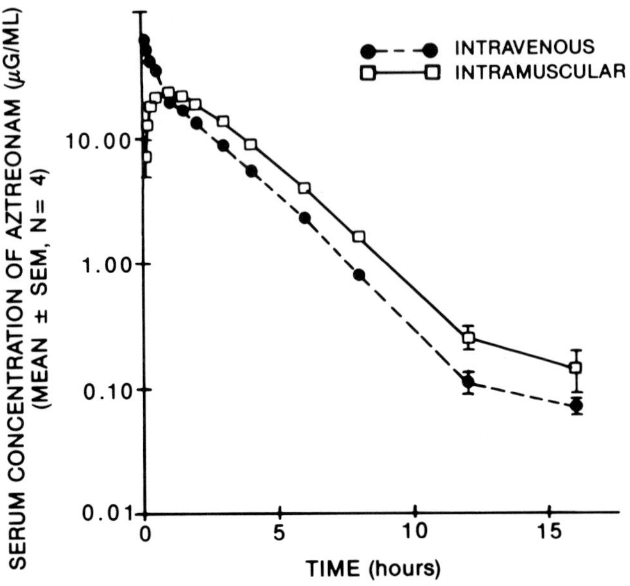

FIGURE 16. Serum concentrations of unchanged aztreonam in healthy volunteers after single intramuscular or intravenous doses (500 mg) of ^{14}C-aztreonam.

Table 10
PHARMACOKINETIC PARAMETERS OF AZTREONAM IN HUMANS

Parameter (units)	i.v. Dose[a]	i.m. Dose[a]
Half-life (hr)	1.6	1.7
Serum clearance (mℓ/min/kg)	1.5	1.5
Renal clearance (mℓ/min/kg)	1.0	0.9
V_1 (ℓ/kg)	0.11	—
Vd_{ss} (ℓ/kg)	0.18	—
Vd_{area} (ℓ/kg)	0.21	0.21

[a] Average of four subjects.

In the 0- to 96-hr feces, unchanged aztreonam accounted for only 7 to 11% of the radioactivity and SQ 26,992 accounted for 21 to 28% of the radioactivity; a large portion of the radioactivity in feces consisted of unknown biotransformation products. However, these unknown biotransformation products in feces accounted for only 8 to 11% of the i.m. or i.v. dose of aztreonam.

The initial concentrations of unchanged aztreonam were higher after the i.v. administration than those after the i.m. administration (Figure 16). However, between 1 and 16 hr, the concentrations of unchanged aztreonam were generally higher after the i.m. administration than after the i.v. administration. Similar results were observed for the concentrations of total radioactivity.[14]

Some of the pharmacokinetic parameters calculated for unchanged aztreonam are summarized in Table 10. Most of the parameters that were calculated for both routes of administration were comparable. The volumes of distribution values indicate that aztreonam is widely distributed in the body of humans. The average half-life values for aztreonam were 1.6 hr after the i.v. dose and 1.7 hr after the i.m. dose.

Table 11
BIOTRANSFORMATION PROFILES OF AZTREONAM IN URINE

Av % Relative Distribution of Radioactivity

	i.v. Dose			s.c. or i.m. Dose		
Species (dose)	**Aztreonam**	**SQ 26,992**	**Others**	**Aztreonam**	**SQ 26,992**	**Others**
Rat (50 mg/kg)	85	13	2	83	14	3
Dog (25 mg/kg)	86	10	4	85	11	4
Monkey (25 mg/kg)	77	13	10	78	14	7
Man (500 mg; ~7 mg/kg)	87	9	4	85	10	5

Plots of urinary excretion rate of aztreonam vs. time after the i.v. dose of aztreonam showed that the average apparent elimination half-life of aztreonam was about 1.8 hr, a value similar to the one obtained from concentrations of aztreonam in serum. However, plots of urinary excretion of SQ 26,992 vs. time showed that the elimination half-life of SQ 26,992 averaged about 26 hr, more than 10 times longer than that of unchanged aztreonam.[14]

IV. SUMMARY AND CONCLUSIONS

In summary, after a parenteral dose of aztreonam, the renal route of elimination was favored in rats, dogs, and humans, whereas in monkeys, the average amounts excreted in urine and feces were about the same. In all four species, unchanged aztreonam was the major component of total radioactivity in urine, with the averages for the different species ranging from 77 to 87% of the radioactivity (Table 11). The prominent biotransformation product (SQ 26,992) resulting from hydrolysis of the β-lactam ring accounted for 9 to 14% of the radioactivity in urine or 5 to 10% of the dose. In monkey and human feces, SQ 26,992 was present in larger amounts than in urine.

For various penicillins, 82 to 100% of an i.m. dose administered to humans is excreted in urine, predominantly as unmetabolized penicillin, and 2 to 15% of the dose as penicilloic acids.[15] Also, it was recently shown that after administration of cefotaxime to humans, rats, and dogs,[16,17] appreciable amounts of β-lactam ring hydrolysis products are excreted in urine. The open-ring metabolites appeared to have longer half-lives than the parent drug. Thus, despite the novel structure and spectrum of activity of aztreonam, its disposition and biotransformation properties are similar to those of other β-lactam antibiotics.

In conclusion, the monobactams are an entirely new class of antibiotics. Their significant departure from the penicillins and cephalosporins is that the central structure feature is a single-ring β-lactam moiety. The original compound in this class was isolated from a bacterium instead of a mold. Also, because of the structural simplicity of these compounds, they can be prepared by synthetic procedures and can be chemically modified for improvement in antimicrobial activity. The first of the monobactams to be selected for clinical studies was aztreonam, which has been shown to have a remarkable specificity for Gram-negative organisms, and excellent resistance to the β-lactamases produced by these bacteria. The relative simplicity of the molecule may actually be a contributing factor for its good pharmacokinetic profile, excellent penetration throughout the body, and a good safety profile.

ACKNOWLEDGMENTS

The author is grateful for technical and scientific contributions of the following individuals: Messrs. P. Egli, C. Ita, F. Meeker, J. Mitroka, J. Shaw, and B. Walker; Drs. M. Bathala, A. Dean, D. Bonner, J. Dreyfuss, D. Everett, K. Kripalani, S. Lan, B. Migdalof, R. Morrison, and S. Weinstein; and Mrs. M. A. Leitz and Ms. S. M. Wind. The author also thanks Ms. T. Mikus for secretarial assistance.

REFERENCES

1. **Sykes, R. B., Cimarusti, C. M., Bonner, D. P., Bush, K., Floyd, D. M., Georgopapadakou, N. H., Koster, W. H., Liu, W. C., Parker, W. L., Principe, P. A., Rathnum, M., Slusarchyk, W. A., Trejo, W. H., and Wells, J. S.,** Monocyclic β-lactam antibiotics produced by bacteria, *Nature (London),* 291, 489, 1981.
2. **Breuer, H., Treuner, U. D., Dentzel, T., Applegate, H. E., Bonner, D. P., Bush, K., Cimarusti, C. M., Koster, W. H., Slusarchyk, W. A., Sykes, R. B., and Young, P.,** Monocyclic β-lactam antibiotics. Activity relationships with α-acyl amino derivatives, 21st Intersci. Conf. Antimicrob. Agents Chemother., Abstr. 878, American Society for Microbiology, Washington, D.C., 1981.
3. **Sykes, R. B., Bonner, D. P., Bush, K., and Georgopapadakou, N. H.,** Aztreonam (SQ 26,776), a synthetic monobactam specifically active against aerobic Gram-negative bacteria, *Antimicrob. Agents Chemother.,* 21, 85, 1982.
4. **Bonner, D. P., Whitney, R. R., Baughn, C. O., Miller, B. H., Olsen, S. J., and Sykes, R. B.,** In vivo properties of SQ 26,776, *J. Antimicrob. Chemother.,* 8(Suppl. E), 123, 1981.
5. **Sykes, R. B. and Bonner, D. P.,** Discovery and development of the monobactams, *Rev. Inf. Dis.,* 7 (Suppl. 4), S579, 1985.
6. **Saxon, A., Hassner, A., Swabb, E. A., Wheeler, B., and Adkinson, F. N.,** Lack of cross-reactivity between aztreonam, a monobactam antibiotic, and pencillin in penicillin-allergic subjects, *J. Infect. Dis.,* 149, 16, 1984.
7. **Adkinson, F. N., Jr., Swabb, E. A., and Sugerman, A. A.,** Immunology of the monobactam aztreonam, *Antimicrob. Agents Chemother.,* 25, 93, 1984.
8. **Migdalof, B. H.,** Aztreonam disposition and pharmacokinetics in rats, dogs, and monkeys, *N. J. Med.* (Spec. Issue, 16—19), January, 1986.
9. **Krinpalani, K. J., Singhvi, S. M., Weinstein, S. H., Everett, D. W., Bathala, M. S., Dean, A. V., Ita, C. E., Lawrence, L., Meeker, F. S., Shaw, J. M., Walker, B. D., and Migdalof, B. H.,** Disposition of ^{14}C-aztreonam in rats, dogs, and monkeys, *Antimicrob. Agents Chemother.,* 26, 119, 1984.
10. **Swabb, E. A., Leitz, M. A., Pilkiewicz, F. G., and Sugerman, A. A.,** Pharmacokinetics of the monobactam SQ 26,776 after single intravenous doses in healthy subjects, *J. Antimicrob. Chemother.,* 8(Suppl. E), 131, 1981.
11. **Singhvi, S. M., Ita, C. E., Shaw, J. M., and Migdalof, B. H.,** Distribution of ^{14}C-aztreonam in rat tissue, *Antimicrob. Agents Chemother.,* 26, 127, 1984.
12. **Singhvi, S. M., Ita, C. E., Shaw, J. M., Keim, G. R., and Migdalof, B. H.,** Distribution of aztreonam into fetuses and milk of rats, *Antimicrob. Agents Chemother.,* 26, 132, 1984.
13. **Rapoport, S. I.,** *Blood-brain Barrier in Physiology and Medicine,* Raven Press, New York, 1976, 79.
14. **Swabb, E. A., Singhvi, S. M., Leitz, M. A., Frantz, M., and Sugerman, A.,** Metabolism and pharmacokinetics of aztreonam in healthy subjects, *Antimicrob. Agents Chemother.,* 24, 394, 1983.
15. **Cole, M., Kenig, M. D., and Hewitt, V. A.,** Metabolism of penicillins to penicilloic acids and 6-aminopenicillanic acid in man and its significance in assessing penicillin absorption, *Antimicrob. Agents Chemother.,* 3, 463, 1973.
16. **Chamberlain, J., Coombes, J. D., Dell, D., Fromson, J. M., Ings, R. J., MacDonald, C. M., and McEeven, J.,** Metabolism of cefotaxine in animals, *J. Antimicrob. Chemother.,* 6 (Suppl. E), 69, 1980.
17. **Reeves, D. S., White, L. W., Holt, H. A., Bahari, D., Bjivater, M. J., and Dax, D. P.,** Human metabolism of cefotaxime, *J. Antimicrob. Chemother.,* 6 (Suppl. A), 93, 1980.

Resistance and Clinical Aspects

Chapter 9

BACTERIAL RESISTANCE TO ANTIMICROBIAL AGENTS

Zafar H. Israili

TABLE OF CONTENTS

I. INTRODUCTION

Antimicrobial agents have played an extremely important role in the treatment and control of bacterial infections. However, resistance to antimicrobial drugs was recognized as a problem since the advent of chemotherapy. The bacteria seem to be extraordinarily adept at becoming resistant, with very few exceptions, to each new antibacterial agent that is discovered, either by isolation from natural sources or by synthesis. The medicinal chemists design drugs to interfere with each critical step in the growth and/or in the production of protective cell walls of bacteria, and the bacteria evolve novel mechanisms to overcome the inhibitory activity of the drug. An understanding of the mechanisms involved in bacterial resistance should be an important step towards developing strategies to overcome this serious problem in human medicine. Several relevant and recent reviews covering many aspects of mechanisms of resistance to antimicrobial agents have been published.[1-18]

Bacterial resistance to sulfonamides was noted as early as the late 1930s, and resistance to penicillin G in *Staphylococcus aureus* and *Escherichia coli* was recognized in the 1940s not long after the introduction of the drug.[19,20] In general, as each new drug has been introduced, resistance to it has emerged within a few years. Occasionally, it may take many years for the appearance of resistance to a new agent; for example, the resistance of pneumococci to penicillin took many years to develop.

Bacterial resistance per se may not be a direct consequence of heavy use of antibiotics. Organisms isolated prior to the antibiotic era have been found to contain resistance factors.[21] Even before penicillin was introduced clinically, penicillin-resistant organisms were discovered. This may be due to the fact that antibiotic-producing microorganisms have always existed in the soil (e.g., Actinomycetales) and intrinsic bacterial mechanisms for resistance were probably always present in bacteria.[22] Therefore, it is likely that natural antibiotic resistance is partly an evolutionary process and partly an acquisition process from organisms that already have resistance factors. However, the extensive use of antimicrobial agents in hospitals and in the community has speeded up the selection of pathogens which are stably resistant to many antibiotics.[6,23] The indiscriminate and extensive use of antibiotics in animal feed has also contributed to the development of bacterial resistance to these drugs in man.[24]

In order to produce the antibiotic action, drugs must be able to penetrate the cell capsule or the external slime layer (glycocalyx) and traverse various membranes to reach their target(s) in the bacteria (Figure 1). These membranes include: (1) outer membrane found only in Gram-negative bacteria, (2) cytoplasmic membrane found in all bacteria, and (3) intracytoplasmic membrane prominent in some classes of bacteria such as *Rhodopseudomonas*. The outer membrane is composed of lipopolysaccharides, phospholipids, lipoproteins, and proteins.[25-29] The hydrophilic lipopolysaccharide units project out of this membrane. The inner core consists mainly of hydrophobic phospholipids and proteins. The majority of drugs cross the outer membrane by passive diffusion through the phospholipid bilayer or across water-filled porin channels that traverse the outer membrane.[27,28] Porin proteins line these channels and control the size of the pores. The inner end of the outer membrane is connected via polysaccharide bridges to the peptidoglycan layer. The peptidoglycan layer in Gram-positive bacteria (which do not have the outer membrane) is thicker than in Gram-negative bacteria. The outer membrane is separated from the cytoplasmic (phospholipid) membrane by periplasmic space. The protein-synthesizing ribosomes are located in the cytoplasm.

In spite of the wide variety of chemical structures of the antimicrobial agents, the mode of action of the majority of the drugs studied involves inhibition of any of the following classical targets: (1) nucleic acid biosynthesis, (2) peptidoglycan biosynthesis, (3) lipid biosynthesis, (4) ribosome function, (5) membrane function, and (6) folic acid metabolism. In addition, some antibiotics inhibit nonclassical targets such as lipoprotein and lipopolysaccharide biosynthesis.[30]

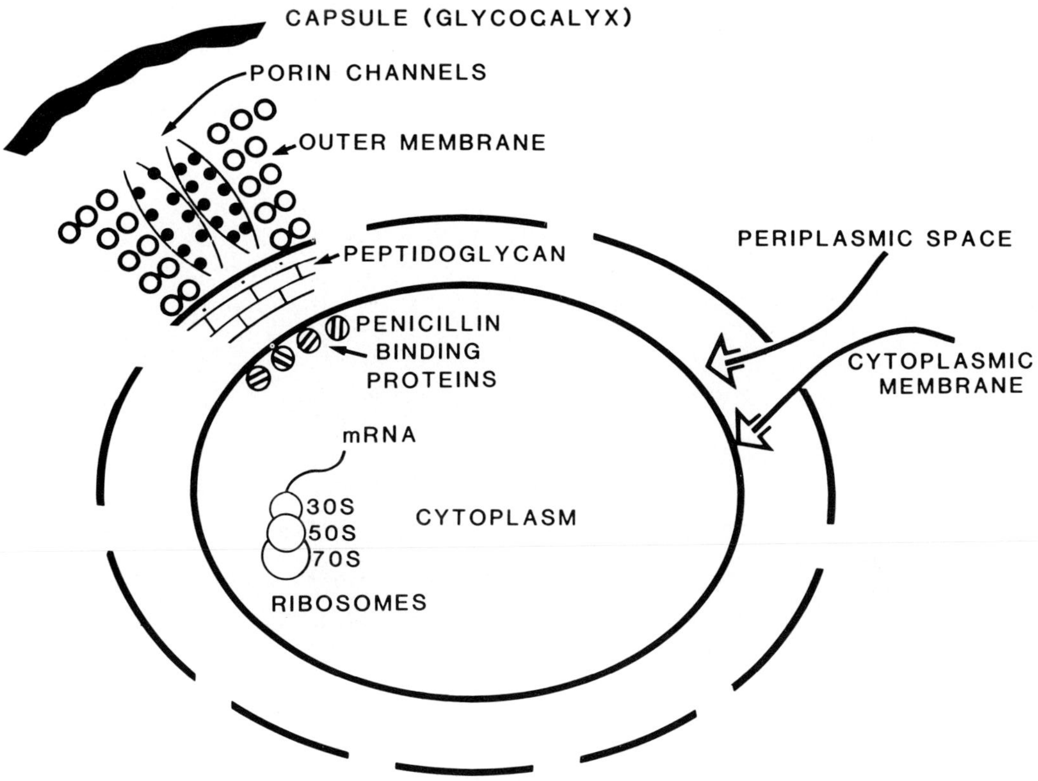

FIGURE 1. Schematic representation of a bacterial cell.

II. GENERAL MECHANISMS OF RESISTANCE

An antimicrobial agent can be rendered ineffective (or less effective) by mechanisms which could be intrinsic in nature or acquired by the bacteria. These may include one or more of the following:

1. An alteration of the drug by the microorganism resulting in changes in its binding, permeability, or inhibitory action. The bacteria may modify the antibacterial agent in such a way that the modified drug (metabolite) either enters the bacterial cell poorly or when it enters the cell it fails to bind adequately to the target, as is the case with aminoglycosides and chloramphenicol. Acetyltransferases in many resistant bacteria acetylate chloramphenicol to a diacetyl derivative which binds poorly to the target 50S ribosomes (see later). Similarly, aminoglycosides are metabolized by resistant bacteria to yield antibiotically inactive products (see later). Alternately, the metabolic activation of a prodrug may be suppressed in resistant bacteria. For example, metronidazole is activated by the bacteria, and resistant bacteria have reduced capacity to convert the drug to its active form (see later).
2. A change in the environment may occur, such as in oxygen tension, the carbon source, and the presence of alternate pathways which may circumvent antibiotic action. For example, resistance to trimethoprim and sulfonamides develops in bacteria which utilize thymidine as an alternate to folate (see later).
3. A change in the microorganism itself is the most important and common cause of resistance. This alteration may involve any or some of the following:

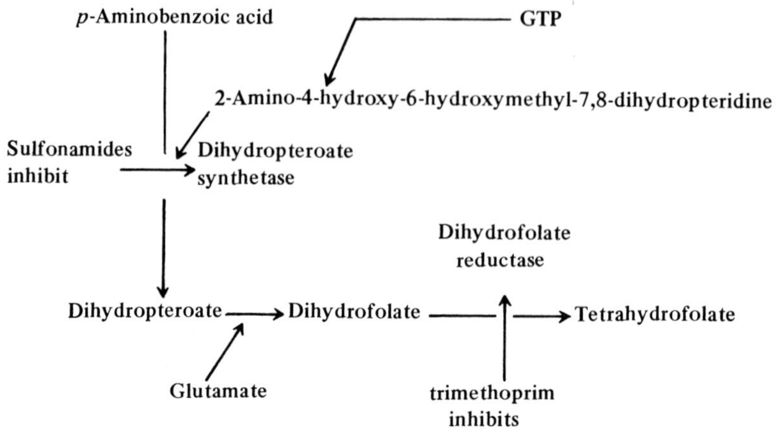

FIGURE 2. Folate pathway in the bacteria and sites of action of sulfonamides and trimethoprim.

Change in permeability barriers (cell wall and membranes) to prevent access of the drug to the target — for example, decreased permeability across the cell wall occurs in *Mycobacteria* and Gram-negative bacteria resistant to rifampin and in *Pseudomonas aeruginosa* resistant to sulfonamides, trimethoprim, and chloramphenicol. Tetracycline resistance is due to decreased drug accumulation as a result of altered membrane permeability in the resistant bacteria.

Altered affinity of the receptor (binding proteins) for the drug in bacteria — Altered penicillin-binding proteins (PBP) have been found in beta-lactam-resistant *Streptococcus pneumoniae, Neisseria gonorrhoeae,* and *Staphylococcus aureus* (see later). Bacterial resistance to macrolide antibiotics is due to an altered 23S RNA (the target) to which the antibiotics bind less efficiently.

Enhanced synthesis of enzyme(s) which deactivate the drug — Increased synthesis of beta-lactamases (which deactivate the penicillins and cephalosporins) is the major mechanism of resistance to beta-lactam antibiotics. The enhanced enzyme synthesis is often the result of antibiotic-induced derepression, e.g., by cefamandole, cefoxitin, etc.[31,32] The induced enzyme may also bind (trap) the antibiotics, thus restricting their activity.

Overproduction of a substrate of a target enzyme — The intrinsic resistance in *Neisseria* to sulfonamides is probably due to hyperproduction of *p*-aminobenzoic acid, a precursor in the folate pathway (Figure 2).

Novel synthesis of target enzymes resistant to the drug — For example, some bacteria become resistant to trimethoprim by synthesizing an altered dihydrofolic reductase (the target) which has a much lower affinity for the drug than the sensitive enzyme (Figure 2).

Modification of a target enzyme so that it becomes insensitive to the drug, yet still functions in bacterial metabolic pathways — The resistance of certain bacteria (e.g., *Pneumococcus*) to sulfonamides is caused by the change in the affinity of dihydropteroate synthetase for sulfonamides compared with the affinity for *p*-aminobenzoic acid (Figure 2).

Reduction in the physiological importance of a target — In certain streptococci and some Gram-negative species the role of peptidoglycan in maintaining bacterial cell wall viability is significantly reduced; thus drugs acting at this site are rendered ineffective.

Use of alternate pathways or precursors to synthesize critical biochemicals required by the bacteria — Resistance of mutant strains of Gram-negative bacilli and *S. aureus* to sulfonamides is because of their ability to bypass folate blockade and utilize exogenous thymine (see later).

Elimination of autolytic enzymes whose activity is stimulated by the drug (see Section IV.C).

The resistance described above in categories 1 and 3 is termed genetic resistance, while that described in category 2 is considered to be phenotypic resistance.

III. GENETIC BASIS OF RESISTANCE

Although mutation must be responsible for the origin of some resistant variants, in the vast majority of cases the acquisition of resistance is due to the transfer of genetic material from a resistant microorganism to a sensitive one. The resistance mechanisms may be conferred by genes on the bacterial chromosome or plasmid. In the case of chromosomally mediated resistance, the genetic information remains with the particular bacterial cell and offspring, while the plasmid-mediated resistance is often transmissible.

The early studies of bacterial resistance focused on single-step mutation which was of chromosomal origin. For example, resistance to the early sulfonamides was due to a change in one amino acid in the enzyme dihydropteroate synthetase (Figure 2), which resulted in lower binding of sulfonamides to the enzyme than *p*-aminobenzoic acid, the normal substrate.[33] Similarly, resistance to streptomycin developed by a single-step mutation that altered the target ribosomal protein.[34] Frequently the resistance may not be due to mutation, but rather be a result of selection of organisms within the bacterial population which intrinsically make an altered target, have an altered cell wall, or produce drug-inactivating enzymes in response to an antimicrobial agent. Thus, the selected organisms survive and replicate.

In the late 1950s, it was found that enteric bacteria such as *Shigella dysenteriae* or *S. flexinari* became resistant not only to sulfonamides but also to several other different types of drugs (streptomycin, tetracyclines, and chloramphenicol). The bacterial resistance was shown not to be chromosome mediated,[35] but due to the presence of extrachromosomal DNA, which was transmissible.[36,37] This resistance, previously called R-factor (resistance factor), is now referred to as plasmid resistance.

Plasmids are self-replicating extrachromosomal pieces of DNA (0.03 to 10% the size of bacterial chromosome) with molecular weights between 1 and 300 Mdaltons. Plasmids are present in virtually all bacteria and are widely dispersed in nature. Plasmids in enterobacteria often determine multiple resistance. It is not unusual for a single plasmid to carry genetic determinants for resistance to eight or more unrelated drugs.[38,39] Plasmids carry not only bacterial resistance genes but also often genes that allow the bacteria to attach to mucosal surfaces, to produce local toxins, and to invade and colonize. Plasmid incompatibility has provided a system for taxonomic classification, and 13 incompatible groups of plasmids have been recognized.[38,40-43]

Plasmid resistance can be transferred from bacteria to bacteria by any of the following processes: (1) conjugation (direct cell to cell contact), (2) transduction (bacteriophage vector or infection with a nonlethal virus), (3) transformation (uptake of DNA from the environment), and (4) transposition, by transfer of resistant determinant from one plasmid to another, via transposons.

Transposons, the so-called jumping genes, are movable DNA elements carrying the resistance genes, which can jump or hop from plasmid to plasmid and from plasmid to chromosome. Transposons are found in many bacterial species such as *Neisseria, Haemophilus, Shigella, Salmonella,* and *Pseudomonas aeruginosa*. Transposon-mediated resistance

Table 1
MECHANISMS OF RESISTANCE TO SULFONAMIDES

Type	Examples
Intrinsic	
Detoxification of sulfonamides	Probably several species, but not well established
Poor penetration through cell envelope	*Pseudomonas aeruginosa* and maybe many other species
No target site	Folate auxotrophs, e.g., *Streptococcus faecium, Lactobacillus* spp.
Acquired	
Mutation to impaired penetration	Probably many species
Plasmid-mediated restricted penetration	Many species
Hyperproduction of *p*-aminobenzoic acid	*Neisseria, Staphylococcus aureus*
Mutation to thymine auxotrophy	Many species (mutant Gram-negative bacteria, *S. aureus,* etc.)
Plasmid-mediated production of additional sulfonamide-resistant dihydropteroate synthetase	Many enterobacteria
Simultaneous operation of many mechanisms	Several R factors, (Su^R) plasmid

Table 2
MECHANISMS OF RESISTANCE TO TRIMETHOPRIM

Type	Examples
Intrinsic	
Detoxification of trimethoprim	Organisms in ruminal flora
Poor penetration through cell envelope	*Pseudomonas aeruginosa*
Insensitive target site (DHFR)	*Neisseria, Nocardia,* many anerobes, *Brucella, Bacteroides*
Acquired	
Mutation to impaired penetration	Probably for several species
Plasmid-mediated restricted penetration	May be many species
Overproduction of DHFR	*E. coli, Klebsiella pneumoniae Salmonella,* enterobacteria
Mutation to production of insensitive DHFR	Many species, including *E. coli,* enterobacteria
Mutation to thymine auxotrophy	Many species (e.g., *E. coli, Staphylococcus,* enterobacteria, *Streptococcus pneumoniae)*
Plasmid-mediated production of novel trimethoprim-resistant plasmid-encoded-DHFR	Many enterobacteria, several plasmids
Simultaneous operation of more than one mechanism (e.g., plasmid-mediated plus mutated chromosomal DHFR production	Many bacteria

to a variety of antibiotics (beta-lactams, aminoglycosides, chloramphenicol, cotrimoxazole [trimethoprim-sulfamethoxazole], and tetracycline) is the main cause of various nosocomial outbreaks of antibiotic-resistant community and hospital infections.[43-45]

Plasmid resistance, from an epidemiologic viewpoint, is the most important resistance since (1) it is transmissible, (2) it is usually highly stable, (3) it confers resistance to many different classes of antibiotics simultaneously, and (4) it is often associated with other characteristics which enable the microorganism to colonize and invade a susceptible host.

IV. SPECIFIC EXAMPLES OF MECHANISMS OF RESISTANCE TO ANTIBIOTICS

A. Sulfonamides and Trimethoprim (Tables 1 and 2)

Sulfonamides and trimethoprim inhibit enzymes in the folate pathway of the bacteria (Figure 2). The sulfonamides block the conversion of *p*-aminobenzoic acid to dihydropteroate by inhibiting dihydropteroate synthetase. Trimethoprim inhibits dihydrofolate reductase (DHFR), thereby blocking the reduction of dihydrofolate to tetrahydrofolate.[1,33,46]

Several intrinsic mechanisms of resistance to sulfonamides and trimethoprim have been found to occur in bacteria. In certain organisms, e.g., *Pseudomonas aeruginosa,* which exhibit resistance to both drugs simultaneously, a permeability barrier to both compounds has been suggested as the mechanism of resistance.[46] Permability barriers may also be present in other insensitive pseudomonads and some Gram-negative bacteria.[5,47]

Resistance to trimethoprim and sulfonamides occurs mostly by target site alteration. In many bacteria, a single mutation results in the chromosomal production of an altered DHFR with properties different from the normal enzyme.[48] The resistance of certain bacteria (e.g., *Neisseria* spp., *Clostridium* spp., *Nocardia restricta, Bacteroides,* and *Caulobacter*) to sulfonamides and trimethoprim is due to the replacement of the chromosomally determined drug-sensitive DHFR by a plasmid-encoded enzyme that may be as much as 20,000 times less susceptible to enzyme inhibition.[5,46] In many bacteria the insensitive and sensitive enzymes (of different susceptibilities) are simultaneously produced in different proportions. The enzymes differ in molecular weight and other physicochemical and kinetic properties. For example, DHFR isolated from trimethoprim-sensitive *E. coli* was found to have a molecular weight of 18,500, while enzymes from resistant *E. coli* had molecular weights of ~35,000 (plasmid type I) and 33,776 (plasmid type II). The concentrations of trimethoprim required to produce 50% inhibition of the enzymes were 0.007, 57, and 70,000 μM, respectively. Further, antibodies produced against a certain resistant DHFR did not cross-react with another plasmid-encoded DHFR. Regardless of the mechanism involved, resistance to one DHFR inhibitor produces resistance to all others. For example, *E. coli* that are resistant to trimethoprim are also resistant to two other DHFR inhibitors, diaminoquinazoline and diaminopyrimidine, as well as to sulfadiazine, which is not a DHFR inhibitor but does act in the folate pathway.[49]

The intrinsic resistance in *Neisseria* and *Staphylococcus* is probably due to hyperproduction of *p*-aminobenzoic acid. Since sulfonamides are both substrates as well as competitive inhibitors of dihydropteroate synthetase, it has been suggested that intrinsic resistance in some bacteria may be due to their capability to degrade the sulfonamides.[5]

In some bacteria the resistance to sulfonamides or trimethoprim is due to the production of a replacement plasmid-encoded, drug-insensitive dihydropteroate synthetase.[33]

The folate auxotrophs, e.g., *Streptococcus fecalis* and *Lactobacillus* spp., which utilize exogenous folate, are intrinsically resistant to both sulfonamides and trimethoprim;[5,50] methotrexate can reverse this resistance.[51]

Bacterial resistance (e.g., of enterococci) to sulfonamides and trimethoprim may be a result of mutation in the thymidylate synthetase gene to thymine auxotrophy such that cells would require an exogenous source of thymine for growth.[52] Long-term use of trimethoprim plus sulfonamides, such as in patients with urinary tract infection, can lead to thymine-dependent organisms resistant to either drug by virtue of bypassing the folate pathway. Mutant strains of Gram-negative bacilli and *Staphylococcus aureus* have been found to bypass folate blockade and utilize thymine.

B. Aminoglycosides (Table 3)

The target of this diverse group of antibiotics, with aminocyclitol as the common structural feature, is the 30S ribosome subunit in the cytoplasm; both the free amino and hydroxyl groups on drug molecule are essential for binding. The binding of aminoglycosides to the ribosomes actively engaged in protein synthesis causes extensive degradation of polysomes and accumulation of aberrant initiation complexes containing 70S ribosomes, leading to misreading of mRNA and resulting ultimately in inhibition of protein synthesis.[9,53] Further, during the transfer (oxygen and energy dependent) of the drug from transporter molecule in

Table 3
MECHANISMS OF RESISTANCE TO AMINOGLYCOSIDE

Type	Examples
Intrinsic	
Absence of transport system	Anaerobic bacteria, e.g., *Bacteroides fragilis, Clostridium perfringens*
Acquired	
Poor penetration through cell envelope (altered porin channels)	Enterobacteriaceae, *Pesudomonas aeruginosa, Neisseria gonorrhoeae, Pseudomonas*
Target site alteration (poor attachment to 30S ribosomal unit)	*N. gonorrhoeae*
Plasmid-mediated enhanced metabolism (by acetylation, phosphorylation, adenylation) to inactive products	*Staphylococcus aureus, Streptococcus faecalis, Streptococcus pyogenes, Streptococcus pneumoniae,* Enterobacteriaceae, *P. aeruginosa*

the cytoplasmic membrane to specific binding sites on ribosomes, disorganization of the cytoplasmic membrane occurs, with resultant loss of permeability control for small molecules leading to cell death.[9]

The most important mechanism of resistance to aminoglycosides is poor uptake and accumulation of the drugs by the bacteria, which may be due to the lack of a transport system[1,9,13,54,55] or alteration in the porin channels.[9,56,57] Other mechanisms of resistance include altered binding site on 30S ribosomes and metabolic inactivation (acetylation of amino groups and adenylation or phosphorylation of the hydroxy group) of the drugs.[13,58,59] The metabolizing enzymes, associated with the cytoplasmic membrane, are plasmid mediated and sometimes present on transposons;[13,60,61] the inactivated drug cannot bind to the ribosomes.

In some bacteria, resistance to aminoglycosides may be overcome by inhibiting the aminoglycoside-deactivating enzymes; for example, 7-hydroxytropolone, a specific inhibitor of 2″-*O*-adenyltransferase, potentiates the action of aminoglycosides.[62,63]

C. Beta-Lactams (Table 4)

The main action of this largest group of antibiotics is on the cell-wall-synthesizing enzymes, transpeptidases, which crosslink peptidoglycan, the rigid, bag-like macromolecule responsible for the mechanical strength of the bacterial envelope.[64-67] The inactivation of the enzymes leads to cessation of production of the peptidoglycan layer and thus bacterial growth. For beta-lactams to be effective they need to bind to specific high-affinity targets, penicillin-binding proteins (PBPs), located in the cytoplasmic membrane of all bacteria with cell walls (Figure 1); this binding is practically irreversible. The number of PBPs varies from three to eight (based on electrophoretic characteristics) in various species. They are numbered by convention in the order of decreasing molecular weight (35,000 to 120,000 daltons). The higher molecular weight PBPs (1a, 1b, 2, and 3) appear to be essential for cell survival. The PBPS of *E. coli* have been thoroughly studied and are associated with cell elongation, maintenance of rod or bacillus shape, and septum formation. These PBPs are also associated with carboxypeptidase, transglycosylase, endopeptidase, and transpeptidase activity.[68-75]

Beta-lactams have variable specificity for binding to PBPs and most of these drugs bind to multiple PBPs. Beta-lactams that preferentially bind to PBP-3, such as some penicillins, monobactams, the ureidopenicillins, and some of the newer cephalosporins, cause the formation of long and filamentous forms of bacteria that lyse only slowly and only over sustained exposure to antibiotic. The beta-lactams that bind to PBP-2, such as ticarcillin, mecillinam, imipenam, and penams, form ovoid spheroplasts which immediately burst, causing cell death. The beta-lactams (cephalosporins) that attach to PBP 1a or 1b are the most potent and rapid inducers of bacterial cell lysis.[74,76,77]

Table 4
MECHANISMS OF RESISTANCE TO BETA-LACTAMS

Type	Examples
Intrinsic	
Permeability barrier	*Bacteroides fragilis, Campylobacter jejuni*
Nonfunctional or deficient porin channels	*Pseudomonas aeruginosa*
Presence of beta-lactamases (constitutive)	*Bacteroides fragilis*, many Gram-positive bacteria, *Streptococcus pneumoniae, Neisseria gonorrhoea, Escherichia coli*
Acquired	
Alteration in one or more penicillin-binding protein (decreased affinity for drug)	Enterococci of group D, Streptococci, *Pseudomonas aeruginosa, Streptococcus pneumoniae, Neisseria gonorrhoea, Staphylococcus epidermis*
Increased production of beta-lactamases (inducible)	AmpA gene of *E. coli, Enterobacter, Bacteroides,*[a] *Clostridium ramosum, Clostridium clostridiforme*
Development of "tolerance" (decreased production of autolysins)	*Streptococcus pneumoniae, Lactobacillus, Staphylococcus aureus*, etc.,
Poor permeability due to changes in:	Some bacteria, e.g.,
The porin channels (by mutation), or	*Neisseria gonorrhoeae, Salmonella typhimurium*
Outer membrane composition (lipopolysaccharides, lipid and/or protein)	Some Gram-negative bacteria, *Pseudomonas aeruginosa*
Trapping of drugs by bacterial enzymes	*Pseudomonas*, enterobacteria
Combination of mechanisms	Several bacteria
Unknown mechanisms	*Bacteroides ureolitious, Fusobacterium mortiferum, Fusobacterium naviforme*

[a] = *Bacteroides fragilis, B. melaninogenicus, B. oralis, B. ruminicola, B. bivius, B. capillosus, B. disiens, B. furcosus, and B. splanchnicus.*

The majority of bacteria resistant to beta-lactam antibiotics possess beta-lactamases which inactivate penicillins, cephalosporins, and cephamycins.[10,78-80] These enzymes hydrolyze the cyclic amide bond on the beta-lactam ring, a process that results in antibiotically inactive products. Inactivation of beta-lactams by beta-lactamases has been known for more than 40 years. Resistance in some bacteria to some beta-lactams may arise by the enzyme binding to the antibiotic without subsequent hydrolysis, thus preventing its transport to the target PBP (see later).

Beta-lactamases are widespread in the microbial world, being produced by both Gram-positive and Gram-negative bacteria, blue-green algae, and yeast. The majority of beta-lactamases produced by Gram-positive bacteria are inducible and exocellular, i.e., secreted into the medium surrounding the cell. The enzymes produced by Gram-negative bacteria are in most cases secreted in the periplasmic space. The beta-lactamases may be inducible, i.e., produced after challenge with antibiotic (e.g., in *P. aeruginosa, Acinetobacter* sp., *Enterobacter* sp., *Serratia* sp., etc.), or constitutive, i.e., present at all times and produced continuously without substrate induction (e.g., in *E. coli, N. gonorrhoeae,* and *Bacteroides fragilis).* These enzymes may be plasmid encoded or chromosomally mediated.[10,31,81-87]

A number of classification schemes for beta-lactamases have been proposed based on the type of bacteria, substrate specificity or molecular size, or amino acid sequence homology of the enzymes.[10,88-92]

Resistance to beta-lactam antimicrobial agents can also be acquired by alteration in the targets of these drugs. Alteration in PBPs results in decreased affinity for the beta-lactams (e.g., many penicillins and cephalosporins). The decrease in affinity (sometime 400- to 500-fold) for a beta-lactam may involve a single PBP or may be generalized; this may explain different susceptibilities to beta-lactams between some species.[17,93-96] Reduction in beta-lactam binding may also be combined with other resistance mechanisms to decrease beta-lactam efficacy. On the other hand, alteration by mutation of the PBP can result in increased binding and thus increased sensitivity to beta-lactams.

The bacteria possess autolysins, enzymes that degrade the bacterial cell walls. The antimicrobial activity of beta-lactams via inhibition of cell wall synthesis is augmented by the continued action of autolysins. However, some bacteria (e.g., *Staphylococcus aureus)* suppress their autolytic system and thus become tolerant to certain concentrations of beta-lactams.[71,76,97]

Beta-lactams utilize the hydrophilic porin channels for permeation into the periplasmic space (Figure 1). Cephalosporins are restricted to this route of permeation but penicillins (at least ampicillin) may also penetrate via solubilization through the phospholipid bilayer. Carbapenems may have additional mechanisms of permeation.[17,98] Porin-deficient mutants (e.g., *Salmonella typhimurium)* or bacteria with nonfunctional porins (e.g., *P. aeruginosa)* are resistant to some beta-lactams.[99-102] Some Gram-negative bacteria develop nonspecific permeability barriers to beta-lactams (and other antibiotics) by alteration in the lipopolysaccharide, lipid, and/or protein composition of their outer membranes.[85,103,104] Resistance to cefoxitin and certain other beta-lactams in resistant *B. fragilis* may be due to the presence of a barrier to diffusion into the periplasmic space.

There are several antibiotics other than beta-lactams which act by inhibiting cell wall synthesis. Bacitracin, vancomycin, and D-cycloserine inhibit the biosynthesis of peptidoglycan at steps different from that for beta-lactams.[105] Clinical resistance to these drugs does not develop very often.[17,106,107]

Resistance of *Pseudomonas* and *Enterobacter* to some of the newer beta-lactamase-resistant beta-lactams may arise by the enzyme binding to the antibiotic without subsequent hydrolysis and thereby blocking its transport to the target PBP.[108]

Several antibiotics, e.g., the cephamycins (cephamycin and cefoxitin) and imipenem, have high intrinsic activity against Gram-negative bacteria and little or no susceptibility to the hydrolyzing enzymes produced by these bacteria.[109,110] However, some of these newer beta-lactamase-stable antibiotics, which tend to be high inducers of beta-lactamases, negatively affect (antagonize) the activity of a wide range of antibiotics including aminoglycosides.[84,111-113] Therefore, widespread inappropriate use of such drugs could result in selection of resistant organisms to the whole class of beta-lactams and/or aminoglycosides.

Since the middle 1970s, several compounds, such as the naturally occurring clavulanic acid, olivanic acids, izumenolide, and panosialin and the semisynthetic-penicillanic acid sulfones and halopenicillinic acids have been found to irreversibly inhibit beta-lactamases.[30,114-117] Clavulanic acid was found to be effective when combined with amoxicillin or ticarcillin in treating resistant infections.[118-124] The combination of benzylpenicillin, ampicillin, or cefoperazone with sulbactam was shown to inhibit many bacteria previously resistant to these antibiotics.[125-127] Penicillanic acid sulfone was reported to act synergistically with cefoperazone, cefoxitin, mecillinam, and pipercillin against a large number of Gram-positive, Gram-negative, and anerobic bacteria.[128,129] 6β-Iodopenicillanic acid, when combined with ampicillin, potentiated its activity both in vitro and in vivo.[130] Potassium 2β-chloromethyl-2α-methyl-penem-3α-carboxylate sulfone was synergistic when combined with amoxicillin or ampicillin.[131] If these inhibitors are relatively inexpensive and nontoxic to the host, the safest and cheapest of available drugs might once again become the agents of choice for treatment of infections.

D. Rifampins (Table 5)

The target of the antibiotic action of rifampins is the DNA-dependent RNA polymerases located in the cytoplasm. The drug forms a stable complex with the enzyme by interacting at its β-subunit site.[53,132-135] Rifampin causes ribosomal damage and fragmentation of the

Table 5
MECHANISM OF RESISTANCE TO RIFAMPICIN

Type	Example
Intrinsic	
Decreased permeability	*Mycobacteria*, some Gram-negative organisms
Acquired	
Altered DNA-dependent RNA polymerases (change in one amino acid in beta subunit of the enzyme alters binding of drug)	Enterobacteriaceae, *Neisseria meningitidis, Haemophilus influenzae, Mycobacterium phlei, Streptococcus faecium, Lactobacillus* spp.

Table 6
MECHANISMS OF RESISTANCE TO CHLORAMPHENICOL

Type	Examples
Acquired	
Reduced permeability	*Pseudomonas aeruginosa, Pseudomonas putida, Neisseria gonorrhoeae, Haemophilus influenzae*
Rapid inactivation by nitroreductases	*Bacteroides fragilis*
Inactivation by acetyltransferase (reduced binding of the resulting diacetyl derivative to 50S ribosomes)	*B. fragilis* (some strains), many Gram-positive and Gram-negative bacteria, *H. influenzae*, staphylococci, pseudomonads, Enterobacteriaceae
Altered ribosomal target	*Bacillus subtilis*
Unknown mechanism	*Flavobacteria*, indole-positive *Proteus, Nocardia asteroides*

mesosomal membrane and subsequent inhibition of protein synthesis.[134] Intrinsically resistant *Mycobacteria* and Gram-negative organisms have decreased permeability to the drug, since polymerases isolated from resistant *E. coli, S. typhimurium,* or *Mycobacteria intracellularae* are sensitive to the drug at concentrations similar to sensitive species.[53,136] In some microorganisms the resistance is due to decreased binding of the drug to RNA polymerase as a result of conformational changes in the enzyme[135] or alteration of an amino acid in its β-subunit. In many cases resistance is due to selection of the microorganism.

E. Chloramphenicol (Table 6)

The bacteriostatic action of chloramphenicol is via its binding to the 50S ribosomal subunit of a wide spectrum of Gram-positive and Gram-negative bacteria.[137] This binding inhibits peptide chain elongation and thus protein synthesis.[53] In order to reach the target, the drug has to pass through the outer and inner membranes of the bacteria. In *E. coli* the drug reaches the periplasm by diffusion through the porin channels (Figure 1). Transfer from the periplasm to the cytoplasm probably occurs via solubilization in and diffusion through the cytoplasmic membrane.[138]

Most microorganisms, in the absence of plasmid-specified degradation or permeability changes, are susceptible to the drug.[1] Many Gram-positive and Gram-negative bacteria become resistant to chloramphenicol by synthesizing (often plasmid-mediated) chloramphenicol acetyltransferase.[139] This intracellular enzyme converts chloramphenicol to its di-acetyl derivative which binds poorly to 50S ribosome, thus allowing protein synthesis to continue.[140-143] At least 15 different variants of the enzyme have been detected; most of these have approximately 22,500 mol wt. The enzymes in Gram-positive bacteria are inducible, while those in Gram-negative bacteria are constitutive.[144,145] It has been reported that chloramphenicol may also be inactivated in certain bacteria (e.g., *B. fragilis)* by the action of nitroreductases. Several other mechanisms of resistance to chloramphenicol may be operating

Table 7
MECHANISMS OF RESISTANCE TO TETRACYCLINES

Type	Examples
Intrinsic	
Poor penetration	*Pseudomonas aeruginosa*
Acquired	
Poor penetration due to plasmid-mediated changes in transport system of cell membrane	Anaerobic bacteria: *Bacteroides fragilis, Neisseria gonorrhoeae, Clostridium perfringens*, Enterobacteriaceae, *Haemophilus* spp.
Active transport of drug out of cell	*E. coli*
Unknown mechanisms	*Clostridium ramosum, Fusobacterium mortiferum, Bacteroides ruminicola*

Table 8
MECHANISMS OF RESISTANCE TO METRONIDAZOLE

Type	Examples
Intrinsic	
Decreased capacity to transform the drug to active intermediate (low levels of nitroreductase)	*Bacteroides fragilis*
Acquired	
Decreased uptake	Some bacteria

in bacteria; these may include plasmid-mediated permeability barriers, e.g., in *P. aeruginosa*[146] and *Haemophilus influenzae*,[147] or the presence of ribosomes (e.g., in *B. subtilis*) with poor binding for the drug.[148,149]

The resistance of certain bacteria may be counteracted by combining chloramphenicol with an inhibitor of acetyltransferase. Such an inhibitor, isolated from a streptomycetes,[150] increased the inhibitory activity of chloramphenicol against several chloramphenicol-resistant bacteria.

F. Tetracyclines (Table 7)

Tetracylines interfere with the normal topology of ribosomes by weakening the bridge formation between 30S and 50S ribosomal subunits by aminoacyl-tRNA-mRNA, a requirement of the bacteria, thus disrupting the protein synthesis.[53,151]

The main mechanism of resistance to tetracycline involves plasmid-mediated changes in transport systems in the cell membrane. These alterations result in reduced accumulation of the drug due to enhanced active efflux[151-153] and/or decreased influx of the drug.[154-158] The resistance of both anaerobic and aerobic bacteria is inducible.[4,159] Intrinsic resistance to tetracycline has also been documented in *P. aeruginosa*.[160]

G. Metronidazole (Table 8)

Metronidazole is converted to the active metabolite, the nitro radical anion, by the reduction of the nitro group by nitroreductases present in the bacteria. The primary target for the antibiotic activity is DNA. The resistance of certain bacteria to metronidazole may be due to the decreased capacity of the bacteria to convert the drug into the active metabolite or a decrease in the uptake of the drug by the bacteria.[161]

H. Macrolides and Lincosamides (Table 9)

The macrolides and lincosamides bind to the 23S ribosomal subunit and inhibit peptidyl transferase activity or "translocation". Lincomycin may also cause a premature detachment of ribosome from mRNA. The antibiotic thus interferes with protein synthesis.

Table 9
MECHANISMS OF RESISTANCE TO MACROLIDES AND LINCOSAMIDES

Type	Examples
Intrinsic	
Poor penetration	Many Gram-negative bacteria
Acquired	
Decreased binding to 23S ribosomal RNA (plasmid-encoded methylating enzymes that modify specific adenine residues of the ribosome)	*Bacteroides fragilis, Staphylococcus aureus, Streptococcus* spp., *S. pneumoniae, Clostridium perfringens,* enterococci
Decreased uptake	Some bacteria
Plasmid-mediated inactivation (lincomycin)	*S. haemolyticus, S. aureus, S. epidermidis*

Modification of the targets located within the cytoplasm constitutes the most significant form of erythromycin resistance, which is generally plasmid mediated and can be constitutive or inducible by the drug.[162] For example, erythromycin-resistant *Staphylococcus aureus, Streptococcus* spp., and *S. pneumoniae* carry plasmid-encoded methylating enzymes that modify specific adenine residues of 23S ribosomal RNA,[163,164] thus decreasing the affinity of the drug for its target. These strains are cross-resistant to both macrolide and lincosamide antibiotics, and resistance to streptogramin class of antibiotics is also included. This type of resistance has been named macrolide-lincosamide-streptogramin phenotype.[165] Many Gram-negative bacteria are generally impermeable to erythromycin because the drug cannot penetrate the outer membrane.[1,103] Plasmid-mediated resistance to clindamycin and erythromycin develops in certain bacteria such as *Bacteroides fragilis.*[5,159,166,167] Recently, inactivation of lincomycin as a mechanism of resistance to the drug has been noted in some bacteria, e.g., *S. haemolyticus, S. aureus,* and *Staphylococcus epidermidis.*[168]

V. PROBLEM MICROORGANISMS

A number of microorganisms have caused and are the cause of serious human problems. The present list includes the following:

1. *Staphylococcus aureus* — It is resistant to beta-lactams, due to altered penicillin-binding proteins, and to many other antibiotics, and continues to be a pathogen of major importance. The organism evolves new pathogenic capabilities and modes of resistance to host defense mechanisms and newer antibiotics. The bacteria presents problems especially in narcotic addicts and chronic care facility patients exposed to many beta-lactam antibiotics.
2. *S. epidermidis* — In patients with indwelling foreign devices, it is resistant to penicillins and cephalosporins due to altered penicillin-binding proteins.
3. *Streptococcus fecalis* — It is becoming increasingly resistant to aminoglycosides, erythromycin, and tetracyclines due to plasmid-mediated enzymes.
4. *Haemophilus influenzae* — Its resistance has been observed to both penicillins and chloramphenicol by virtue of beta-lactamase and chloramphenicol transacetylase.
5. *Enterobacteriaceae, Escherichia coli, and Klebsiella pneumoniae* — These organisms have developed beta-lactamase-mediated resistance to beta-lactam antibiotics and are on the increase.
6. *Serratia marcescens* and *Pseudomonas aeruginosa* — These organisms are resistant to aminoglycosides and penicillins and are widespread.

VI. MECHANISMS TO REDUCE ANTIBIOTIC RESISTANCE

Prevention of widespread, disseminated infection due to a variety of microorganisms may be accomplished by the following mechanisms:

1. By paying careful attention to hygienic practices, i.e., improvement of hygiene in hospitals and among hospital personnel. The hospital staff members can transmit organisms from other patients or are carriers themselves. Hand washing continues to be the primary way of preventing the spread of most bacteria. Also, adherence to meticulous techniques of placing and maintaining intravascular lines and other prosthetic devices will reduce the incidence of subsequent infection.
2. Reduced movement of patients to decrease the dissemination of resistant organisms within hospitals. Patients with or without clinical symptoms can transmit infection.
3. More rational use of antibiotics by selecting the agent based upon identity of the organism, its probable sensitivity, the site and seriousness of infection, the environment in which the infection developed, and the type of patient. The rational use of antibiotics may also include controlling, decreasing, or cycling the use of these drugs.
4. Discovery of new antibiotics from nature or by chemical synthesis, or molecular modification of presently used agents to produce compounds which would be inert to known mechanisms of resistance.
5. Enhancement of antibiotic susceptibility by development of inhibitors of antibiotic-deactivating bacterial enzymes, e.g., beta-lactamases (beta-lactams), adenyltransferase (aminoglycosides), acetyltransferase (chloramphenicol), methyltransferase (macrolides), etc., found in bacteria.
6. Decrease in the indiscriminate usage of antibiotics, especially subtherapeutic use in farm animals, in order to diminish human exposure to resistant bacteria and antibiotic residue in tissues. It has been shown that the antibiotic-resistant organisms from antibiotic-fed animals are capable of colonizing in the human intestinal tract, and salmonellosis in humans resulting from the consumption of infected animal products has been documented.[24,169]

ACKNOWLEDGMENT

The author would like to thank Dr. F. Bharmal and Dr. Peter G. Dayton for their helpful comments.

REFERENCES

1. **Bryan, L. E.,** *Bacterial Resistance and Susceptibility to Chemotherapeutic Agents,* Cambridge University Press, Cambridge, 1982.
2. **Levy, S. B.,** Microbial resistance to antibiotics: an evolving and persistent problem, *Lancet,* 2, 83, 1982.
3. **Lowe, J. A., III,** Mechanisms of antibiotic resistance, *Annu. Rep. Med. Chem.,* 17, 119, 1982.
4. **Tally, F. P. and Malamy, M. H.,** Mechanism of antimicrobial resistance and resistance transfer in anaerobic bacteria, *Scand. J. Infect. Dis. Suppl.* 35, 37, 1982.
5. **Then, R. L.,** Mechanisms of resistance to trimethoprim, the sulfonamides, and trimethoprim-sulfamethoxazole, *Rev. Infect. Dis.,* 4, 261, 1982.
6. **McGowan, J. E., Jr.,** Antimicrobial resistance in hospital organisms and its relation to antibiotic use, *Rev. Infect. Dis.,* 5, 1033, 1983.
7. **Russell, A. D. and Quesnel, L. B.,** *Antibiotics, Assessment of Antimicrobial Activity and Resistance,* Academic Press, New York, 1983.
8. **Bryan, L. E.,** *Antimicrobial Drug Resistance,* Academic Press, New York, 1984.

9. **Bryan, L. E.,** Aminoglycoside resistance, in *Antimicrobial Drug Resistance,* Bryan, L. E., Ed., Academic Press, New York, 1984, 242.
10. **Bush, K. and Sykes, R. B.,** Interaction of beta-lactam antibiotics with beta-lactamases as a cause for resistance, in *Antimicrobial Drug Resistance,* Bryan, L. E., Ed., Academic Press, New York, 1984, 1.
11. **Datta, N.,** Bacterial resistance to antibiotics, *Ciba Found. Symp.,* 102, 204, 1984.
12. **Datta, N.,** *Antibiotic Resistance in Bacteria,* Churchill Livingstone, London, 1984.
13. **Davies, J. E.,** Resistance to aminoglycosides: mechanisms and frequency, *Rev. Infect. Dis.,* 5(Suppl. 2), 261, 1983.
14. **Gangadharam, P. R. J.,** *Drug Resistance in Mycobacteria,* CRC Press, Boca Raton, Fla., 1984.
15. **Neu, H. C.,** Current mechanisms of resistance to antimicrobial agents in microorganisms causing infection in the patient at risk for infection, *Am. J. Med.,* 76, 11, 1984.
16. **Neu, H. C.,** Changing mechanisms of bacterial resistance, *Am. J. Med.,* 77, 11, 1984.
17. **Parr, T. R. and Bryan, L. E.,** Nonenzymatic resistance to beta-lactam antibiotics and resistance to other cell wall synthesis inhibitors, in *Antimicrobial Drug Resistance,* Bryan, L. E., Ed., Academic Press, New York, 1984, 81.
18. **Washington, J. A., 2nd,** Emerging patterns of microbial resistance, *Orthop. Clin. N. Am.,* 15, 417, 1984.
19. **Abraham, E. P. and Chain, E.,** An enzyme from bacteria able to destroy penicillin [Letter], *Nature (London),* 146, 837, 1940.
20. **Kirby, W. M. M.,** Extraction of highly potent penicillin inactivator from penicillin-resistant staphylococci, *Science,* 99, 452, 1944.
21. **Datta, N. and Hughes, V.,** Plasmids of the same Inc group in Enterobacteria before and after the medical use of antibiotics, *Nature (London),* 306, 616, 1983.
22. **Murray, B. E. and Moellering, R. C., Jr.,** Patterns and mechanisms of antibiotic resistance, *Med. Clin. N. Am.,* 62, 899, 1978.
23. **Phillips, I.,** Environmental factors contributing to antibiotic resistance, *Infect. Control,* 4, 448, 1983.
24. **Schell, O.,** *Modern Meat: Antibiotics, Hormones, and the Pharmaceutical "Farm",* Random House, New York, 1984.
25. **Leive, L.,** The barrier function of the Gram-negative envelope, *Ann. N.Y. Acad. Sci.,* 235, 109, 1974.
26. **Salton, M. J. R. and Owen, P.,** Bacterial membrane structure, *Annu. Rev. Microbiol.,* 30, 451, 1976.
27. **Nikaido, H.,** Nonspecific transport through the outer membrane, in *Bacterial Outer Membranes,* Inouye, M., Ed., John Wiley & Sons, New York, 1979, 361.
28. **Nikaido, H. and Nakae, T.,** The outer membrane of Gram-negative bacteria, *Adv. Microbiol. Physiol.,* 20, 163, 1979.
29. **Osborn, M. J. and Wu, H. C. P.,** Proteins of the outer membrane of Gram-negative bacteria, *Annu. Rev. Microbiol.,* 34, 369, 1980.
30. **Allen, N. E.,** Nonclassical targets for antibacterial agents, *Annu. Rep. Med. Chem.,* 20, 155, 1985.
31. **Sanders, C. C.,** Novel resistance selected by the new expanded-spectrum cephalosporins: a concern, *J. Infect. Dis.,* 147, 585, 1983.
32. **Sanders, C. C., Moellring, R. C., Jr., Martin, R. R., et al.,** Resistance to cefamandole: a collaborative study of emerging clinical problems, *J. Infect. Dis.,* 145, 118, 1982.
33. **Anand, N.,** Sulfonamides: structure-activity relationships and mechanisms of action, in *Inhibition of Folate Metabolism in Chemotherapy,* Hitchings, G. H., Ed., Springer-Verlag, New York, 1983, 25.
34. **Mitsuhashi, S. and Kawabe, H.,** Aminoglycoside antibiotic resistance in bacteria, in *Aminoglycosides,* Welton, A. and Neu, H. C., Eds., Marcel Dekker, New York, 1982, 97.
35. **Watanabe, T.,** Infective heredity of multiple drug resistance in bacteria, *Bacteriol. Rev.,* 27, 87, 1963.
36. **Falkow, S.,** *Infectious Multiple Drug Resistance,* Pion Ltd., London, 1975.
37. **Foster, T. J.,** Plasmid-mediated resistance to antimicrobial drugs and toxic metal ions in bacteria, *Microbiol. Rev.,* 47, 361, 1983.
38. **Levy, S. B., Clowes, R., and Koenig, E.,** *Molecular Biology, Pathology, and Ecology of Bacterial Plasmids,* Plenum Press, New York, 1981.
39. **Ling, V., Gerlach, J., and Kartner, N.,** Multidrug resistance, *Breast Cancer Res. Treat.,* 4, 89, 1984.
40. **Mitsuhashi, S.,** *Drug Resistance Plasmids,* University of Tokyo Press, Tokyo, 1977.
41. **Albritton, W. L.,** Resistance plasmids of *Haemophilus* and *Neisseria,* in *Antimicrobial Drug Resistance,* Bryan, L. E., Ed., Academic Press, New York, 1984, 515.
42. **Brunton, J.,** Antibiotic resistance plasmids of *Streptococci, Staphylococci,* and *Bacteroides,* in *Antimicrobial Drug Resistance,* Bryan, L. E., Ed., Academic Press, New York, 1984, 530.
43. **Jacoby, G. A.,** Resistance plasmids of *Pseudomonas aeruginosa,* in *Antimicrobial Drug Resistance,* Bryan, L. E., Ed., Academic Press, New York, 1984, 497.
44. **Sundstrom, L., Vinayagamoorthy, T., and Skold, O.,** Novel type of plasmid borne resistance to trimethoprium, *Antimicrob. Agents Chemother.,* 31, 60, 1987.
45. **Datta, N., Nugent, M., and Richards, H.,** Transposons encoding trimethoprim or gentamicin resistance in medically important bacteria, *Cold Spring Harbor Symp. Quant. Biol.,* 45, 45, 1981.

46. **Hamilton-Miller, J. M. T.,** Resistance to antibacterial agents acting on antifolate metabolism, in *Antimicrobial Drug Resistance,* Bryan, L. E., Ed., Academic Press, New York, 1984, 173.
47. **Then, R. L. and Angehrn, P.,** Low trimethoprim susceptibility of anaerobic bacteria due to insensitive dihydrofolate reductases, *Antimicrob. Agents Chemother.,* 15, 1—6, 1979.
48. **Then, R. L. and Hermann, F.,** Mechanisms of trimethoprim resistance in enterobacteria isolated in Finland, *Chemotherapy,* 27, 192, 1981.
49. **Burchall, J. J., Elwell, L. P., and Fling, M. E.,** Molecular mechanisms of resistance to trimethoprim, *Rev. Infect. Dis.,* 4, 246, 1982.
50. **Bushby, S. R. M. and Hitchings, G. H.,** Trimethoprim, a sulfonamide potentiator, *Br. J. Pharmacol. Chemother.,* 33, 72, 1968.
51. **Zervos, M. J. and Schaberg, D. R.,** Reversal of the *in vitro* susceptibility of enterococci to trimethoprim-sulfamethoxazole by folinic acid, *Antimicrob. Agents Chemother.,* 28, 446, 1985.
52. **Maskell, R. O., Kabudejo, D. A., Payne, R. H., and Pead, L.,** Human infection with thymidine requiring bacteria, *J. Med. Microbiol.,* 11, 33, 1978.
53. **Gale, E. F., Cundliffe, E., Reynolds, P. E., Richmond, M. H., and Waring, M. J.,** *The Molecular Basis of Antibiotic Action,* John Wiley & Sons, London, 1981.
54. **Bryan, L. E., Koward, S. K., and Van Den Elzen, H. M.,** Mechanism of aminoglycoside antibiotic resistance in anaerobic bacteria: *Clostridium perfringens* and *Bacteroides fragilis, Antimicrob. Agents Chemother.,* 15, 7, 1979.
55. **Hancock, R. E. W.,** Aminoglycosides uptake and mode of action, with special reference to streptomycin and gentamicin. I. Antagonists and mutants, *J. Antimicrob. Chemother.,* 8, 249, 1981.
56. **John, J. F., McNeill, W. F., Kresel, P., and Price, K. E.,** Evidence for a chromosomal site specifying amikacin-resistance in multiply-resistant *Serratia marcescens, Antimicrob. Agents Chemother.,* 21, 587, 1982.
57. **Miller, G. H., Sabatelli, F. J., Hare, R. S., and Waitz, J. A.,** Survey of aminoglycoside resistance patterns, *Dev. Ind. Microbiol.,* 21, 91, 1980.
58. **Dickie, P., Bryan, L. E., and Pickard, M. A.,** Effect of enzymatic adenylation on dihydrostreptomycin accumulation in *Escherichia coli* carrying an R-factor: model explaining aminoglycoside resistance by inactivating mechanisms, *Antimicrob. Agents Chemother.,* 14, 569, 1978.
59. **Davies, J. E.,** Aminoglycoside-aminocyclitol antibiotics and their modifying enzymes, in *Antibiotics in Laboratory Medicine,* Lorain, V., Ed., Williams & Wilkins, Baltimore, 1980, 487.
60. **Miller, G. H., Hare, R. S., Sabatelli, F. J., and Waitz, J. A.,** Survey of aminoglycoside resistance patterns, in *Bacterial Resistance to Aminoglycoside Antibiotics,* Weidemann, B. and Reeves, D. S., Eds., Byr-Essex, Munich, 1981, 148.
61. **Shannon, K. and Phillips, I.,** Mechanisms of resistance to aminoglycosides in clinical isolates, *J. Antimicrob. Chemother.,* 9, 91, 1982.
62. **Allen, N. E., Alborn, W. E., Jr., Hobbs, J. N., Jr., and Kirst, H. A.,** 7-Hydroxytropolone: an inhibitor of aminoglycoside-2′-O-adenyltransferase, *Antimicrob. Agents Chemother.,* 22, 824, 1982.
63. **Kirst, H. A., Marconi, G. G., Counter, F. T., Ensminger, P. W., Jones, N. D., Chaney, M. O., Toth, J. E., and Allen, N. E.,** Synthesis and characterization of a novel inhibitor of an aminoglycoside-inactivating enzyme, *J. Antibiot. (Tokyo),* 35, 1651, 1982.
64. **Tipper, D. J. and Strominger, J. L.,** Mechanism of action of penicillins. A proposal based on their structural similarity to acyl-D-alanyl-D-alanine, *Proc. Natl. Acad. Sci. U.S.A.,* 54, 1133, 1965.
65. **Strominger, J. L.,** Penicillin-sensitive enzymatic reactions in bacterial cell wall synthesis, *Harvey Lect.,* 64, 170, 1970.
66. **Ghuysen, J.-M. and Shockman, G. D.,** Biosynthesis of peptidoglycan, in *Bacterial Membranes and Walls,* Leive, L., Ed., Marcel Dekker, New York, 1973, 37—130.
67. **Spratt, B. G.,** Biochemical and genetical approaches to the mechanism of action of penicillin, *Philos. Trans. R. Soc. London Ser. B,* 289, 273—283, 1980.
68. **Blumberg, P. M. and Strominger, J. L.,** Interaction of penicillin with the bacterial cell in penicillin binding proteins and penicillin sensitive enzymes, *Bacteriol. Rev.,* 38, 291, 1974.
69. **Nikaido, H.,** Outer membrane permeability of bacteria: resistance and accessibility of targets, in *Beta-Lactam Antibiotics: Mode of Action, New Developments and Future Prospects,* Salton, M. R. J. and Shockman, G. D., Eds., Academic Press, New York, 1981, 249.
70. **Sykes, R. B. and Georgopapadakou, N. H.,** Bacterial resistance to beta-lactam antibiotics: an overview, in *Beta-Lactam Antibiotics: Mode of Action, New Developments and Future Prospects,* Salton, M. R. J. and Shockman, G. D., Eds., Academic Press, New York, 1981, 199.
71. **Tomasz, A.,** Penicillin tolerance and the control of murein hydrolases, in *Beta-Lactam Antibiotics: Mode of Action, New Developments and Future Prospects,* Salton, M. R. J. and Shockman, G. D., Eds., Academic Press, New York, 1981, 227.
72. **Greenwood, D.,** *Antibiotics of the Beta-Lactam Group,* John Wiley & Sons, New York, 1982.
73. **Mitsuhashi, S.,** *Beta-Lactam Antibiotics,* Springer-Verlag, New York, 1981.

74. **Georgopapadakou, N.,** Bacterial penicillin-binding proteins, *Annu. Rep. Med. Chem.*, 18, 119, 1983.
75. **Sykes, R. B. and Bush, K.,** Physiology, biochemistry and inactivation of beta-lactamases, in *Beta-Lactam Antibiotics, Chemistry and Biology,* Vol. 3, Gorman, M. and Morin, R. B., Eds., Academic Press, New York, 1983, 155.
76. **Tomasz, A.,** From penicillin-binding proteins to lysis and death of bacteria: a 1979 view, *Rev. Infect. Dis.*, 1, 434, 1979.
77. **Waxman, D. J. and Strominger, J. L.,** Penicillin-binding proteins and the mechanism of action of beta-lactam antibiotics, *Annu. Rev. Biochem.*, 52, 825, 1983.
78. **Sykes, R. B. and Matthew, M. H.,** The beta-lactamases of Gram-negative bacteria and their role in resistance to beta-lactam antibiotics, *J. Antimicrob. Chemother.*, 2, 115, 1976.
79. **Hamilton-Miller, J. M. T. and Smith, J. T.,** *Beta-Lactamases,* Academic Press, London, 1979.
80. **Hamilton-Miller, J. M. T.,** Beta-lactamases and their clinical significance, *J. Antimicrob. Chemother.*, 9(Suppl. B), 11, 1982.
81. **Neu, H. C.,** The role of beta-lactamases in the resistance of Gram-negative bacteria to penicillin and cephalosporin derivatives, *Infect. Dis. Rev.*, 11, 133, 1974.
82. **Malamy, M. H. and Tally, F. P.,** Mechanisms of drug-resistance transfer in *Bacteroides fragilis, J. Antimicrob. Ther.*, 8(Suppl. D), 59, 1981.
83. **Sanders, C. C. and Sanders, W. E., Jr.,** Emergence of resistance to cefamandole: possible role of cefoxitin inducible beta-lactamases, *Antimicrob. Agents Chemother.*, 15, 792, 1979.
84. **Sanders, C. C. and Sanders, W. E., Jr.,** Emergence of resistance during therapy with newer beta-lactam antibiotics: role of inducible beta-lactamases and implication for the future, *Rev. Infect. Dis.*, 5, 639, 1983.
85. **Sanders, C. C., Sanders, W. E., Jr., Goering, R. V., and Werner, V.,** Selection of multiple antibiotic resistance by quinolones, beta-lactams and aminoglycosides with special reference to cross-resistance between unrelated drug classes, *Antimicrob. Agents Chemother.*, 26, 797, 1984.
86. **Seeberg, A. H., Tolxdorff-Neutzling, R. M., and Wiedemann, B.,** Chromosomal beta-lactamases of *Enterobacter cloacae* are responsible for resistance to third generation cephalosporins, *Antimicrob. Agents Chemother.*, 23, 918, 1983.
87. **Davies, J. E. and Smith, D. I.,** Plasmid-mediated resistance to antimicrobial agents, *Annu. Rev. Microbiol.*, 32, 469, 1978.
88. **Medeiros, A. A.,** Beta-lactamases, *Br. Med. Bull.*, 40, 18, 1984.
89. **Sykes, R. B.,** The classification and terminology of enzymes that hydrolyze beta-lactam antibiotics, *J. Infect. Dis.*, 145, 762, 1982.
90. **Ambler, R. P.,** The structure of beta-lactamases, *Philos. Trans. R. Soc. London, Ser. B,* 289, 321, 1980.
91. **Jaurin, B. and Grundstrom, T.,** AmpC cephalosporinase of *Escherichia coli* K-12 has a different evolutionary origin from that of beta-lactamases of penicillinase type, *Proc. Natl. Acad. Sci. U.S.A.*, 78, 4897, 1981.
92. **Medeiros, A. A.,** Classification and distribution of beta-lactamases, in *Microbiology,* Leive, L. and Schlessinger, D., Eds., American Society for Microbiology, Washington, D.C., 1984, 385.
93. **Dougherty, T. J.,** Intrinsic resistance: penicillin target alteration and effects on cell wall synthesis, in *Microbiology,* Leive, L. and Schlessinger, D., Eds., American Society for Microbiology, Washington, D.C., 1984, 398.
94. **Malovin, F. and Bryan, L. E.,** Modification of penicillin-binding proteins as mechanisms of β-lactam resistance, *Antimicrob. Agents Chemother.*, 30, 1, 1986.
95. **Hartman, B. J. and Tomasz, A.,** Low-affinity penicillin-binding protein associated with beta-lactam resistance in *Staphylococcus aureus, J. Bacteriol.*, 158, 513, 1984.
96. **Utsui, Y. and Yokota, T.,** Role of an altered penicillin-binding protein in methicillin and cephem-resistant *Staphylococcus aureus, Antimicrob. Agents Chemother.*, 28, 397, 1985.
97. **Sabath, L. D., Wheeler, N., Laverdiere, M., Blazevic, D., and Wilkinson, B. J.,** A new type of penicillin resistance of *Staphylococcus aureus, Lancet,* 1, 443, 1977.
98. **Jaffe, A., Chabbert, Y. A., and Derlot, E.,** Selection and characterization of beta-lactam-resistant *Escherichia coli* K-12 mutants, *Antimicrob. Agents Chemother.*, 23, 622, 1983.
99. **Zimmermann, W.,** Penetration through the Gram-negative cell wall: a codeterminant of the efficacy of beta-lactam antibiotics, *Int. J. Clin. Pharmacol. Biopharm.*, 17, 131, 1979.
100. **Angus, B. L., Carey, A. M., Caron, D. A., Kropinski, A. B. M., and Hancock, R. E. W.,** Outer membrane permeability in *Pseudomonas aeruginosa:* comparison of a wild-type with an antibiotic-supersusceptible mutant, *J. Bacteriol.*, 21, 299, 1982.
101. **Nikaido, H.,** Outer membrane permeability and beta-lactam resistance, in *Microbiology,* Leive, L. and Schlessinger, D., Eds., American Society for Microbiology, Washington, D.C., 1984, 381.
102. **Yoshimura, F. and Nikaido, H.,** Diffusion of beta-lactam antibiotics through the porin channels of *Escherichia coli* K-12, *Antimicrob. Agents Chemother.*, 28, 84, 1985.

103. **Scudamore, R. A., Beveridge, T. J., and Goldner, M.,** Outer membrane penetration barriers as components of intrinsic resistance to beta-lactams and other antibiotics in *Escherichia coli* K-12, *Antimicrob. Agents Chemother.*, 15, 182, 1979.
104. **Godfrey, A. J., Hatlelid, L., and Bryan, L. E.,** Correlation between lipopolysaccharide structure and permeability resistance in beta-lactamase-resistant *Pseudomonas aeruginosa, Antimicrob. Agents Chemother.*, 26, 181, 1984.
105. **Watanakunakorn, C.,** The antibacterial action of vancomycin, *Rev. Infect. Dis.*, 3, S210, 1981.
106. **Shlaes, D. M., Marino, J., and Jacobs, M. R.,** Infection caused by vancomycin-resistant *Streptococcus sanguis*. II, *Antimicrob. Agents Chemother.*, 25, 527, 1984.
107. **Buu-Hoi, A., Branger, C., and Acar, J. F.,** Vancomycin-resistant Streptococci or *Leuconostoc* sp., *Antimicrob. Agents Chemother.*, 28, 458, 1985.
108. **Then, R. L. and Angehrn, P.,** Trapping of nonhydrolyzable cephalosporins by cephalosporinases in *Enterobacter cloacae* and *Pseudomonas aeruginosa* as a possible resistance mechanism, *Antimicrob. Agents Chemother.*, 21, 711, 1982.
109. **Neu, H. C.,** The new beta-lactamase-stable cephalosporins, *Ann. Intern. Med.*, 97, 408, 1982.
110. **Fisher, J.,** Beta-lactams resistant to hydrolysis by the beta-lactamases, in *Antimicrobial Drug Resistance,* Bryan, L. E., Ed., Academic Press, New York, 1984, 33.
111. **Fu, K. P. and Neu, H. C.,** The role of inducible beta-lactamase in the antagonism seen with certain cephalosporin antibiotics, *J. Antimicrob. Chemother.*, 7, 104, 1981.
112. **Gootz, T. D. and Sanders, C. C.,** Characterization of beta-lactamase induction in *Enterobacter cloacae, Antimicrob. Agents Chemother.*, 23, 91, 1983.
113. **Tausk, F., Evans, M. E., Patterson, L. S., Federspiel, C. F., and Stratton, C. W.,** Imipenem-induced resistance to antipseudomonal beta-lactams in *Pseudomonas aeruginosa, Antimicrob. Agents Chemother.*, 28, 41, 1985.
114. **Cole, M.,** *Drugs of the Future*, 6, 697, 1981.
115. **Cartwright, S. J. and Waley, S. G.,** Beta-lactamase inhibitors, *Med. Res. Rev.*, 3, 341, 1983.
116. **Hamanaka, E. S. and Kellogg, M. S.,** Antibacterial agents, *Annu. Rev. Med. Chem.*, 18, 109, 1983.
117. **Hamanaka, E. S. and Kellogg, M. S.,** Antibacterial agents, *Annu. Rev. Med. Chem.*, 19, 107, 1984.
118. **Leigh, D. A. and Robinson, O. P. W.,** *Augmentin,* Excerpta Medica, Amsterdam, 1982.
119. **Martinell, R., Da Silva-Lopes, A. A., De Oliveira, M. M. M. G., and Rocha, H.,** Amoxicillin-clavulinic acid in treatment of urinary tract infection due to gram-negative bacteria resistant to penicillin, *Antimicrob. Agents Chemother.*, 20, 800, 1981.
120. **Mehtar, S. and Watson, A.,** Augmentin. Oral treatment of beta-lactamase positive respiratory tract infections, 21st Intersci. Conf. Antimicrob. Agents Chemother., Chicago, November 4th to 6th, 1981, Abstr. No. 445.
121. **Gurwith, M. J., Stein, G. E., Gurwith, D., Vanderlaan, B. A., and Patterson, M. J.,** Amoxycillin/clavulanate in the treatment of urinary tract infection in college age women, 21st Intersci. Conf. Antimicrob. Agents Chemother., Chicago, November 4th to 6th, 1981, Abstr. No. 446.
122. **Watson, A., and Price, J. D.,** Augmentin: a review of clinical trials in the UK including bacterial and clinical efficacy and safety, 21st Intersci. Conf. Antimicrob. Agents Chemother., Chicago, November 4th to 6th, 1981, Abstr. No. 444.
123. **Jacobs, R., Warren, R., Kluza, R., Trang, J., Underwood, F., Brown, A., and Dorman, D.,** Single dose pharmacokinetics of Timentin (ticarcillin-clavulanic acid) in children with cystic fibrosis, 22nd Intersci. Conf. Antimicrob. Agents Chemother., Abstr. 884, Las Vegas, October 24 to 26, 1983.
124. Abstracts of the Symposium, Beta-lactamase inhibition: therapeutic advances-ticarcillin disodium/clavulanate potassium, Timentin, Ft. Lauderdale, Florida, March 22 to 23, 1985.
125. **Aronoff, S., O'Brien, C., Murdall, D., and Blumer, J.,** A double blind comparative study of the safety and efficacy of Sultamicillin *vs* phenoxymethyl penicillin in the treatment of group A streptococcal pharyngitis in children, 22nd Intersci. Conf. Antimicrob. Agents Chemother., Abstr. 79, Las Vegas, October 24 to 26, 1983.
126. **Lynch, J. E., Retesma, D. A., Pitts, N. E., and Girard, A. E.,** Influence of sulbactam (CP-45,899) on the activity of cefoperazone under conditions simulating human pharmacokinetics, 21st Intersci. Conf. Antimicrob. Agents Chemother., Abstr. 439, Chicago, November 4, 1981.
127. **Lin, T. Y., Ginsburg, C. M., and McCracken, G. H., Jr.,** Pharmacokinetics and serum bactericidal activity of Sultamicillin in infants and children, 22nd Intersci. Conf. Antimicrob. Agents Chemother., Abstr. 841, Las Vegas, October 24 to 26, 1983.
128. **Fu, K. P. and Neu, H. C.,** Synergistic activity of clavulanic acid and penicillanic acid sulfone with pipercillin and cefoperazone, 20th Intersci. Conf. Antimicrob. Agents Chemother., Abstr. 600, New Orleans, September 22 to 24, 1980.

129. **Neu, H. C.,** The effect of beta-lactamase inhibitor, clavulanic acid and penicillanic acid sulfone upon the *in vitro* activity of mecillinam, 20th Intersci. Conf. Antimicrob. Agents Chemother., Abstr. 601, New Orleans, September 22 to 24, 1980.
130. **Moore, B. A. and Brammer, K. W.,** 6β-Iodopenicillanic acid (UK-38,006), a beta-lactamase inhibitor that extends the antibacterial spectrum of beta-lactam compounds: initial bacteriological characterization, *Antimicrob. Agents Chemother.,* 20, 327, 1981.
131. **Pursiano, T. A., Buck, R. E., Randolph, J. A., Misiek, M., Price, K. E., and Leitner, F.,** *In vitro* evaluation of BL-P 2013, a new beta-lactamase inhibitor, 21st Intersci. Conf. Antimicrob. Agents Chemother., Abstr. 435, Chicago, November 4 to 6, 1981.
132. **Lill, U., Santo, R., Sippel, A., and Hartman, G. R.,** Inhibitors of the RNA polymerase reaction, *Colloq. Ges. Biol. Chem.,* 20, 48, 1970.
133. **Rabussay, D. and Zillig, W.,** A rifampin resistant RNA-polymerase from *E. coli* altered in the β-subunit, *FEBS Lett.,* 5, 104, 1969.
134. **Konno, K., Oizumi, K., Ariji, S., Yamaguchi, J., and Oka, S.,** Mode of action of rifampin on mycobacteria. I. Electron microscopic study of the effect of rifampin on *Mycobacterium tuberculosis, Am. Rev. Resp. Dis.,* 107, 1002, 1973.
135. **Konno, K., Oizumi, K., and Oka, S.,** Mode of action of rifampin on mycobacteria. II. Biosynthetic studies on the inhibition of ribonucleic acid polymerase of *Mycobacterium bovis* BCG by rifampin and uptake of rifampin-^{14}C by *Mycobacterium phlei, Am. Rev. Resp. Dis.,* 107, 1006, 1973.
136. **Hui, J., Gordon, N., and Kajioka, R.,** Permeability barrier to rifampin in mycobacteria, *Antimicrob. Agents Chemother.,* 11, 773, 1977.
137. **Nierhaus, D., and Nierhaus, K. H.,** Identification of the chloramphenicol binding proteins in *E. coli* ribosomes by partial reconstitution, *Proc. Natl. Acad. Sci. U.S.A.,* 70, 2224, 1973.
138. **Chopra, I. and Ball, P. R.,** Transport of antibiotics into bacteria, *Adv. Microbiol. Physiol.,* 23, 183, 1982.
139. **Smith, A. L. and Burns, J. L.,** Resistance to chloramphenicol and fusidic acid, in *Antimicrobial Drug Resistance,* Bryan, L. E., Ed., Academic Press, New York, 1984, 293.
140. **Britz, M. L. and Wilkinson, R. G.,** Chloramphenicol acetyltransferase of *Bacteroides fragilis, Antimicrob. Agents Chemother.,* 14, 105, 1978.
141. **Shaw, W. V.,** Chloramphenicol acetyltransferase from chloramphenicol-resistant bacteria, *Methods Enzymol.,* 43, 735, 1975.
142. **Dang-Van, A., Tiraby, G., Acar, J. F., Shaw, W. V., and Bouanchand, D. H.,** Chloramphenicol resistance in *Streptococcus pneumoniae:* enzymatic acetylation and possible plasmid linkage, *Antimicrob. Agents Chemother.,* 13, 557, 1978.
143. **Shaw, W. V.,** Chloramphenicol acetyltransferase: enzymology and molecular biology, *CRC Crit. Rev. Biochem.,* 14, 1, 1983.
144. **Thibault, G., Guitard, M., and Daigneault, R.,** A study of the enzymatic inactivation of chloramphenicol by highly purified chloramphenicol acetyltransferase, *Biochim. Biophys. Acta,* 614, 339, 1980.
145. **Fritton, J. E., Packman, L. C., Harford, S., Zaidenzaig, Y., and Shaw, W. V.,** Chloramphenicol plasmids and the evolution of chloramphenicol resistance, in *Microbiology,* Schlessinger, D., Ed., American Society for Microbiology, Washington, D.C., 1978, 249.
146. **Irvin, J. E. and Ingram, J. M.,** Chloramphenicol-resistant variants of *Pseudomonas aeruginosa* defective in amino acid transport, *Can. J. Biochem.,* 58, 1165, 1980.
147. **Burns, J. L., Mendelman, P. M., Levy, J., Stull, T. L., and Smith, A. L.,** A permeability barrier as a mechanism of chloramphenicol resistance in *Haemophilus influenzae, Antimicrob. Agents Chemother.,* 28, 46, 1985.
148. **Osawa, S., Takata, R., Tanaka, K., and Tamaki, M.,** Chloramphenicol resistant mutants of *Bacillus subtilis, Mol. Gen. Genet.,* 127, 163, 1973.
149. **Baugham, G. A. and Fahnestock, S. R.,** Chloramphenicol resistance mutation in *Escherichia coli* which maps in the major ribosomal protein gene cluster, *J. Bacteriol.,* 137, 1315, 1979.
150. **Miyamura, S., Koizumi, K., and Nakagawa, K.,** An inhibitor of chloramphenicol acetyltransferase produced by Streptomyces, *J. Antibiot. (Tokyo),* 32, 1217, 1979.
151. **Levy, S. B.,** Resistance to the tetracyclines, in *Antimicrobial Drug Resistance,* Bryan, L. E., Ed., Academic Press, New York, 1984, 192.
152. **Chopra, I. and Howe, T. G. B.,** Bacterial resistance to tetracyclines, *Microbiol. Rev.,* 42, 707, 1978.
153. **Chopra, I., Shales, S. W., Ward, J. M., and Wallace, L. J.,** Reduced expression of Tn 10-tetracycline resistance, in *Escherichia coli* containing more than one copy of the transposon, *J. Gen. Microbiol.,* 126, 45, 1981.
154. **McMurry, L. M. and Levy, S. B.,** Two transport systems for tetracycline in *Escherichia coli:* critical role for an initial rapid uptake system insensitive to energy inhibitors, *Antimicrob. Agents Chemother.,* 14, 201, 1978.

155. **Ball, P. R., Shales, S. W., and Chopra, I.,** Plasmid mediated tetracycline resistance involves increased efflux of the antibiotic, *Biochem. Biophys. Res. Commun.*, 93, 74, 1980.
156. **McMurry, L. M., Petrucci, R. E., Jr., and Levy, S. B.,** Active efflux of tetracycline encoded by four genetically different tetracycline-resistant determinants in *Escherichia coli, Proc. Natl. Acad. Sci. U.S.A.*, 77, 3974, 1980.
157. **McMurry, L. M., Aronson, D., and Levy, S. B.,** Susceptible *Escherichia coli* cells can actively excrete tetracyclines, *Antimicrob. Agents Chemother.*, 24, 544, 1983.
158. **Zupancic, T. J., King, S. R., Pogue-Geile, K. L., and Jaskunas, S. R.,** Identification of a second tetracycline-inducible polypeptide encoded by Tn 10, *J. Bacteriol.*, 144, 346, 1980.
159. **Privitera, G., Fayolle, F., and Sebald, M.,** Resistance to tetracycline, erythromycin, and clindamycin in the *Bacteroides fragilis* group: inducible versus constitutive tetracycline resistance, *Antimicrob. Agents Chemother.*, 20, 314, 1981.
160. **Bryan, L. E.,** Resistance to antimicrobial agents: the general nature of the problem and the basis of resistance, in *Pseudomonas aeruginosa, Clinical Manifestations of Infection and Current Therapy,* Doggett, R. G., Ed., Academic Press, New York, 1979, 219.
161. **Rabin, H. R. and Lockerby, D. L.,** Resistance to nitrofurans and nitroimidazoles, in *Antimicrobial Drug Resistance,* Bryan, L. E., Ed., Academic Press, New York, 1984, 317.
162. **Mitsuhashi, S. and Inoue, M.,** Resistance to macrolides and lincomycins, in *Antimicrobial Drug Resistance,* Bryan, L. E., Ed., Academic Press, New York, 1984, 279.
163. **Weisblum, B.,** Altered methylation of ribosomal ribonucleic acid in erythromycin-resistant *Staphylococcus aureus,* in *Microbiology,* Schlessinger, D., Ed., American Society for Microbiology, Washington, D.C., 1974, 199.
164. **Weisblum, B.,** Inducible erythromycin resistance in bacteria, *Br. Med. Bull.*, 40, 47, 1984.
165. **Courvalin, P., Ouniss, H., and Arthur, M.,** Multiplicity of macrolide-lincosamide-streptogramin antibiotic resistance determinants, *J. Antimicrob. Ther., 16 (Suppl. A),* 91, 1985.
166. **Tally, F. P., Snydman, D. R., Gorbach, S. L., and Malamy, M. H.,** Plasmid mediated transferable resistance to clindamycin and erythromycin in *Bacteroides fragilis, J. Infect. Dis.*, 139, 83, 1979.
167. **Sosa, A., Tally, F. P., Jacobus, N. V., and Gorbach, S. L.,** *Bacteroides fragilis* resistance to clindamycin *in vitro, Antimicrob. Agents Chemother.*, 22, 771, 1982.
168. **Leclercq, R., Carlier, C., Duval, J., and Courvalin, P.,** Plasmid-mediated resistance to lincomycin by inactivation in *Staphylococcus haemolyticus, Antimicrob. Agents Chemother.*, 28, 421, 1985.
169. **Levy, S. B.,** Playing antibiotic pool. Time to tally the score, *N. Eng. J. Med.*, 311, 663, 1984.
170. National Research Council, Committee to Study the Human Health Effects of Subtherapeutic Antibiotic Use in Animal Feeds, National Academy of Sciences, Washington, D.C. 1980.

Chapter 10

NOSOCOMIAL INFECTIONS

Hugh N. Lunan

TABLE OF CONTENTS

I. DEFINITION AND HISTORY

A nosocomial infection is one which is acquired in-house or during the period of institutionalization of a patient, due to some defect in the patient's resistance. An infection, by its broadest definition, is the interplay between organism and host under favorable conditions to the organism, but unfavorable to the host.

Two important concepts in the definition are "infecting agent" and host.[1,2] Biological agents, such as bacteria, rickettsia, virus, and fungi are important causes of tissue injury and disease. There is no question that the newer therapeutic agents have modified the status of biological disease, yet on the other hand, their overutilization has increased the number of infections occurring annually. Microorganisms cause disease only incidentally in their struggle for existence in the competitive biology of life. By no means are all of these microorganisms injurious to man. For instance, to a considerable extent, the normal *Escherichia coli* flora of the intestinal tract is responsible for the production of vitamin K. Clinical disease, in fact, results only when the invading agent evokes anatomic as well as functional damage in the course of seeking to obtain the necessary requirements for its survival. Important, also, is the fact that there are many other factors which determine the disease-producing potential of the organism: (1) the invading organism must have a port of entry; (2) the invasion must be of such a magnitude for the agent to survive in its new environment; (3) the organism must be able to obtain in the new environment the appropriate conditions to foster its survival; and (4) the organism must be capable of having a continuum of nourishment.

In order to appreciate the progress that has been made in the area of infection and its control, it is appropriate to take a brief look at the history of infectious diseases. Emperor Constantine, thinking that the church was the most humane organization of his time, decided and convinced the Bishops of the Council of Nicea to provide a hospital in every cathedral city (A.D. 325). It is very likely, however, that in those days these places of health care were gruesome and often fearful places by contemporary standards. Patients were grouped together irrespective of whether their diseases were infectious. Highly communicable diseases such as typhoid, measles, chicken pox, small pox, and plague were probably introduced directly into the wards, killing many patients who were not already dying from other diseases. Surgical patients, on the other hand, faced particular peril as infections were inevitable.

More than a century ago, in what must be recognized as one of the first studies in hospital epidemiology, James Young Simpson at the University of Edinburgh collected data on 2000 amputees who had remained at home. He found that the mortality among those amputees who had been hospitalized was much higher than in those who remained at home. For this reason, he coined the term "hospitalism" — suggesting that some factor related to hospital care had conferred certain risks on these patients. In 1843, Oliver Wendell Holmes in his essay on "Puerperal Fever" suggested that physicians unwittingly were playing major roles in the occurrence of that dreaded complication of childbirth. In 1848, Ignaz Phillip Semelweis, through meticulous observation and recordkeeping, demonstrated a reduction in maternal mortality following the institution of calcium chloride solution for handwashing before attending a delivery. For his research, he received nothing but rebukes and abuses from the medical profession of his time.[9]

Several decades later, however, Pasteur and Lister were to lay the groundwork for the modern practice of surgery using the aseptic technique. But the advent of the antibiotic era has generated another tide of complacency which, to varying degrees, is still with us despite the lessons learned by the emergence of antibiotic resistance. Unquestionably, there has been increased awareness of the determinitive role of the host in the development of nosocomial infections. The newer concepts of immunocompetence and immunononcompetence are indeed more clearly defined today than they were a decade or two ago. In addition, there is a greater appreciation by health care personnel for the kinds of infections that are likely to be acquired in the hospital environment in immunoincompetent patients.[45] For example, the severely granulocytopenic patient undergoing chemotherapy could fall prey to a whole host of bacterial or fungal infections; in addition the patient who needs an organ transplant but has to take immunosuppressive therapy may develop an increased susceptibility to in-house infections. Further, the advent of intravascular pressure-monitoring devices, catheters, and numerous invasive procedures have brought about problems of in-house contamination.

The fact is that medicine has failed somewhat in its mission to eradicate or decrease the incidence of in-house infections, possibly by not providing clear evidence of the methods that are known to control the increase in nosocomial infections. The scientific basis for nosocomial infection control as a variable discipline within hospitals is outrageously deficient. Recommendations have been made and at least given some credence by the Joint Commission on Accreditation of Hospitals, yet there is still inability to define what is important in the control of infection or, indeed, doubt whether infection control ever worked at all in accomplishing its stated purpose.[6]

Hospital-acquired infections occur in a little over 5% of all patients admitted to general hospitals, prevalence being approximately 12% in large tertiary care hospitals and 6% in community hospitals, with Gram-negative bacteria predominating.[14,15]

Prevalence	Incidence
12% In large tertiary care institutions	6%
6% In community hospitals	35%

In terms of figures, there are over 2.5 million hospital-acquired infections every year despite the increase in the use of antimicrobial agents.[7]

The commonly encountered infections are as follows:

Infections	Occurrence (%)
Urinary tract infections	40
Surgical wound infections	25
Upper respiratory tract infections, skin infections (burns included)	15
Septic infections, bacteremias, and mixed infections	20

Certain infections occur with greater frequency in certain diseases, for example, patients with diabetes mellitus are prone to develop such infections as cutaneous infections, monilial infections, bacteruria, and tuberculosis; and if ketoacidosis ensues and is uncontrolled, there is the development of mucorymycosis. Patients with hyperthyroidism develop *Candida albicans,* alveolar proteinosis, and *Nocardia.* Very young children who have undergone spleenectomy have a predilection for developing schistosomiasis, malaria, bartonellosis, and pneumococcal infections. Sickle-cell patients develop salmonella infections; patients with Hodgkins disease and sarcoidosis develop cryptococcosis, and patients with granulocytopenia will develop a host of infections due to various organisms.[8] Reasons for the coexistence of the infections with these disorders are largely unknown. One thing seems certain, and that is in all of these disorders there is a reduction of host response. The predominance of many of these microorganisms is seen in iatrogenic alterations of host resistance following the use of corticosteroids, antimetabolites, radiomimetic procedures, and immunosuppressive and antimicrobial agents. Chronic therapy with corticosteroids not only poses the risk of introducing new infections, but may also reactivate and disseminate a latent disease, e.g., tuberculosis.

II. MECHANISMS PROPOSED

These include: (1) depression of the reticuloendothelial system causing the impairment of immunologic development by decreasing antibody formation; and (2) suppression of other host defense mechanisms: e.g., with cytotoxic agents, there is an increased incidence of infections probably due to the ability of those agents to induce a reduction in granulocytes and lymphocytes, thereby decreasing antibody production.

Routes of spread of microorganisms include:

1. Person-to-person contact, directly or indirectly, via contaminated equipment or bed linen
2. Air
3. Vehicles such as contaminated food, water, injectable drugs, or blood
4. Vectors, such as insects

III. OFFENDERS AND CLINICAL MANIFESTATIONS

Penicillinase-producing staphylococci have been the main offenders; but more recently the Gram-negative enteric bacilli and fungi have emerged. Reappearance of fever plus other signs at the site of initial infection are the major clinical manifestations of an infection. Site may vary from one organism to the other. Manifestations are fever at about the 4th or 5th day of hospitalization, local inflammation, phlebitis, cellulitis or unexplained hyperventilation, and confusion or disorientation.

Catheterization is a major cause of infection; e.g., if a patient is subjected to a variety of diagnostic tests and/or procedures using intravenous catheters or urethral catheters (in particular, the indwelling type left in place for a few days), the risks are greater. Less common causes of infection are injections, thoracentesis, joint aspirations, lumbar punctures, tissue biopsies, and paracentesis. Two to three percent of all surgical procedures are complicated by infections. Respiratory equipment, diagnostic manipulations, and the concomitant indiscriminate use of antibiotics are also some of the causes of infections.

The most frequently encountered organisms are *E. coli, Klebsiella, Proteus, Pseudomonas,* mini-*Herellea, Serratia,* and less commonly, the penicillinase-producing staphylococci, enterococci, and *Clostridia.* Fungi include *Candida, Mucor, Nocardia,* and *Aspergillus.*[9,10] Patients with depressed host resistance are prone to infections by cytomegalovirus-disseminated herpes zoster and *Pneumocystis carinii.*[16]

IV. POSTOPERATIVE, SOFT TISSUE, AND OTHER INFECTIONS

A. Postoperative Infections

Postoperative infections consist of wound infections or the collection of pus formed in or around the oeprative site. Commonly, infections are seen when long operations are done or when extensive resections are performed, e.g., abdominal perineal resection, wounds involving arterial bypass, grafts, or prostheses replacement. Organisms present are staphylococci and Gram-negative enteric bacteria. Although postoperative infections do not, to a large extent, respond very well to chemoprophylaxis, administration of 10 million units of penicillin every 2 hours immediately preceding and after the operation has somehow been shown to reduce postoperative wound sepsis.

B. Cutaneous and Subcutaneous Soft Tissue Infections

These infections result from lesions caused by limited or reduced arterial blood flow such as abscesses in the skin, cellulitis, decubitus ulcers, and vascular stasis ulcers. Organisms are staphylococci; Group A and anaerobic streptococci, Gram-negative enteric bacilli, and even *Clostridia* may be pathogenic under these circumstances. In certain disease conditions, infection of certain areas is common: e.g., in leukemia there is a perineal and perianal infection. In burn patients, the site for infection occurs after Gram-positive flora has been replaced by Gram-negative flora. Organisms are *Pseudomonas,* which may cause a sudden development of shock — bacteremic or septic shock.

C. Urinary Tract Infections

Urinary tract infections may be due to the instrumentation of the urethra, bladder, or ureters, especially the placement of indwelling catheters, which permit entry of bacteria from the external environment. Organisms are *E. coli, Klebsiella, Enterobacter, Serratia, Proteus, Pseudomonas,* and enterococci. In managing urinary tract infections, treatment should be reserved for patients who are symptomatic and those suspected of having bacteremia. Asymptomatic bacteriuria in patients with indwelling catheters should not be treated with antibiotics since the chances of superinfection become rather high and it is unlikely that they would be eradicated.[11,12,17]

D. Pneumonia

Pulmonary infection is very common, occuring many times as a complication of other procedures such as cardiac surgery. It is seen also in the elderly, debilitated patients, and patients with compromised immune response.[13] Nebulizers carry *Pseudomonas, Herella,* and *Flavobacteria;* these are capable of causing severe pulmonary infection. The major therapeutic problem in pulmonary infections is mechanical, and if this is overcome, treatment becomes routine. Methods of treatment include: positive pressure breathing, postural drainage, frequent suctioning, and sometimes bronchoscopy. These procedures at times may even be more important than antibiotic therapy.

E. Bacterial Endocarditis

Open heart patients are at high risk in developing endocarditis. Signs in this disorder are generally absent and the most likely diagnostic tool is isolation of bacteria from blood specimen. Antimicrobial therapy to patients undergoing open heart surgery is routine.

F. Bacteremia

Urinary tract infections predominate, but bacteria can reach the blood stream from nearly any source, for example, intravenous sets or hyperalimentation. Staphylococci and corynybacteria have been implicated.

V. LABORATORY FINDINGS AND THERAPEUTIC MEASURES

Pus or appropriate body fluid culture and sensitivity testing form the backbone for rational therapy.

In soft tissue infections, the use of a penicillinase-resistant penicillin is advocated and those with bacteriuria should be treated with a drug active against enteric bacteria or enterococci. Prevention of the spread of infection can be achieved by surveillance, isolation of infected patients, control of personnel, control of environment, patient care procedures, and antimicrobial prophylaxis.

The concept of developing an infection control team is highly desirable, with the functions of such a team being to:

1. Develop and enforce policies necessary for the appropriate management of patients with communicable diseases;
2. Develop surveillance systems which identify patients at high risk;

3. Quantitate the incidence and prevalence of hospital-acquired infections;
4. Liaise with personnel from all areas of the hospital with a view to ensuring that an appropriate infection-control environment is maintained;
5. Develop educational techniques for employees in the prevention of the spread of infectious agents within the hospital; and
6. Monitor antibiotic utilization.

VI. CONCLUSION

Nosocomial infections are undoubtedly related to medical progress, and the futuristic plans of relying to a great extent on modern scientific technology, will undoubtedly support the increase in hospital-acquired infections. The data that are available regarding nosocomial infections suggest that a substantial number of the incidences of these diseases could be prevented. What is needed, then, is an all-out effort by hospital personnel to reform their behavior with regard to infection-control practices. If success can be achieved by adhering to these practices, then infection control will evolve as a rapid scientific discipline and indeed a vital part of hospital care.

REFERENCES

1. **Feingold, D. S.,** Hospital acquired infections, *N. Engl. J. Med.*, 283, 2384, 1970.
2. **Thoburn, R., et al.,** Infections acquired by hospitalized patients, *Arch. Intern. Med.*, 121, 1, 1968.
3. **Eickhoff, T. C.,** Nosocomial infections, a 1980 view: progress priorities and prognosis, *Am. J. Med.*, 70, 381, 1981.
4. **Williams, R. E. O.,** Changing perspectives in hospital infection, in *Proc. Int. Conf. on Nosocomial Infections,* Brachman, P. S. and Eickhoff, T. C., Eds., American Hospital Association, Chicago, 1971, 1.
5. **Wenzell, R. P.,** Surveillance, in *Principles and Practice of Infectious Diseases,* Mandell, G. L., Douglas, R. G., and Bennett, J. E., Eds., John Wiley & Sons, 1979, 2213.
6. *Accreditation Manual for Hospitals,* 1978 ed., Joint Commission on Accreditation of Hospitals, Chicago, 1978.
7. Special issue, The Senic Project, *Am. J. Epidemiol.*, 11, 465, 1980.
8. **Freid, M. A. and Vosti, K. L.,** The importance of underlying diseases in patients with Gram negative bacteremia, *Arch. Int. Med.*, 121, 418, 1968.
9. **Reinarz, J. A., et al.,** Potential role of inhalation therapy requipment in nosocomial pulmonary infection, *J. Clin. Invest.*, 44, 831, 1965.
10. **Duma, R. J., et al.,** Septicemia from intravenous infusions, *N. Engl. J. Med.*, 284, 257, 1971.
11. **Kunin, C. M.,** *Detection, Prevention and Treatment of Urinary Tract Infection,* 3rd ed., Lea & Febiger, Philadelphia, 1979.
12. **Kass, E. H. and Schneiderman, L. J.,** Entry of bacteria into the urinary tract of patients with inlaying catheters, *N. Engl. J. Med.*, 256, 556, 1957.
13. **Wightman, J. A. K.,** A prospective survey of the incidence of post operative pulmonary complications, *Br. J. Med.*, 55, 85, 1968.
14. **Scheckler, W. E.,** Nosocomial infections in a community hospital, 1972—76, *Arch. Intern. Med.*, 138, 1792, 1978.
15. **McCabe, W. R. and Jackson, C. C.,** Gram negative bacteremia, etiology and ecology, *Arch. Intern. Med.*, 110, 847, 1962.
16. **Weinstein, R. A., Stamm, W. E., Kramer, L., and Corey, L.,** Pressure monitoring devices — overlooked sources of nosocomial infections, *JAMA*, 236, 936.
17. **Warren, J. W., Plat, R., Thomas, R. J., Rosner, B., and Kass, E. H.,** Antibiotic irrigation and catheter-associated urinary tract infections, *N. Engl. J. Med.*, 299, 570, 1978.
18. **Moses, R. H.,** *Diseases of Medical Progress,* Charles C Thomas, Springfield, Ill., 1969.

Chapter 11

URINARY TRACT INFECTIONS

Michael Thompson

TABLE OF CONTENTS

I. INTRODUCTION

Urinary tract infections (UTIs) are among the most common infections encountered in clinical practice. Although the vast majority of these infections can be successfully managed with antimicrobial agents, serious kidney damage can occur when infection is associated with urinary tract obstruction or renal papillary damage. Further, these infections serve as a main cause of nosocomial infections in the U.S.

Improper prescribing can contribute to the growth of multiply-resistant strains of microorganisms. Since the number of effective agents has grown considerably in recent years, trying to decide which agent to use for specific forms of UTIs can be confusing. Although researchers continue to improve upon existing antimicrobial agents in general, there are still specific issues dealing with the management of UTIs that will require further study in order to improve upon existing methods of treating this disorder.

This manuscript will highlight various aspects concerning the etiology, pathogenesis, clinical manifestations, and treatment of bacterial UTIs. Areas in need of further research will be discussed in order to bridge the gap between prescribers and researchers. This has become crucial in todays climate of health care cost containment and in terms of what is required for continued patient safety.

II. DEFINITIONS

Before proceeding with the discussion, a brief section in which commonly used terms are defined is in order. Considerable confusion exists concerning terminology associated with this disorder.[1,2] In fact, the British Medical Council published an official statement recently, in order to define commonly used terms in an effort to promote consistency and to provide clarity.[2]

Urinary tract infection is defined as the presence of microorganisms in a properly collected urine specimen. Bacteriuria is considered significant if 100,000 colony-forming units of bacteria per milliliter or more are discovered upon examination of infected urine. In contrast, suprapubic aspirates containing any growth of bacteria are considered significant.[1] Recent studies have challenged these diagnostic criteria in women with acute dysuria and frequency,[3] and for patients with indwelling catheters.[4,5] These criteria, however, are widely used and established in medical practice.

UTIs may be complicated or uncomplicated. In addition, they may or may not be associated with symptoms. When significant bacteriuria exists and there are no additional abnormalities associated with the urinary tract or factors that would predispose the patient to recurrence, the infection is considered to be uncomplicated. The term "complicated" UTIs is used when there is an abnormality in the path of urine flow (e.g., intrarenal obstruction due to stones, metabolic disorders, prostatic hypertrophy, neurogenic dysfunction, etc.). Catheters serving as the focus of infection in UTIs and debilitated patients are factors often associated with complicated infections.

Infection may occur in any of the structures that comprise the urinary tract. Acute pyelonephritis is the term referring to infection of the kidney and defines the syndrome in which patients complain of fever, flank pain, and tenderness combined with significant bacteriuria. The definition for chronic pyelonephritis varies.[1,2] Glauser[1] defines chronic pyelonephritis as the presence of scarring in the kidney parenchyma due to infection. The interstitial inflammation and scarring caused by infection is indistinguishable from renal inflammation of other origins. Interstitial nephritis is then defined as renal interstitial inflammation and scarring from noninfectious causes such as chronic obstruction, analgesic abuse, hyperuricemia, diabetes, sickle cell anemia, and other disease states. In a report issued by the British Medical Research Council, chronic pyelonephritis is not recommended as a term because it was felt that its use has led a considerable confusion in the medical community.[2] These investigators prefer to use the terms ''chronic interstitial nephritis'' or ''chronic tubulointerstitial disease''. Thus, bacterial infection is included as a cause of interstitial nephritis rather than being seen as a separate disease process, as defined by Glauser.

Bacterial cystitis (bladder infection) is defined as a syndrome consisting of dysuria and frequency which is often accompanied with significant bacteriuria pyuria and sometimes hematuria.[2] Abacterial cystitis (commonly called the urethral syndrome), is defined as a syndrome consisting of frequency and dysuria in the absence of significant bacteriuria.[2] The term ''urethral syndrome'' is not recommended by the British Medical Council, since no evidence of actual uretheral disease is present in most patients. This term is used frequently, however, in the U.S.

The term prostatitis is used to describe a symptom complex rather than a specific disease entity. Considerable confusion surrounding this term existed since some patients diagnosed as having an infected prostate had normal prostatic secretions, while others exhibited marked inflammatory responses. Currently, prostatitis is defined as being acute bacterial, chronic bacterial, nonbacterial, and prostatodynic.[6] Patients with acute-bacterial prostatitis have systemic symptoms associated with infected urine. Specific symptoms will be duscussed later in the manuscript when the clinical manifestations of UTIs will be discussed in further detail. The term ''chronic prostatitis'' is used in patients with continuous prostatic infections as caused by a specific microorganism. Between infectious episodes the patient is asymptomatic. Prostatic secretions in these patients may appear normal, but will usually reveal leukocytes, macrophages, and a reduced number of ''lecithin bodies'' usually abundant in normal prostatic fluid. Nonbacterial prostatitis is a term used when the diagnosis of bacterial prostatitis is excluded, but examination of prostatic fluid reveals: (1) greater than 10 to 15 leukocytes per high power field, (2) an increase in oval fat bodies, and (3) a decrease in or the absence of lecithin granules usually seen in normal prostatic secretions. Finally, the term ''prostatodynia'' is reserved for patients with symptoms of prostatitis but without objective evidence of prostatic inflammation.

III. ETIOLOGIC AND EPIDEMIOLOGIC CONSIDERATIONS

The majority of UTIs encountered in clinical practice are caused by the Enterobacteriaceae from the gut. *Escherichia coli* is the most common organism implicated in these infections in ambulatory patients. Hospital-acquired infections as well as repeated UTIs may be due to *E. coli*, but are also likely to be due to other organisms such as *Proteus, Pseudomonas aeruginosa,* and *Klebsiella* spp.[1] The use of repeated courses of antibiotics will lead to infections caused by multiply resistant organisms as well as the development of mixed infections.[7,8] Gram-positive organisms rarely cause infection of the urinary tract, but infection with *Streptococcus faecalis* is not uncommon. When *Staphylococcus aureus* is found repeatedly in the urine, renal microabcesses secondary to bacteremia and a primary source of infection (e.g., endocarditis or osteomyelitis) should be sought. Although *Chlamydia trachomatis, Trichomonas vaginalis,* and *Ureaplasma urealyticum* commonly cause urethritis, they do not generally cause infection in other portions of the urinary tract. Fungi such as *Candida* spp. may occur in patients with indwelling catheters and diabetes. However, the role of viruses in these infections is poorly understood at present.

UTIs occur most often in women and girls. Approximately 1.2% of schoolgirls less than 18 years of age will have a UTI, and by 18 approximately 5% will have had at least one infection.[7] During the childbearing years, women are particularly predisposed to develop UTIs. This may in part be due to increased sexual activity and the recognized fact that UTIs commonly occur in pregnancy.[8]

Infections of the urinary tract in males are seen on opposite ends of the age spectrum. During the first year of life, the incidence of these infections is high in males, secondary to congenital abnormalities affecting the urinary tract. Except for infection seen due to sexually transmitted disease, the incidence of UTIs in males less than 50 years old is extremely low, and the presence of an infection should prompt the physician to conduct a thorough examination of the patient. Above the age of 50, the incidence increases due to prostatic hypertrophy which impedes proper urinary flow, thereby predisposing to secondary infection.[8]

Although UTIs occur frequently in diabetic patients, the development of these infections is not necessarily correlated with age, degree of glucosuria, or use of instruments. Predisposing factors remain to be adequately defined.

Finally, urologic manipulation (i.e., use of indwelling catheters, cystoscopy, etc.) accounts for more than 400,000 UTIs annually in the U.S.[8] In addition, infections of the urinary tract are cited as the most common form of nosocomial infection in institutional settings.[10]

IV. PATHOGENESIS

Although bacteria can reach the urinary tract hematogenously as a result of "seeding" from other primary sites of infection (e.g., osteomyelitis, endocarditis, and empyema), ascension via the urethra serves as the major route of infection for most patients. In the majority of patients, the bacteria commonly implicated in UTIs originate from the intestinal aerobic Gram-negative bacterial flora. Females who have poor personal hygiene habits can introduce these bacteria into the vaginal and urethral areas if they wipe the rectal area towards the urethra while cleaning. Thus, poor perianal hygiene is a major contributing factor. In addition, the trauma of sexual intercourse can also be responsible for the introduction of these bacteria into the urethra.[1] Once these bacteria gain access to the urethra, infection can readily occur since the female urethra is short and these bacteria can ascend quicker to the bladder and upper portions of the urinary tract.

Whenever urinary flow is reduced or obstructed (e.g., secondary to tumors or prostatic hypertrophy), the likelihood of a UTI developing increases.[8] The reason for this lies in the fact that the "washing out" of bacteria that occurs with each urinary void is considered a major intrinsic defense mechanism against the development of UTIs.[11] Post-void residual urine occurs in patients with neurologic lesions affecting the bladder or sphincter musculature, through effects of drugs (e.g., anticholinergics), or due to poor micturition habits.[8] Urinary stasis will lead to eventual infection since the urine is a good culture medium. If proper voiding is therefore interfered with, infections will occur.

Vesicoureteral reflux also contributes to the development of UTIs. This reflux is the result of an incompetent valve at the vesicoureteral junction and occurs as a result of congenital abnormalities or during pregnancy. A normal vesicoureteral valve is an important mechanism of defense against ascending bacterial infection. However, reflux may not be required for bacteria to reach the renal pelvis since the motile bacteria are able to ascend the ureter, even against a descending flow in some cases.[1]

Urologic manipulation is responsible for a significant number of UTIs in hospitalized patients.[10] Since the bacteria will normally inhabit the distal third of the urethra, manipulation with instruments such as catheters can introduce bacteria into the bladder. Studies have shown that nosocomial UTIs can prolong hospital stay anywhere from 2 to 10 days.[12] This may prove to have disastrous effects on maintaining hospital costs, since hospitals which receive a fair portion of their revenue from third-party systems under new prospective payment plans may not be fully reimbursed for expenses of nosocomial infections. This is a problem of increasing concern in the health care community.[12]

V. CLINICAL MANIFESTATIONS AND DIAGNOSIS

Clinical signs and symptoms as well as associated laboratory findings are listed in Table 1. UTIs are often classified as lower tract infections (i.e., urethritis or cystitis) or upper

Table 1
SIGNS AND SYMPTOMS OF VARIOUS FORMS OF URINARY TRACT INFECTION

Type of infection	Common symptoms	Laboratory findings
Urethritis	Dysuria, urgency, frequent urination; may be associated with cystitis	Bacteriuria may not be present in 1/3 to 1/2 of patients (urethral syndrome); if associated with cystitis >100,000 bacteria/mℓ possible
Cystitis	Dysuria, urgency, frequent urination; suprapubic pain, pyuria, hematuria	>100,000 bacteria/mℓ; pyuria (5 to 10 white blood cells per hpf); red blood cells may be present in urine
Pyelonephritis	Fever, chills, flank pain; nausea and vomiting; costovertebral angle tenderness; symptoms of cystitis; hematuria	>100,000 bacteria/mℓ; leukocyte casts; pyuria; red blood cells in urine; proteinuria; leukocytosis may be seen on peripheral blood smears
Acute bacterial prostatitis	Fever, chills, urethral discharge, frequent urination, perineal pain; arthralgias and myalgias; enlarged tender prostate on rectal examination	Urine and EPS[a] should be analyzed in order to detect bacteria; urine sample analyzed following prostate massage should have more bacteria present than before prostate massage (at least 10-fold difference)
Chronic prostatitis	Dysuria, frequent urination, urgency; fever, perineal discomfort/lumbosacral pain	EPS should contain some bacteria (usually less than 10,000/mℓ), as well as leukocytes, oval fat bodies, and low amount of lecithin bodies
Nonbacterial prostatitis	Same general symptoms as listed above	EPS contains no bacteria; leukocytes, oval fat bodies may be present with low amount lecithin bodies
Prostatodynia	Same general symptoms as above	EPS normal with no cells or bacteria noted

[a] EPS = expressed prostatis secretion. Procedure for expressing prostate and comparing bacteria amounts in EPS to urine amounts in order to determine if the prostate is truly infected is discussed in Reference 6.

tract infections (i.e., acute or chronic pyelonephritis). Patients with upper tract infections will often have symptoms of lower tract infections as well. The symptoms and lab findings of prostatitis are sometimes different from other forms of UTI, as may be seen in Table 1.

Urethritis may be associated with cystitis or it may occur alone as a separate entity. When it is associated with cystitis, the patient will complain of significant dysuria, frequency and urgency as well as suprapubic pain, and sometimes hematuria. When urethritis occurs in the absence of bladder involvement, dysuria, urgency, and frequency may or may not be associated with a urethral discharge. It is not uncommon for these patients to present with less than 10,000 bacteria per mℓ on urinalysis.

Although pyelonephritis can be associated with symptoms of cystitis, as previously mentioned, systemic involvement is often associated with this infection and patients will present with fever, chills, nausea, and vomiting, in addition to flank pain. Costovertebral angle tenderness is also common. These patients will generally appear to be more ill appearing than those with cystitis and may require hospitalization if they have become dehydrated from vomiting or if there are other complications. Laboratory abnormalities are highlighted in Table 1.

The varied symptomatology and lab findings of prostatis are also presented in Table 1. As may be seen, some patients will have no laboratory findings (prostatodynia), while others will be symptomatic with urine specimens containing smaller amounts of bacteria than typically seen in cystitis and pyelonephritis. A diagnosis of prostatitis is best made by expressing prostatic secretions and then observing thoese secretions for bacteria, cells, and fat-laden macrophages (oval fat bodies).

In order to properly diagnose a urinary tract infection, a midstream urine sample should be obtained using standard precautions to ensure against contamination. A more detailed discussion of the various methods of obtaining a urine sample is presented elsewhere.[8] A Gram stain of the urine should be performed in order to make an initial diagnosis. Although this will not delineate which organism is causing the infection, it will provide a clue as to the general class infecting the tract, which will aid in initial empiric selection of antimicrobial therapy. This is important since many patients will be in pain and waiting for final culture, and sensitivities will be undesirable.

VI. ERADICATIVE THERAPY

The treatment of a UTI varies according to the site of infection within the tract, the microorganism isolated, whether the patient is ambulatory or hospitalized, whether there are complicating factors such as obstruction or catheter placement, as well as the extent of infection. The toxicity of the drug selected, its pharmacokinetics, and costs are additional factors to be considered.

Patients diagnosed as having the acute urethral syndrome can usually be successfully managed with tetracycline 500 mg q.i.d. or doxycycline 100 mg b.i.d. for at least 7 days. Erythromycin can be prescribed for pregnant patients due to the well-known adverse effects of tetracycline on the developing fetus. Tetracycline is recommended in these patients because urethritis of this nature (i.e., not associated with cystitis and $<10^3$ bacteria per milliliter) is usually caused by *Chlamydia trachomatis*. As mentioned earlier, this organism commonly causes urethritis, but rarely infects other urinary tract structures. If an exudate is produced, it should be cultured in order to be sure that the patient does not have genococcal urethritis.

Uncomplicated cystitis can usually be successfully managed with an oral sulfonamide (e.g., sulfisoxazole), a sulfonamide/trimethoprim combination (e.g., Bactrim), ampicillin, amoxicillin, or nitrofurantoin. These agents are clearly effective against the common Gram-negative enteric bacilli that frequently infect the urinary tract. These same agents are also effective when used in uncomplicated pyelonephritis as well as uncomplicated cystitis when *E. coli* and *Proteus mirabilis* are the offending organisms. *Klebsiella* spp. are generally resistant to ampicillin therapy but are often sensitive to 1st-, 2nd-, or 3rd-generation cephalosporins. In cases of cephalosporin-resistant *Klebsiella, Enterobacter,* and *Pseudomonas,* aminoglycosides will be effective. Currently, third-generation cephalosporins (e.g., cefoperazone) and newer penicillins (e.g., piperacillin) may also be successful in eradicating some of these organisms. Although they have not been used clinically as widely as the aminoglycosides, they do offer the advantage of a much lower incidence of nephrotoxicity. Costs may be a considerable drawback for some of these newer agents. *Streptococcus faecalis* generally responds to ampicillin therapy while a penicillinase-resistant penicillin (e.g., nafcillin) is recommended for staphylococcal infections. Carbenicillin and ticarcillin may also be effective against *Pseudomonas* and some *Enterobacter* spp.[13]

It is important to note that community-acquired infections are most likely due to *E. coli, Proteus,* or *Klebsiella*. Other more serious Gram-negative infections such as *Pseudomonas, Enterobacter,* indole + *Proteus,* etc., are usually associated with complicated or hospital-acquired infections. The synergistic effect of a β-lactam antibiotic with an aminoglycoside is usually recommended in these cases.

For cystitis and pyelonephritis, antibiotics can be started once a urine sample has been obtained and bacteria and leukocytes are seen. If a repeat culture performed 24 to 48 hours after initiating therapy is sterile, then further identification or antibiotic sensitivity testing is generally not required.[1] However, the choice of drug should be confirmed through further sensitivity testing if repeat urines at 24 to 48 hr are still positive or if there is any doubt concerning the appropriateness of treatment. It is important to note that symptomatic improvement in pyelonephritis does not necessarily indicate successful treatment. Approximately one third of patients receiving ineffective treatment will experience clinical improvement.[13] Therefore, the importance of reculturing urine to determine the effectiveness of antimicrobial therapy is of paramount importance.

Response to therapy can be categorized into four classifications: (1) persistence, (2) relapse, (3) reinfection, and (4) cure.[13] Persistence indicates that the quantity of bacteria has not decreased at 24 to 48 hr after starting drug therapy and in spite of the fact that urinary concentrations were adequate. These patients should technically be considered treatment failures and another regimen should be selected as indicated by culture and sensitivity. Relapse describes the situation in which bacteria disappeared during drug therapy but returned once therapy was discontinued. In patients with obstruction or other underlying complicating factors, complete eradication may not be possible and continued antimicrobial therapy may be required in order to suppress the infection. However, when relapse occurs in patients without obstructive lesions or other complications, a second but longer course of the same antibiotic therapy may be indicated if the organism has maintained its susceptibility. Reinfection means that the patient has contracted a new infection (with bacteriuria) with a different organism a month or more after treatment. Complete cure means that the organism has been effectively eradicated from the urinary tract.

Treating infections of the prostate gland can pose problems in certain instances. Although the bacteria that cause acute and chronic bacterial prostatitis are the same organisms that are responsible for the majority of UTIs, diffusion of the antibiotic into prostatic secretions can be hindered due to the inability of the drug to penetrate into the organ.[6] In acute bacterial prostatitis, patients respond dramatically to antibiotics, possibly because the prostate is inflamed and the membrane permeability is altered. Most patients respond rapidly if therapy is initiated with Bactrim 2 tablets administered b.i.d. or ampicillin 250 to 500 mg q.i.d. or one of the other penicillins or cephalosporins discussed. Those individuals with severe systemic symptoms should be treated with a combination of an aminoglycoside plus ampicillin or a cephalosporin.[6] The problem of antibiotic penetration into prostatic secretions is most evident in chronic bacterial prostatitis. Although trimethoprim concentrates in the prostate of normal dogs, in clinical practice it is often difficult to completely cure this infection.[6] Trimethoprim is generally considered the drug of choice. From Gram-positive infections, erythromycin is recommended. Acute prostatitis is generally treated for 14 days, while chronic bacterial prostatitis can be treated for as long as 3 months if necessary.

The question of how long antibiotics should be administered in UTIs is controversial. Uncomplicated cystitis can usually be treated for 10 to 14 days with complete cure.[8] However, noncompliance with prescribed therapy and high costs of prescribed medications are important contributors to therapeutic failures. Some investigators have attempted to circumvent this problem by studying the therapeutic efficacy of single dose antimicrobial therapy for the management of UTIs.[14-21] This would have the obvious advantage of lower patient costs and improved compliance because patients would have only to take the medication one time in a single dose. Results using amoxacillin 3.0 g orally demonstrated clearly that this dosing method was equally as effective in lower UTIs when compared to conventional therapy when 250 mg q.i.d. was given (efficacy rates for both groups were 92%).[15] Trimethoprim-Sulfamethoxazole, cefaloridine, cefamandole, kanamycin, and nitrofurantoin have all been studied with variable results.[15-21]

Although 10 to 14 days is usually cited as the recommended duration of therapy for upper UTIs (pyelonephritis), some investigators have reported failure rates as high as 50% when these infections were treated for 10 days.[15] Other investigators have reported that pyelonephritis may have to be treated for as long as 6 weeks or more in order to reduce relapse.[22,23]

However, studies dealing with this subject will have to be sure to analyze data from patients with uncomplicated UTIs separately from those with complicated infections, since the latter subgroup of patients will usually require longer durations of treatment and removal of any underlying obstructive problems in order for these infections to be effectively cured.

VII. PREVENTIVE AND SUPPRESSIVE THERAPY

UTIs can be prevented in those individuals who are known to develop these infections frequently at closely spaced intervals. Preventive therapy should only be instituted once any existing infection has already been completely eradicated. Preventive therapy is frequently used in women who have frequent infections possibly related to sexual activity. Low doses are used in an effort to minimize adverse effects but yet prevent overgrowth of bacteria. An existing infection should first be eradicated with effective fulldose therapy. Trimethoprim-Sulfamethoxazole, 1/2 to 1 tablet given at bedtime, nitrofurantoin, 50 to 100 mg at bedtime, and methenamine mendelate, 1 g q.i.d. given with ascorbic acid 6 g/day, are all agents that can be used to prevent these infections in susceptible patients.

In patients who have a persistent focus of infection, suppressive therapy can be instituted in order to decrease flair-up and exacerbations of infections. These are patients with chronic infections in whom suppressive therapy manages to keep the number of bacteria reduced, thereby reducing the number of acute symptomatic infections. Patient examples would include individuals with chronic indwelling catheters, patients with chronic bacterial prostatitis, or individuals with other obstructive defects. Although it is best to remove or treat the underlying problem in order to prevent any level of infection, this is not always clinically possible. However, suppressing the rapid overgrowth of bacteria present will aid in reducing the number of active infections. This is important since these patients are those who have "complicated" UTIs which will often require more than one antimicrobial and longer durations of therapy. Methenamine mandelate with ascorbic acid (dosages as previously mentioned) is usually preferred over the other drugs used in preventive therapy in patients who will require long periods of suppressive therapy because it will not cause development of bacterial resistance to drugs frequently used to eradicate acute infections.[8]

VIII. RECOMMENDED AREAS FOR FUTURE RESEARCH

Future research is needed in the area of single dose therapy for uncomplicated infections. Although some progress has been made in this area, larger patient populations as well as different dosing regimens should be thoroughly examined before recommending widespread use. This area is particularly appealing, since it will reduce the amount of money required by a patient and since it has the potential to increase compliance. Since most patients are not likely to follow-up for repeat cultures to be sure that the infection is completely eradicated, we must be sure at least, from a purely research perspective, that success rates are consistently high enough on large numbers of patients to alleviate fears of relapse and increasing bacterial resistance through the use of subtherapeutic dosage regimens.

Additional research is also needed in order to develop broad-spectrum antimicrobial agents capable of effectively penetrating into prostatic secretions. Although trimethoprim is often recommended as an effective agent in this regard, complete cure is often not possible for certain individuals with chronic forms of this infection. Further work can then be conducted in order to determine if the effectiveness of an antimicrobial capable of penetrating and completely eradicating bacteria in the prostate varies between the different prostatic syndromes as previously outlined in this manuscript.

IX. SUMMARY

Urinary tract infections represent an area in which the development of effective antimicrobial agents has had a tremendous impact on patient cure. The armamentarium of effective antimicrobials is vast, and the problem presented to the prescribing clinician centers around being sure to select the most appropriate form of therapy for the specific patient being treated. In spite of high success rates seen with lower tract infections, complicated infections of the upper tract as well as chronic infections of the prostate and kidney are areas where a complete cure is often not possible unless underlying problems are corrected. However, the judicious use of antibiotics even in these cases can prevent acute symptomatic exacerbations, thereby decreasing hospitalization and improving patient comfort.

REFERENCES

1. **Glauser, M. P.,** Urinary tract infection and pyelonephritis, in *Medical Microbiology and Infectious Diseases,* Braude, A. I., Davis, C. E., and Fierer, J., Eds., W. B. Saunders, Philadelphia, 1981, chap. 146.
2. **Blowers, R., Asschar, A. W., Brunfitt, W., Colley, J. R. T., Saxton, H. M., Scott, J. E. S., Selkon, J. B., Smellie, R. H. R.,** Recommended terminology of urinary tract infection, *Br. Med. J.,* 2, 717, 1979.
3. **Stamm, W. E., Wagner, K. F., Amsel, R., et al.,** Causes of the acute urethral syndrome in women, *N. Engl. J. Med.,* 303, 409, 1980.
4. **Platt, R.,** Quantitative definition of bacteriuria, *Am. J. Med.,* 75 (Suppl. 1B), 44, 1983.
5. **Stark, R. P. and Maki, D. G.,** Bacteriuria in the catheterized patient: what quantitative level of bacteriuria is relevant? *N. Engl. J. Med.,* 311, 560, 1984.
6. **Fair, W. R.,** Prostatitis, in *Medical Microbiology and Infectious Diseases,* Braude, A. I., Davis, C. E., and Fierer, J., Eds., W.B. Saunders, Philadelphia, 1981, chap. 147.
7. **Kunin, C. M.,** Emergence of bacteriuria, proteinuria, and symptomatic urinary tract infections among a population of school girls followed for 7 years, *Pediatrics,* 41, 968, 1968.
8. **Kishi, D. T.,** Urinary tract infections, in *Clinical Pharmacy and Therapeutics,* Herfindal, E. T. and Hirschman, J. L., Eds., Williams & Wilkins, Baltimore, 1984, chap. 7.
9. **Forland, M., Thomas, V., and Shelokov, A.,** Urinary tract infections in patients with diabetes mellitus, *JAMA,* 238, 1924, 1977.
10. **Stamm, W. E.,** Guidelines for prevention of catheter associated urinary tract infections, *Ann. Int. Med.,* 82, 386, 1975.
11. **Fass, R. J., Klainer, A. S., and Perkins, R. L.,** Urinary tract infection: practical aspects of diagnosis and treatment, *JAMA,* 225, 1509, 1973.
12. **Wenzel, R. P.,** Nosocomial infections, diagnosis related groups, and study on the efficacy of nosocomial infection control, *Am. J. Med.,* 78(Suppl. 6B), 3, 1985.
13. **Pazin, G. J. and Braude, A. I.,** Pyelonephritis, in *Infectious Diseases,* Hoeprich, P. D., Ed., Harper and Row, Hagerstown, 1977, chap. 2.
14. **Rubin, R. H., Fang, L. S. T., Jones, S. R., Munford, R. S., Slepack, J. M., Varga, M. A., Onheiber, L., Hall, C. L., Tolkoff-Rubin, N. E.,** Single dose amoxicillin therapy for urinary tract infections. Multicenter trial using antibody coated bacteria localization technique, *JAMA,* 244, 561, 1980.
15. **Fang, L. S. T., Tolkoff-Rubin, N. E., and Rubin, R. H.,** Efficacy of single-dose and conventional amoxicillin therapy in urinary tract infection localized by the antibody-coated bacteria, *N. Engl. J. Med.,* 298, 413, 1978.
16. **Shapiro, E. D. and Wald, E. R.,** Single-dose amoxicillin treatment of urinary tract infections, *J. Pediatr.,* 99, 989, 1981.
17. **Brumfitt, W., Faiers, M. C., and Franklin, I. N. S.,** The treatment of urinary tract infection by means of a single dose of cephaloridine, *Postgrad. Med. J.,* 46(Suppl.), 65, 1970.
18. **Shaw, P. G., Fairley, K. F., and Whitworth, J. A.,** Treatment of urinary tract infection with a single dose of intramuscular administration of cephamandole, *Med. J. Aust.,* 1, 489, 1980.
19. **Ronald, A. R., Boutros, P., and Mourtada, H.,** Bacteriuria localization and response to single-dose therapy in women, *JAMA,* 235, 1854, 1976.
20. **Lohr, J. A., Hayden, G. F., Kesler, R. W., Gleason, C. H., Wood, J. B., Ford, R. F., Perriello, V. A., Benjamin, J. T., Dickens, M.D.,** Three day therapy of lower urinary tract infections with nitrofurantoin macrocrystals: a randomized clinical trial, *J. Pediatr.,* 99, 980, 1981.

21. **Bailey, R. R. and Abbott, G. D.,** Treatment of urinary tract infection with a single dose of trimethoprim-sulfamethoxazole, *J. Can. Med. Assoc.*, 118, 551, 1978.
22. **Turck, M., Anderson, K. N., and Petersdorf, R. G.,** Relapse and reinfection in chronic bacteriuria, *N. Engl. J. Med.*, 275, 70, 1966.
23. **Turck, M., Anderson, K. N., and Petersdorf, R. G.,** Relapse and reinfection in chronic bacteriuria. II, *N. Engl. J. Med.*, 278, 422, 1968.

Index

INDEX

A

B

C

D

E

F

G

H

N

O

P